Flow Shop Lot Streaming

Flow Shop Lot Streaming

Subhash C. Sarin
Virginia Tech

Puneet Jaiprakash
Virginia Tech

Subhash C. Sarin
Paul T. Norton Endowed Professor
Virginia Tech
Grado Department of Industrial and Systems Engineering
250 Durham Hall
Blacksburg, Virginia 24061
USA

Puneet Jaiprakash
Virginia Tech
Grado Department of Industrial and Systems Engineering
250 Durham Hall
Blacksburg, Virginia 24061
USA

ISBN 978-1-4419-4298-2 e-ISBN 978-0-387-47688-9

Printed on acid-free paper.

Printed on acid-free paper.

9 8 7 6 5 4 3 2 1

springer.com

Dedicated to

Prakash Vati and Sat Dev Verma, my parents.

Preface

It is imperative for a manufacturing company to run an efficient operation now-a-days to stay competitive in the world market. The advent of new technologies, a continuous improvement in product quality and changing customer requirements have all lead to shorter production runs, which demand effective methodologies for their execution on the shop floor – the ones that minimize work-in-process and cycle time while meeting customer demands. Due to the batch production nature of such an environment, the use of an appropriate production lot size (or sizes) on the shop floor is central to achieving these objectives. One technique that can effectively influence the flow of a batch (or a lot) of jobs over the machines by appropriately determining the size of production lots (also called *sublots* or *transfer lots*) is lot streaming. By splitting a lot of jobs into smaller-size sublots and processing them in an overlapping fashion over the machines, it tends to achieve the above objective. In this book, we present this technique for the flow shop machine configuration, which constitutes the brunt of its development that, thus, also comprises of the core of the related theoretical contributions made in this field of study.

The material presented in the book has been divided into five chapters, while the last chapter, Chap. 6, contains concluding remarks. Chapter 1 introduces the relevant concepts and definitions that are essential for a clear understanding of the material presented in subsequent chapters. To give the reader an appreciation of the potential benefits of lot streaming, analytical expressions to that end are derived. A historical perspective of this technique is given to put the subject matter on lot streaming in proper perspective and to provide the motivation behind the development of this technique. Some application areas that lend themselves to the use of lot streaming are then presented. A glimpse of the material contained in subsequent chapters is also provided to give the reader an idea of what to expect in these chapters. Chapter 2 presents new and generic mathematical models for the lot streaming problems that contain a variety of relevant features. A mathematical model of a problem, in general, can aid in its analysis, and also, in the development of an appropriate mathematical programming-based methodology for its solution. Chapter 2 is written with this intent in mind. Chapters 3–5 present material in the increasing order of difficulty of the lot streaming problems, namely, for two-machine, three-machine, and m-machine problems. Each of these chapters addresses a variety of problems while presenting for each the requisite analytical development leading up to the algorithm for its solution. These algorithms are illustrated through numerical examples to further aid in their understanding.

The material in this book can be used as a supplement to a course in sequencing and scheduling, production planning and control, production management, supply chain management, or to courses in related areas at graduate or advanced undergraduate levels. As background, it requires mathematical maturity and introductory knowledge of optimization concepts and methodologies. The book provides useful ideas and algorithms for practitioners, and it can serve as a useful research reference.

My first and foremost thanks go to one of my graduate students, Puneet Jaiprakash. For his many direct contributions, I consider him to be a coauthor of this book. I used the first draft of this book in my graduate-level course on sequencing and scheduling taught at Virginia Tech during the 2005 Spring semester. The students from this class provided valuable feedback that assisted in improving the exposition of the material presented in the book. In particular, I would like to recognize the contributions made in this regard by Ming Chen and Liming Yao, two of my doctoral students. I would also like to extend my sincere thanks to the anonymous reviewers for their careful reading of the manuscript and insightful comments.

A project of this magnitude cannot be accomplished without the unconditional support, encouragement, and love of the family. For this, I would like to thank my wife, Veena, and our sons, Sumeet and Shivan. Finally, I would like to thank Sandy Dalton for her help in typing the manuscript.

Subhash C. Sarin
Blacksburg, VA
December 15, 2006

Contents

1 Introduction to the Lot Streaming Problem

1.1 Introduction

We consider the scenario where lots consisting of several discrete and identical items are to be processed on several machines configured as a flow shop. Instead of transferring the entire lot after all of its items have been processed on a machine, we consider transferring the items of the lot in smaller batches called *sublots*. This technique of splitting a lot into sublots (also termed *transfer lots*), and processing different sublots simultaneously over different machines, albeit still maintaining their movement over the machines in accordance with their flow shop configuration, is called *lot streaming*. As an illustration of this concept, consider the following example. Suppose a lot consists of 120 identical items and it is to be processed on two machines. The processing times per item of the lot are $1(= p_1)$ and $2(= p_2)$ time-units on machines 1 and 2, respectively. If the lot were not to be split, it will be processed as shown in Fig. 1.1. However, if the lot were split into sublots of sizes, say, 50, 40, and 30 items and these sublots were processed in an overlapping fashion, the distribution of the lot for processing over the machines will be as shown in Fig. 1.2.

One apparent advantage of lot streaming is reduction in the makespan value (290 time-units against 360 time-units, see Figs. 1.1 and 1.2). However, this advantage may not be that obvious if setup and/or transfer times are encountered during the handling of individual sublots. The problem gets even more interesting depending on whether a setup can be performed at the time of the transfer of a sublot or whether it can be performed a priori, i.e., before the arrival of a sublot on a machine. Also, if more than one lot is to be processed on the machines, the makespan value will depend on whether or not the sublots from different lots are intermingled. The sequence in which the lots themselves are processed can impact the makespan value as well. We address all of these issues in this book.

Before delving into the analysis of various lot streaming problem scenarios, we first present, in the remainder of this chapter, relevant terminology, assumptions, notation, a lot streaming problem classification scheme that helps in organizing the

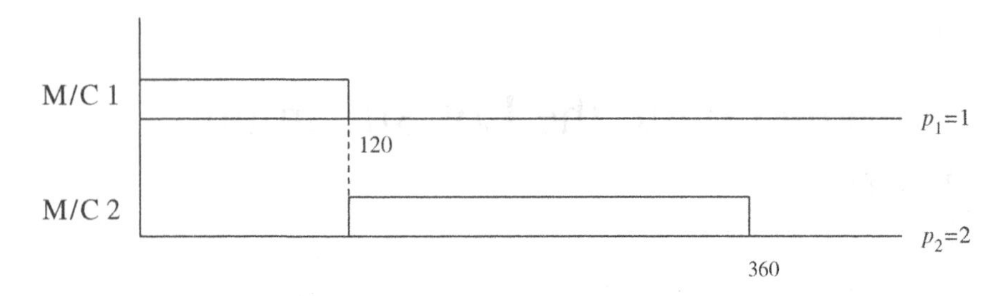

FIGURE 1.1. Processing of a lot without lot streaming

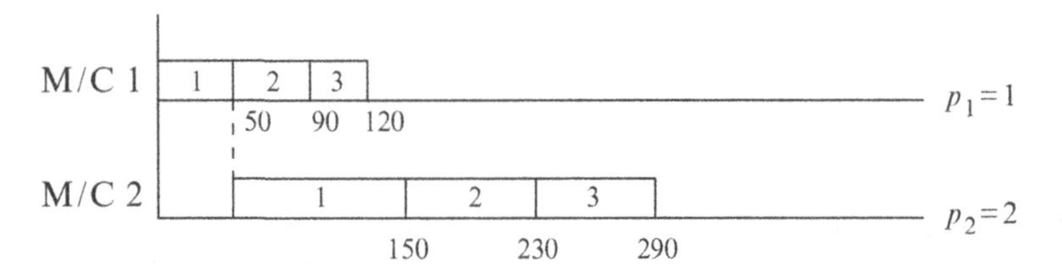

FIGURE 1.2. Processing of a lot with lot streaming

material to be presented in subsequent chapters, a concept of dominance among lot streaming models, potential benefits of lot streaming, as well as a brief historical perspective and some application areas of lot streaming.

1.2 Terminology

1.2.1 *Consistent and Variable Sublots*

When identical sublot sizes are used for transferring a lot between every pair of consecutive machines, the sublots are termed *consistent.* However, if the sublot sizes used for transferring a lot among the machines vary, the sublots are called *variable.* An example of consistent sublots is shown in Fig. 1.3, where sublot sizes of 50, 40, and 20 items are maintained while transferring a lot between machines 1 and 2 as well as between machines 2 and 3. Figure 1.4 depicts an example of variable sublot sizes where the sublots used to transfer a lot from machine 1 to machine 2 consist of 50, 40, and 20 items while those from machine 2 to machine 3 consist of 40, 60, and 10 items. The resulting makespan values are 370 and 390 units, respectively, for these two types of sublots.

1.2.2 *Equal Sublots*

This refers to the case when all the sublots of a lot are of the same size. This is a special case of consistent sublots. An example of equal sublots is shown in Fig. 1.5 where a lot consisting of 120 items is split into three equal sublots of 40 items each.

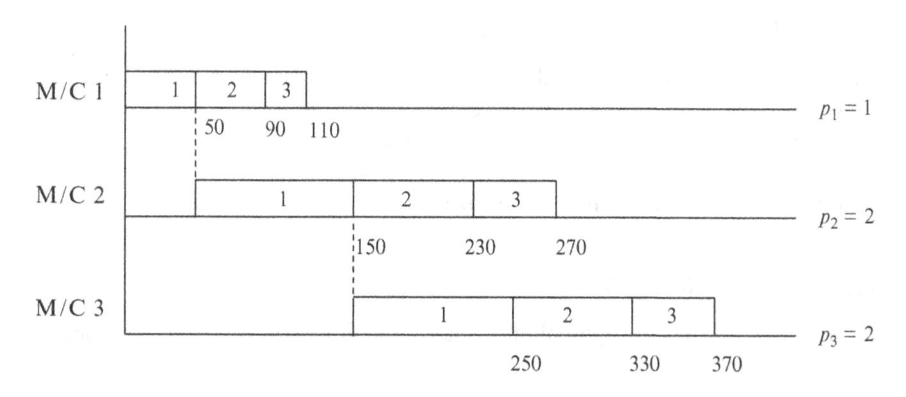

FIGURE 1.3. Lot streaming with consistent sublots of sizes 50, 40, and 20 items

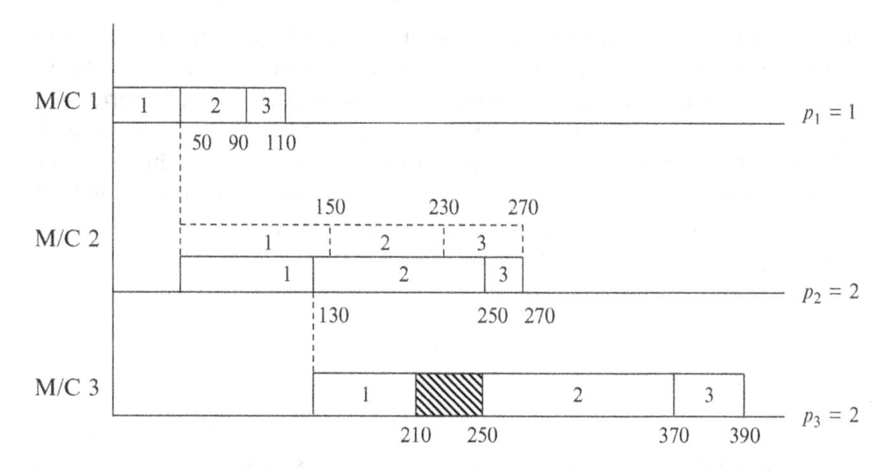

FIGURE 1.4. Lot streaming with variable sublot sizes of 50, 40, and 20 items between machines 1 and 2, and 40, 60, and 10 items between machines 2 and 3

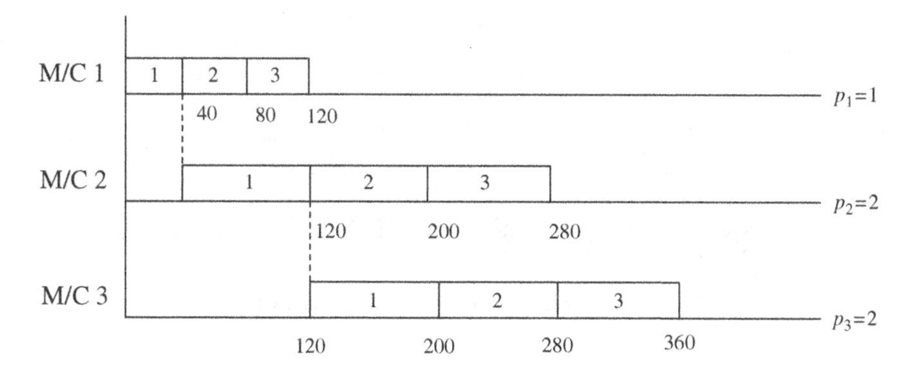

FIGURE 1.5. Lot streaming with equal sublots of size 40

1.2.3 *Continuous and Discrete Sublot Sizes*

The sublots of a lot are, in general, assumed to take real-valued (continuous) sizes. However, integer sublot sizes are more relevant for the manufacturing facilities involved in the production of discrete parts. The integrality restriction, typically, makes a lot streaming problem difficult to solve. We consider scenarios pertaining to both continuous and discrete sublot sizes.

1.2.4 *Critical Sublot*

A sublot that impacts the optimal makespan value, if its size is varied, is termed a *critical sublot*.

1.2.5 *Lot-Attached and Lot-Detached Setups*

The lot-attached setup refers to the case when the setup for a lot on a machine can be started only after the first sublot of that lot has arrived at that machine. However, under lot-detached setup, the setup for a lot on a machine can be performed even when the first sublot of that lot is being processed on a previous machine. In this case, the setup on a machine is performed as soon as that machine finishes processing the previous lot assigned to it. These are illustrated in Figs. 1.6 and 1.7 where the setup time for lot j on machine k is represented by t_{jk}.

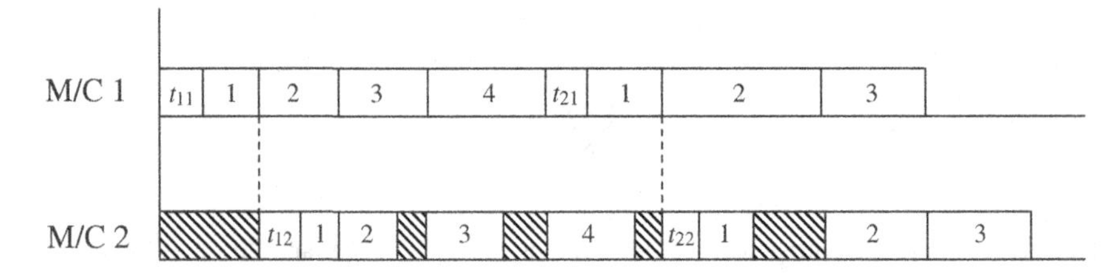

FIGURE 1.6. Lot streaming with lot-attached setups

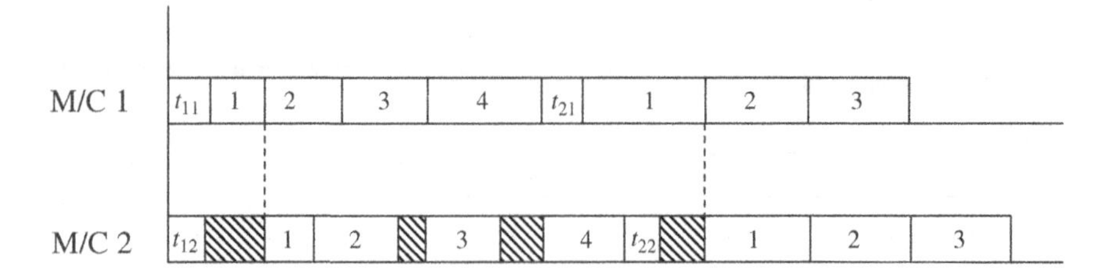

FIGURE 1.7. Lot streaming with lot-detached setups

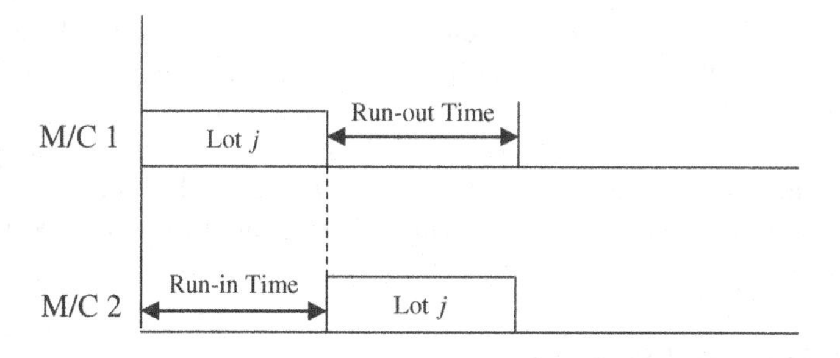

FIGURE 1.8. Run-in and run-out times for lot j

1.2.6 *Run-In and Run-Out Times for a Lot j*

These terms are generally defined for lot streaming problems that involve two machines. The run-in time for a lot j is defined as the time elapsed after its start on machine 1 and before its processing begins on machine 2. The run-out time for a lot j is the time elapsed between the end of its processing on machine 1 and that on machine 2. These are shown in Fig. 1.8.

As a lot consists of a number of items, under lot streaming and to minimize the makespan, these items are processed as soon as possible and with no intermittent idle time on machine 1. However, depending upon their processing times on both the machines, it is possible to delay the start of these items on machine 2 without increasing the makespan value (so that the items of the lot are processed consecutively on machine 2). The elapsed time between the starts of the lot on machines 1 and 2, after this shift in the start time of the items on machine 2, is called run-in time. We designate this by RI_{j1} for lot j, which can be expressed as

$$\mathrm{RI}_{j1} = \max\left(p_{j1}, p_{j1} + \left(n_j - 1\right) p_{j1} - \left(n_j - 1\right) p_{j2}\right),$$

where p_{ji} is the processing of an item of lot j on machine i, and n_j the number of items in lot j. In the same vein, run-out time, namely the difference between the completion times of the last item on machines 1 and 2, designated by RO_{j2}, can be expressed as:

$$\mathrm{RO}_{j2} = \max\left(p_{j2}, p_{j2} + \left(n_j - 1\right) p_{j2} - \left(n_j - 1\right) p_{j1}\right).$$

1.2.7 *Run-In and Run-Out Times for a Sequence of Lots*

Consider the streaming of N lots on two machines. Assume that the sequence in which these lots are to be processed on the machines is given. As before, we schedule the start of every item of a lot as early as possible on machine 1 to minimize the makespan. Hence, the lots are processed continuously on machine 1. However, depending on the time required to process an item of a lot on the machines, idle times might be encountered in between the processing of the sublots of a lot on machine 2. If we were to right shift the start times of the sublots of all the lots on machine 2 just enough to eliminate the idle times encountered on that machine, the makespan would still remain the same. The total idle time on machine 2 (now appearing before the first lot on machine 2), i.e., the time elapsed between the starts of the first lot on machines 1 and 2, is termed the *run-in time*. The idle time on machine 1 after it has finished processing the last lot and before the completion of the last lot on machine 2 is called the *run-out time*. Both of these run-in and run-out times are defined for a given sequence in which the lots are processed on the machines.

1.2.8 *Start and Stop Lags*

A start lag is the minimum interval of time required between the start times of a lot on two successive machines. Similarly, a stop lag is the minimum time interval required between the completion times of a lot on two successive machines. Note that if start and stop lags are given, then they constitute lower bounds of RI and RO, respectively.

1.2.9 *Transfer Lag*

This term is also, usually, defined only for flow shop problems involving two machines. A transfer lag (see Fig. 1.9) is the difference between the start time of a job on machine 2 and its completion time on machine 1 in the presence of start and stop lags. Thus, for a job j, with processing times p_1 and p_2 on machines 1 and 2, respectively, the transfer lag d_j is given by

$$d_j = \max\left\{u_j - p_1, v_j - p_2\right\},$$

where u_j and v_j are its start and stop lags. A negative value of d_j represents an overlap between the processing of a job j on the two machines (see Fig. 1.9).

For a lot streaming problem, suppose a lot j consists of n_j items. The inherent lot streaming process gives rise to the start lag $u_j = p_{j1}$, and the stop lag $v_j = p_{j2}$, while the processing times of the lot on machines 1 and 2 are $p_1 = n_j p_{j1}$ and $p_2 = n_j p_{j2}$, respectively. The transfer lag for lot j, under lot streaming, can then be calculated using the above expression.

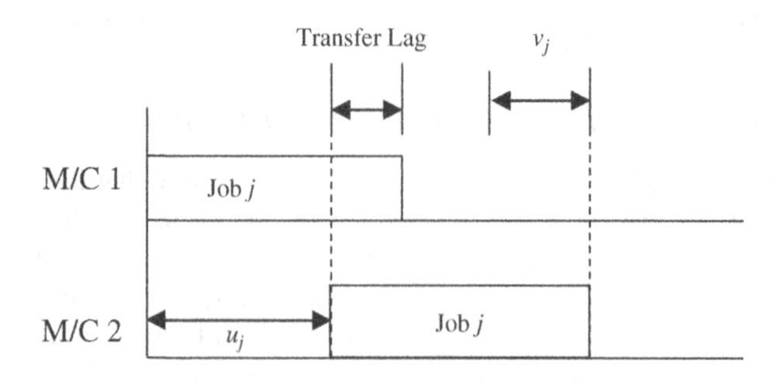

FIGURE 1.9. Transfer lag for job j

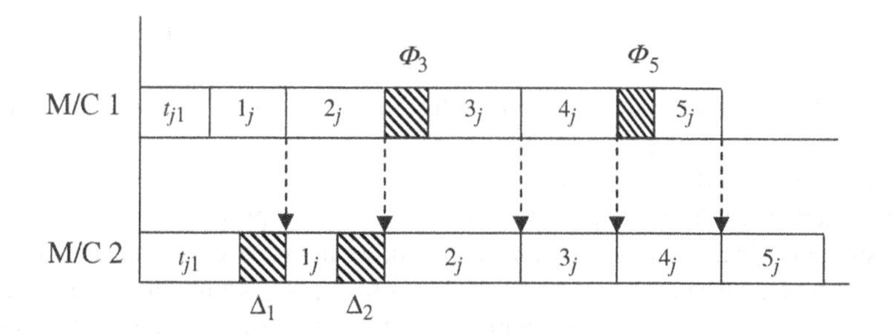

FIGURE 1.10. No-wait flow shop with detached setup

1.2.10 *No Idling*

This refers to the situation where no idle time is permitted in between the processing of the sublots of a lot on a machine, i.e., once a machine starts processing a lot, it must do so until all its sublots have been processed.

1.2.11 *Intermittent Idling*

Under intermittent idling, an idle time may be present between the processing of two successive sublots of a lot on a machine.

1.2.12 *No-Wait Flow Shop*

In a no-wait flow shop, each sublot of a lot is processed continuously on all the machines, i.e., the start time of a sublot j on machine k should equal the completion time of that sublot on machine $(k-1)$. This could lead to an inserted idle time before the processing of a sublot of a lot on a machine. This is depicted in Fig. 1.10 by Φ_3 and Φ_5 for sublots 3 and 5 on machine 1 and by Δ_1 and Δ_2 for sublots 1 and 2 on machine 1.

1.3 Assumptions, Notation, and Classification Scheme

1.3.1 *Assumptions*

1. All lots are available at time zero.
2. Unless stated otherwise, the machine configuration that we consider constitutes a flow shop.
3. Also, unless stated otherwise, the performance measure that we consider is *makespan*, i.e., the time elapsed between the start of the first operation of the first sublot on machine 1 and the completion of the last operation of the last sublot on the last machine. The objective is to minimize the makespan.
4. All sublots of a lot are processed together, i.e., intermingling among the sublots belonging to different lots is not permitted. Furthermore, preemption of a sublot

is not permitted, i.e., once the processing of a sublot is started, it cannot be interrupted.
5. Sublot transfer times are assumed negligible unless they are considered explicitly in a model being discussed.
6. The lot setup and removal times as well as the sublot setup and transfer times are independent of the sequence in which the lots are processed.
7. A sublot is assumed to be transferred in entirety in the case of equal and consistent sublots, while in the case of variable sublots, we assume that a sublot can be reconfigured as soon as the items constituting that sublot become available.

1.3.2 *Notation*

We use the following notation:
Parameters

- m Number of machines constituting a flow shop
- N Total number of lots available
- p_{jk} Processing time of an item in the jth lot on machine k; a single subscript is used if the processing time is only machine dependent
- n_j Number of sublots for the jth lot; the subscript j is omitted for problems involving a single lot
- U_j Number of items in lot j; the subscript j is omitted for problems involving a single lot
- t_{jk} Setup time for the jth lot on machine k; the subscript j is omitted for problems involving a single lot
- r_{jk} Removal time for the jth lot on machine k; the subscript j is omitted for problems involving a single lot

Variables

- x_{ij} Proportion of work allotted to the ith sublot of lot j, $\sum_{i=1}^{n_j} x_{ij} = 1$
- s_{ij} Number of items in the ith sublot of lot j, $s_{ij} = x_{ij}U_j$. Note that, in general, s_{ij} can take on continuous or discrete values. We consider both situations in the sequel

1.3.3 *Classification Scheme*

We use an eight-field classification scheme as shown below to represent the various lot streaming models:

{No. of machines}{machine configuration}/{no. of lots}/{sublot type}/{idling}/
{sublot sizes}/{objective function}/{special features}.

Machine configuration refers to the arrangement of the machines, e.g., as a flow shop (F), job shop (J), or parallel machines (P). Sublot type refers to the type of sublots, namely consistent (C), variable (V), or equal (E). Intermittent idling may or may not be permissible between two sublots. This is denoted by II for intermittent idling and by NI for no idling. Sublot sizes may be continuous (CV) or discrete (DV). As mentioned earlier, we assume minimizing the makespan as the objective for all the problems presented unless otherwise specified. Consequently, we state the objective function only if it is different from that of minimizing the makespan. The last field specifies special features that may be present. These pertain to setup, transfer or removal times, which can be lot and sublot – attached or detached, and intermingling of the sublots belonging to different lots, among others.

Thus, 3F/N/C/II/DV refers to a three-machine, flow shop system involving N lots where sublot sizes remain the same on all the machines, idling is permitted in between the processing of sublots, each sublot consists of integer number of items, and no special features are involved. Since we are dealing only with flow shop systems in this book, we will drop F from the classification scheme in the sequel.

1.4 Dominance in Lot Streaming Models

Dominance among lot streaming problems is discussed in [35]. We say that a model a dominates model b if the following relationship holds: $C_{\max}(a) \leq C_{\max}(b)$, where $C_{\max}(x)$ represents the makespan of model x. To understand the concept of dominance amongst lot streaming models, we make the following observations:

1. The no-idling requirement is a special case of the intermittent-idling requirement, since a schedule obtained by permitting intermittent idling might end up with no idle time in between the processing of the sublots.
2. Integer-size sublots are a special case of continuous-size sublots.
3. Consistent sublots are a special case of variable sublots, and equal sublots are a special case of consistent sublots.

Hence, the minimum possible makespan will be achieved by the most general/least restrictive case of V/II. This model dominates both the C/II and V/NI models, whereas all of these dominate the C/NI model. For the case of equal sublots, the E/II model dominates the E/NI model. Since equal sublots are a special case of consistent sublots, the C/NI model dominates the E/NI model and the C/II dominates E/II. However, no clear relationship exists between the E/II and C/NI as well as between the C/II and V/NI models. This dominance relationship among lot streaming problems is shown in Fig. 1.11.

1.5 Potential Benefits of Lot Streaming

Although the industry-based and simulation-based studies confirm that lot streaming is indeed beneficial in several batch production environments, the latter

$C_{\max}(\{V/II\}) \le C_{\max}(\{C/II\})$ and $C_{\max}(\{V/NI\})$

$C_{\max}(\{C/II\}) \le C_{\max}(\{E/II\})$ and $C_{\max}(\{C/NI\})$

$C_{\max}(\{V/NI\}) \le C_{\max}(\{C/NI\})$

$C_{\max}(\{E/II\})$ and $C_{\max}(\{C/NI\}) \le C_{\max}(\{E/NI\})$

FIGURE 1.11. Dominance among lot streaming models

suffers from the drawback that it is data specific. Analytical results, on the other hand, have a general appeal. In this section, we discuss the potential benefits of lot streaming in flow shop systems with respect to three commonly used performance measures [24], i.e.,

- Makespan ($C_{\max}$)
- Mean flow time (MFT)
- Average WIP level ($\overline{\text{WIP}}$)

For each of these criteria, an expression of the ratio of the value of a measure under lot streaming to its value without lot streaming is developed, which can then be used to evaluate the benefits of lot streaming. For ease of analysis, we develop these results for the m-machine flow shop configuration involving equal and continuous sublot sizes. We address both single- and multiple-lot cases in which transfer and setup times are negligible. Thus, in view of our classification scheme, the lot streaming problem that we consider is $m/\{1, \text{N}\}/\text{E}/\text{II}/\text{CV}/\{C_{\max}, \text{MFT}, \overline{\text{WIP}}\}$.

1.5.1 *m/1/E/II/CV/*{$C_{\max}$, *MFT,* $\overline{WIP}$}

1.5.1.1 Makespan ($C_{\max}$)

In the presence of equal sublot sizes, the makespan under lot streaming is the sum of the processing times of the first sublot on machines 1 to $(\upsilon - 1)$, processing of the lot on machine υ, and, the processing times of the last sublot on machines $(\upsilon + 1)$ to m (see Fig. 1.12), i.e.,

$$C_{\max}^{\text{LS}} = \frac{U}{n}\sum_{i=1}^{\upsilon-1} p_i + U p_{\max} + \frac{U}{n}\sum_{i=\upsilon+1}^{m} p_i,$$

where $\upsilon = \arg\left\{\max_{1\le i\le m}(p_i)\right\}$.

Note that υ need not be unique, which would imply the existence of more than one path to obtain the optimal makespan value. Also, υ can be the first machine in which case the first term of the $C_{\max}^{\text{LS}}$ expression would not exist. After rearranging, the above expression can be written as:

$$C_{\max}^{\text{LS}} = \frac{U}{n}\left\{\sum_{i=1}^{m} p_i + (n-1)p_{\max}\right\}. \tag{1.1}$$

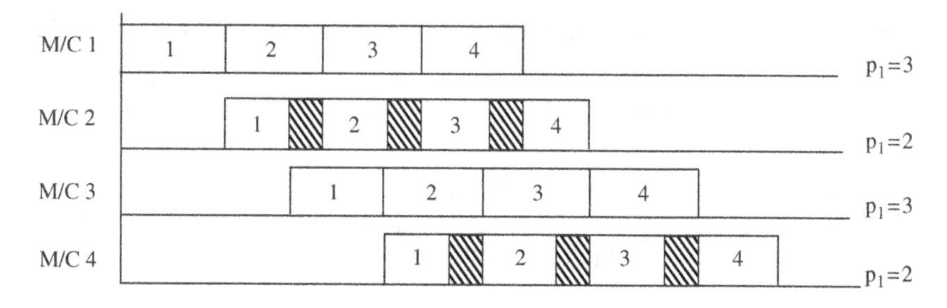

FIGURE 1.12. Lot streaming with equal sublots

In the absence of lot streaming, the makespan is given by

$$C_{\max} = U \sum_{i=1}^{m} p_i. \tag{1.2}$$

It is obvious that when transfer and setup times are not present, the optimal number of sublots, n^*, is given by U, i.e., unit-size sublots will be optimal. Hence, the ratio of (1.1) and (1.2) becomes

$$\frac{C_{\max}^{\mathrm{LS}}}{C_{\max}} = \frac{\sum_{i=1}^{m} p_i + (U-1)p_{\max}}{U \sum_{i=1}^{m} p_i}. \tag{1.3}$$

In case $(U-1)p_{\max} \ll \sum_{i=1}^{m} p_i$, i.e., we have a small-size lot to be processed on a large number of machines, the above ratio approaches $1/U$. However, when U is large, $(U-1)p_{\max} \gg \sum_{i=1}^{m} p_i$, the above ratio approaches $p_{\max}/\sum_{i=1}^{m} p_i$. When all the machines require identical processing times, the ratio reduces to $1/m$, its lower bound. Hence, the maximum makespan reduction (MMR) is given by

$$\mathrm{MMR} = 1 - \frac{1}{m} = \frac{m-1}{m}.$$

Note that the MMR is achieved when the lot size is sufficiently large and all machines require equal processing times. Also, in that case, the makespan reduction depends only on the number of machines.

1.5.1.2 Mean Flow Time (MFT)

Under lot streaming, portions (sublots) of a lot become available for processing earlier at the next machine when compared with the case when no lot streaming is used. Hence, one would expect MFT of the sublots under lot streaming to be lower than that when no lot streaming is used. The flow time $\left(\mathrm{FT}_i^{\mathrm{LS}}\right)$ of a sublot i, in the presence of lot streaming, is the sum of the processing times of the first sublot on machines 1 to $\upsilon - 1$, processing time of sublots 1 to i on machine υ, and the processing times of the ith sublot on machines $(\upsilon + 1)$ to m, i.e.,

$$\mathrm{FT}_i^{\mathrm{LS}} = s \sum_{k=1}^{\upsilon-1} p_k + isp_{\max} + s \sum_{k=\upsilon+1}^{m} p_k,$$

where $s = U/n$, and v is as defined earlier. After simplification, this can be written as:

$$\mathrm{FT}_i^{\mathrm{LS}} = s\sum_{k=1}^{m} p_k + (i-1)sp_{\max}.$$

Hence, the MFT of the entire lot is given as:

$$\mathrm{MFT}^{\mathrm{LS}} = \frac{\sum_{i=1}^{n} \mathrm{FT}_i^{\mathrm{LS}}}{n}.$$

Consider the numerator of the above expression

$$\begin{aligned}\sum_{i=1}^{n} \mathrm{FT}_i^{\mathrm{LS}} &= s\sum_{i=1}^{n}\left(\sum_{k=1}^{m} p_k + (i-1)p_{\max}\right)\\ &= s\left(n\sum_{k=1}^{m} p_k + \frac{n(n-1)}{2}p_{\max}\right)\\ &= U\left(\sum_{k=1}^{m} p_k + \frac{(n-1)}{2}p_{\max}\right).\end{aligned}$$

Hence, the MFT can be written as:

$$\mathrm{MFT}^{\mathrm{LS}} = \frac{U\left(\sum_{k=1}^{m} p_k + \frac{(n-1)}{2}p_{\max}\right)}{n}.$$

After rearranging, we have

$$\mathrm{MFT}^{\mathrm{LS}} = \frac{U\left(\sum_{k=1}^{m} p_k - \frac{p_{\max}}{2}\right)}{n} + \frac{Up_{\max}}{2}. \tag{1.4}$$

Note that the above expression is valid for both continuous and integer values of n. As in the case of makespan, the optimal MFT is obtained when n is as large as possible, i.e., $n^* = U$, implying that unit-size sublots are optimal. Hence, the above expression can be written as

$$\mathrm{MFT}^{\mathrm{LS}} = \frac{(U-1)p_{\max}}{2} + \sum_{k=1}^{m} p_k.$$

When lot streaming is not used, the MFT is identical to the makespan, i.e.,

$$\mathrm{MFT} = U\sum_{k=1}^{m} p_k. \tag{1.5}$$

Hence, the ratio becomes

$$\frac{\mathrm{MFT}^{\mathrm{LS}}}{\mathrm{MFT}} = \frac{\frac{(U-1)p_{\max}}{2} + \sum_{k=1}^{m} p_k}{U\sum_{k=1}^{m} p_k}. \tag{1.6}$$

When the lot size is relatively small when compared with the number of machines, the above ratio approaches $1/U$, which is identical to that obtained for the makespan case. When the lot size is large, the ratio approaches $p_{\max}/2\sum_{i=1}^{m} p_i$, which is half of what was obtained earlier. This ratio becomes $1/2m$ when the processing times are identical for all the machines.

1.5.1.3 Average WIP Level ($\overline{\text{WIP}}$)

The reduction of WIP in the system, under lot streaming, is shown in Fig. 1.13. As can be seen in the figure, during the interval $[0, \text{FT}_1]$, the entire lot is in the system. At time FT_1, the first sublot is completed and leaves the system, thereby decreasing the WIP level to $(U - s)$. This process continues as the sublots get completed one by one. The average WIP, denoted by $\overline{\text{WIP}}$, is the area under the above curve weighted by the completion time of the last sublot. This can be expressed as follows:

$$\begin{aligned}\text{WIP} = U\,(\text{FT}_1) + (U - s)\,(\text{FT}_2 - \text{FT}_1) + (U - 2s)\,(\text{FT}_3 - \text{FT}_2) + \cdots \\ + (U - (n-1)s)\,(\text{FT}_n - \text{FT}_{n-1})\,.\end{aligned}$$

But, $\text{FT}_i - \text{FT}_{i-1} = sp_{\max}, \forall i = 1, \ldots, n$. Therefore,

$$\begin{aligned}\text{WIP}^{\text{LS}} &= Us\sum_{i=1}^{m} p_k + sp_{\max}[U(n-1) - s - 2s - \cdots - (n-1)s] \\ &= Us\sum_{i=1}^{m} p_k + sp_{\max}\left[U(n-1) - \frac{sn(n-1)}{2}\right] \\ &= Us\left[\sum_{i=1}^{m} p_k + p_{\max}\frac{(n-1)}{2}\right].\end{aligned}$$

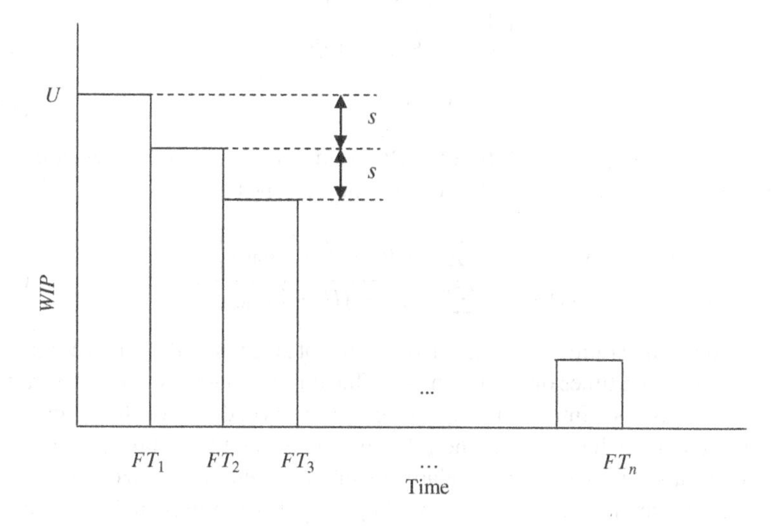

FIGURE 1.13. WIP in a flow shop under lot streaming

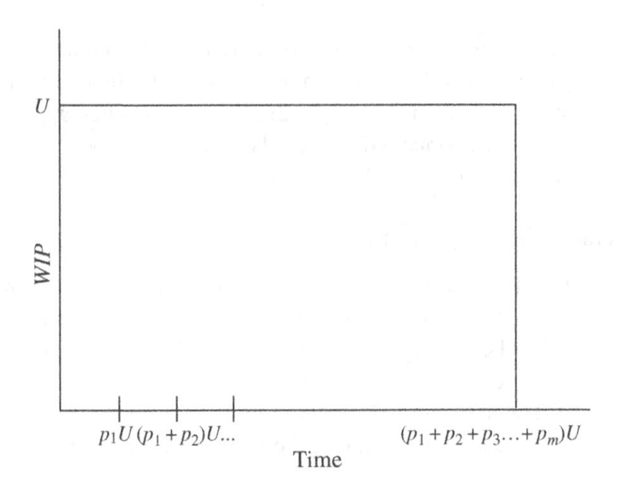

FIGURE 1.14. WIP in a flow shop under no lot streaming

FT_n is nothing but the makespan under lot streaming, given by (1.1). Dividing WIP by this expression, we have

$$\overline{\text{WIP}}^{\text{LS}} = \frac{U\left\{\sum_{k=1}^{m} p_k + \frac{(n-1)}{2} p_{\max}\right\}}{\sum_{k=1}^{m} p_k + (n-1)p_{\max}}. \tag{1.7}$$

When lot streaming in not used, the WIP remains at U throughout, until the last sublot has left the system. The graph of WIP vs. time in this case is as shown in Fig. 1.14. Therefore,

$$\overline{\text{WIP}} = \frac{U \sum_{k=1}^{m} p_k U}{U \sum_{k=1}^{m} p_k},$$

or,

$$\overline{\text{WIP}} = U. \tag{1.8}$$

As in the previous cases, the ratio of $\overline{\text{WIP}}^{\text{LS}}$ and $\overline{\text{WIP}}$ is minimized when n is as large as possible, i.e., $n^* = U$. Hence, the ratio becomes

$$\frac{\overline{\text{WIP}}^{\text{LS}}}{\overline{\text{WIP}}} = \frac{\left\{\sum_{k=1}^{m} p_k + \frac{(U-1)}{2} p_{\max}\right\}}{\sum_{k=1}^{m} p_k + (U-1)p_{\max}}. \tag{1.9}$$

Note that the above ratio approaches 1 when the lot size is smaller when compared with the processing times on the machines. This implies that lot streaming is not as beneficial in this case. Intuitively, this makes sense since, due to the larger processing times and a smaller lot size, the difference between the completion times of the first and last sublots becomes relatively small. When the lot size is relatively large, the ratio approaches 1/2, thereby, implying that the maximal reduction in the average WIP, in the presence of lot streaming, is 50%.

1.5.2 $m/N/E/II/CV/\{C_{\max}, MFT, \overline{WIP}$

We now consider a more complicated case of multiple lots. We still retain the assumption of negligible transfer and setup times. We use the following additional notation

$$\mathrm{BN} \equiv \arg\left\{\max_{1\le k\le m}\left(\sum_{i=1}^{N} U_i\, p_{ik}\right)\right\},$$

designates the bottleneck machine, and $[i]$ denotes the position of lot i in the sequence.

1.5.2.1 Makespan ($C_{\max}$)

For the sake of simplicity, we assume that all the lots have a common bottleneck and this bottleneck is dominant, i.e.,

$$p_{i\mathrm{BN}} \ge p_{ij}, \quad \forall j \ne \mathrm{BN},\ \forall i = 1,\ldots,N.$$

When intermingling of the lots is not permitted, and unit-size sublots are used, the expression to determine makespan resulting from the processing of N lots is as follows:

$$C_{\max}^{\mathrm{LS}} = \sum_{k=1}^{\mathrm{BN}-1} p_{[1]k} + \sum_{i=1}^{N} U_{[i]}\, p_{[i]\mathrm{BN}} + \sum_{k=\mathrm{BN}+1}^{m} p_{[N]k} + I^{\mathrm{LS}},$$

where I^{LS} is the sum of idle times appearing in between the lots on the bottleneck machine and the transfer lags of the last sublot on machines $\mathrm{BN}+1$ to m, under lot streaming.

Similarly, for the case when lot streaming is not used, the makespan expression can be written as

$$C_{\max} = \sum_{k=1}^{\mathrm{BN}-1} U_{[1]}\, p_{[1]k} + \sum_{i=1}^{N} U_{[i]}\, p_{[i]\mathrm{BN}} + \sum_{k=\mathrm{BN}+1}^{m} U_{[N]}\, p_{[N]k} + I,$$

where I is the total of idle times appearing in between the lots on the bottleneck machine when lot streaming is not used. Hence, the ratio becomes

$$\frac{C_{\max}^{\mathrm{LS}}}{C_{\max}} = \frac{\sum_{k=1}^{\mathrm{BN}-1} p_{[1]k} + \sum_{i=1}^{N} U_{[i]}\, p_{[i]\mathrm{BN}} + \sum_{k=\mathrm{BN}+1}^{m} p_{[N]k} + I^{\mathrm{LS}}}{\sum_{k=1}^{\mathrm{BN}-1} U_{[1]}\, p_{[1]k} + \sum_{i=1}^{N} U_{[i]}\, p_{[i]\mathrm{BN}} + \sum_{k=\mathrm{BN}+1}^{m} U_{[N]}\, p_{[N]k} + I}.$$

If we further assume that:

1. The lots are sequenced under lot dominance, such that no idle time is present in between the processing of lots on the bottleneck machine, i.e.,

$$\begin{aligned} &p_{[i]k} \ge p_{[i+1]k-1}, \quad \forall i,\ i = 1,\ldots,N \text{ and } k = 1,\ldots,\mathrm{BN}\\ &\text{and}\\ &p_{[i]k} \le p_{[i+1]k-1}, \quad \forall i,\ i = 1,\ldots,N \text{ and } k = \mathrm{BN}+1,\ldots,m. \end{aligned} \tag{1.10}$$

2. Lot sizes are equal, i.e.,

$$U_i = U, \quad \forall i = 1, \ldots, N. \tag{1.11}$$

Then, the above ratio reduces to

$$\frac{C_{\max}^{\mathrm{LS}}}{C_{\max}} = \frac{U \sum_{i=1}^{N} p_{[i]\mathrm{BN}} + \sum_{k=1}^{\mathrm{BN}-1} p_{[1]k} + \sum_{k=\mathrm{BN}+1}^{m} p_{[N]k}}{U\left(\sum_{k=1}^{\mathrm{BN}-1} p_{[1]k} + \sum_{i=1}^{N} p_{[i]\mathrm{BN}} + \sum_{k=\mathrm{BN}+1}^{m} p_{[N]k}\right)}. \tag{1.12}$$

Note that condition (1) results in $I^{\mathrm{LS}} = I = 0$. When the lot sizes are sufficiently large, such that

$$U \sum_{i=1}^{N} p_{[i],\mathrm{BN}} \gg \left(\sum_{k=1}^{\mathrm{BN}-1} p_{[1]k} + \sum_{k=\mathrm{BN}+1}^{m} p_{[N]k} \right),$$

the ratio becomes

$$\frac{C_{\max}^{\mathrm{LS}}}{C_{\max}} = \frac{U \sum_{i=1}^{N} p_{[i]\mathrm{BN}}}{U\left(\sum_{k=1}^{\mathrm{BN}-1} p_{[1]k} + \sum_{i=1}^{N} p_{[i]\mathrm{BN}} + \sum_{k=\mathrm{BN}+1}^{m} p_{[N]k}\right)},$$

which, under the assumption of equal processing times for all items of all lots, i.e., $p_{i,k} = p, \forall i, k$, becomes

$$\frac{C_{\max}^{\mathrm{LS}}}{C_{\max}} = \frac{N}{[N + (m-1)]}.$$

Once again, note that this expression is purely dependent on problem parameters. Although we have assumed that $I = 0$, it is not unusual to have $I^{\mathrm{LS}} \ll I$, which would imply that significant reduction in makespan can be achieved via lot streaming. Also, the above expression indicates that the gains from lot streaming reduce as the number of lots, N, increases.

When the lot sizes and number of lots are relatively small when compared with the number of machines, i.e.,

$$U \sum_{i=1}^{N} p_{[i],\mathrm{BN}} \ll \left(\sum_{k=1}^{\mathrm{BN}-1} p_{[1]k} + \sum_{k=\mathrm{BN}+1}^{m} p_{[N]k} \right),$$

the ratio becomes

$$\frac{C_{\max}^{\mathrm{LS}}}{C_{\max}} = \frac{\sum_{k=1}^{\mathrm{BN}-1} p_{[1]k} + \sum_{k=\mathrm{BN}+1}^{m} p_{[N]k}}{U\left(\sum_{k=1}^{\mathrm{BN}-1} p_{[1]k} + \sum_{i=1}^{N} p_{[i]\mathrm{BN}} + \sum_{k=\mathrm{BN}+1}^{m} p_{[N]k}\right)}.$$

When all the processing times are equal, this reduces to

$$\frac{C_{\max}^{\mathrm{LS}}}{C_{\max}} = \frac{m-1}{U[N + (m-1)]}.$$

This expression suggests even greater improvements due to lot streaming when compared with the earlier expression. The above expressions reduce to $1/m$ and $1/U$, when $N = 1$, which is consistent with the expressions derived earlier.

1.5.2.2 Mean Flow Time (MFT)

The mean flow time of the lots can be obtained by dividing the sum of sublot mean flow times, derived earlier, by the number of lots, i.e.,

$$\mathrm{MFT}^{\mathrm{LS}} = \frac{\sum_{i=1}^{N} \mathrm{MFT}_i^{\mathrm{LS}}}{N}. \tag{1.13}$$

The (sublot) mean flow time of the lot scheduled first in the sequence, with unit-size sublots, is given as

$$\mathrm{MFT}_{[1]}^{\mathrm{LS}} = \frac{\left(U_{[1]} - 1\right) p_{[1]\mathrm{BN}}}{2} + \sum_{k=1}^{m} p_{[1]k}.$$

If we assume that the bottleneck is dominant and the lots are sequenced under lot dominance (condition in (1.10) above), then the MFT of a lot scheduled in position i can be obtained as

$$\mathrm{MFT}_{[i]}^{\mathrm{LS}} = \frac{\left(U_{[i]} - 1\right) p_{[i]\mathrm{BN}}}{2} + \sum_{k=1}^{m} p_{[i]k}.$$

However, the above expression does not account for the fact that lot i starts processing at $\sum_{j=1}^{i-1} U_{[j]} p_{[j]1}$. Modifying the above expressions accordingly, we have

$$\mathrm{MFT}_{[i]}^{\mathrm{LS}} = \sum_{j=1}^{i-1} U_{[j]} p_{[j]1} + \frac{\left(U_{[i]} - 1\right) p_{[i]\mathrm{BN}}}{2} + \sum_{k=1}^{m} p_{[i]k} + I_{[i]}^{\mathrm{LS}},$$

where $I_{[i]}^{\mathrm{LS}}$ is the slack time present in the processing of lot i on the machines before BN.

Substituting the above expression in (1.13), we have

$$\mathrm{MFT}^{\mathrm{LS}} = \frac{\sum_{i=2}^{N} \sum_{j=1}^{i-1} U_{[j]} p_{[j]1} + \sum_{i=1}^{N} \sum_{k=1}^{m} p_{[i]k} + \sum_{i=1}^{N} \frac{\left(U_{[i]}-1\right)}{2} p_{[i]\mathrm{BN}} + \sum_{i=1}^{N} I_{[i]}^{\mathrm{LS}}}{N}. \tag{1.14}$$

Similarly, the sublot MFT, when lot streaming is not used, can be expressed as

$$\mathrm{MFT}_{[i]}^{\mathrm{LS}} = \sum_{j=1}^{i-1} U_{[j]} p_{[j]1} + \sum_{k=1}^{m} U_{[i]} p_{[i]k} + I_{[i]},$$

where $I_{[i]}$ is the counterpart of $I_{[i]}^{\mathrm{LS}}$.

Hence, we have the lot MFT as follows:

$$\begin{aligned}\mathrm{MFT} &= \frac{\sum_{i=1}^{N} \mathrm{MFT}_i^{\mathrm{LS}}}{N} \\ &= \frac{\sum_{i=2}^{N} \sum_{j=1}^{i-1} U_{[j]} p_{[j]1} + \sum_{i=1}^{N} \sum_{k=1}^{m} U_{[i]} p_{[i]k} + \sum_{i=1}^{N} I_{[i]}}{N}.\end{aligned} \tag{1.15}$$

After taking the ratio of (1.14) and (1.15),

$$\frac{\mathrm{MFT}^{\mathrm{LS}}}{\mathrm{MFT}} = \frac{\sum_{i=2}^{N}\sum_{j=1}^{i-1} U_{[j]}p_{[j],1}+\sum_{i=1}^{N}\sum_{k=1}^{m} p_{[i],k}+\sum_{i=1}^{N}\frac{(U_{[i]}-1)}{2}p_{[i],\mathrm{BN}}+\sum_{i=1}^{N} I_{[i]}^{\mathrm{LS}}}{\sum_{i=2}^{N}\sum_{j=1}^{i-1} U_{[j]}p_{[j],1}+\sum_{i=1}^{N}\sum_{k=1}^{m} U_{[i]}p_{[i],k}+\sum_{i=1}^{N} I_{[i]}}. \tag{1.16}$$

When all lot sizes are the same and the processing time per item is identical on all the machines for lots, then (1.16) reduces to

$$\frac{\mathrm{MFT}^{\mathrm{LS}}}{\mathrm{MFT}} = \frac{\left(\frac{N(N-1)}{2}Up\right)+(Nmp)+\left(N\frac{(U-1)}{2}p\right)}{\left(\frac{N(N-1)}{2}Up\right)+(UNmp)} = \frac{(UN)+2m-1}{U(N+2m-1)}.$$

As the lot size increases, the above ratio approaches $N/(N+2m-1)$, which is significant as long as N is relatively small. However, with an increment in the value of N, the value of the ratio approaches 1, as in the case for the makespan criterion.

1.5.2.3 Average WIP Level ($\overline{\mathrm{WIP}}$)

The depletion of the WIP in this case is as shown in Fig. 1.15.

In the analysis that follows, we restrict ourselves to the situation in which the sublots are of unit size, lots are of the same size and every item of all the lots requires the same processing time on all the machines. As can be seen from

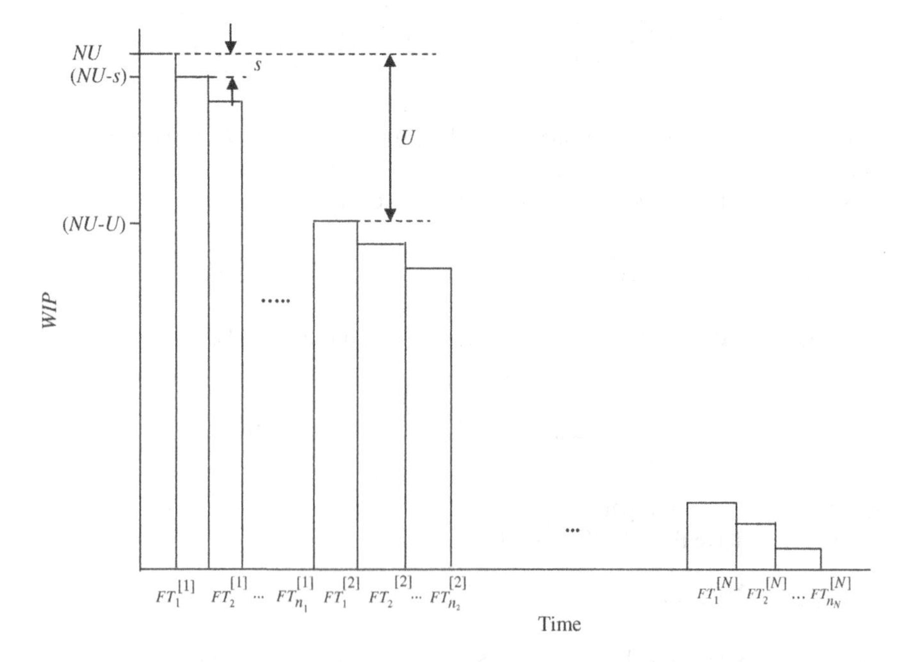

FIGURE 1.15. Depletion of WIP in the presence of multiple lots and lot streaming

Fig. 1.15, at time 0, there are N lots (or NU items) waiting to be processed. At time $\mathrm{FT}_1^{[1]}$, the first sublot of the lot sequenced first is completed and leaves the system, thereby, reducing the WIP to $(NU - s)$ items. As the sublots of the first lot are completed, they leave the system, and at time $\mathrm{FT}_{n_1}^{[1]}$, the sublot n_1 of lot 1 leaves the system. The WIP at this point is $(NU - U)$, since the first lot has been completed in entirety. This process is repeated for all the other lots. The WIP, which is represented by the area under the curve, can be expressed as

$$\begin{aligned}\mathrm{WIP}^{\mathrm{LS}} &= NU\left(\mathrm{FT}_1^{[1]}\right) + (NU - s)\left(\mathrm{FT}_2^{[1]} - \mathrm{FT}_1^{[1]}\right) + (NU - 2s)\left(\mathrm{FT}_3^{[1]} - \mathrm{FT}_2^{[1]}\right)\\ &\quad + \cdots + (NU - (n-1)s)\left(\mathrm{FT}_{n_1}^{[1]} - \mathrm{FT}_{n_1-1}^{[1]}\right)\\ &\quad + \cdots + (NU - (N-1)U - s)\left(\mathrm{FT}_2^{[N]} - \mathrm{FT}_1^{[N]}\right)\\ &\quad + \cdots + (NU - (N-1)U - (n-1)s)\left(\mathrm{FT}_{n_N}^{[N]} - \mathrm{FT}_{n_N-1}^{[N]}\right),\end{aligned}$$

where $\mathrm{FT}_i^{[k]} - \mathrm{FT}_{i-1}^{[k]} = sp, \forall i = 1, \ldots, n_k$ and $\forall k, k = 1, \ldots, N$. Proceeding in a manner similar to the single-lot case (see Sect. 1.5.1.3), this expression can be reduced to

$$NUs\left\{mp + p\frac{(NU-1)}{2}\right\}.$$

Again, the makespan for this case, using (1.1), is $C_{\max}^{\mathrm{LS}} = ps(m + NU - 1)$. Therefore,

$$\overline{\mathrm{WIP}}^{\mathrm{LS}} = \frac{\mathrm{WIP}^{\mathrm{LS}}}{C_{\max}^{\mathrm{LS}}} = \frac{NU\left(m + \frac{(NU-1)}{2}\right)}{m + (NU-1)}.$$

When lot streaming is not used, the depletion of WIP is as shown in Fig. 1.16.

$$\begin{aligned}\mathrm{WIP} &= NU(pmU) + (NU - U)(pU) + \cdots + (NU - (N-1)U)(pU)\\ &= NU^2 p\left\{m + \frac{(N-1)}{2}\right\}.\\ C_{\max} &= Upm + (N-1)Up\\ &= Up(m + (N-1)).\end{aligned}$$

Therefore,

$$\overline{\mathrm{WIP}} = \frac{\mathrm{WIP}}{C_{\max}} = \frac{NU(2m + N - 1)}{2(m + N - 1)}.$$

Hence, the ratio becomes

$$\frac{\overline{\mathrm{WIP}}^{\mathrm{LS}}}{\overline{\mathrm{WIP}}} = \frac{\left(m + \frac{(NU-1)}{2}\right)(m + N - 1)}{\left(m + \frac{(N-1)}{2}\right)(m + NU - 1)}.$$

For a reasonable N and large U, the ratio reduces to

$$\frac{\overline{\mathrm{WIP}}^{\mathrm{LS}}}{\overline{\mathrm{WIP}}} = \frac{1}{2}\frac{m + N - 1}{\left(m + \frac{(N-1)}{2}\right)}.$$

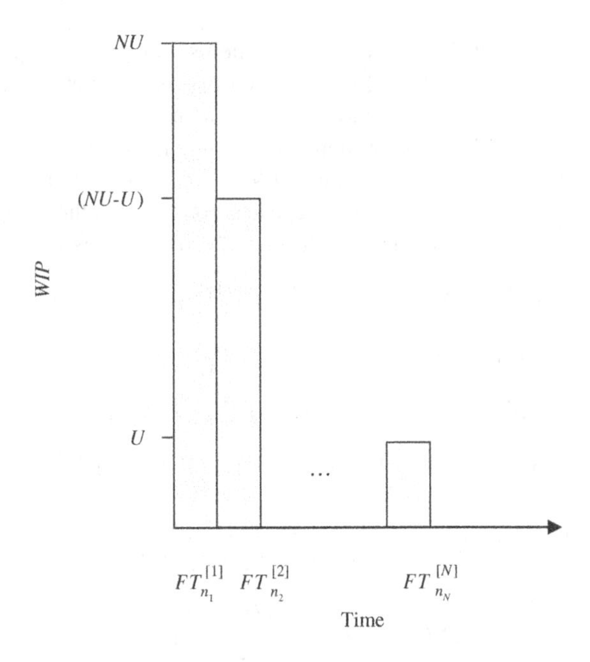

FIGURE 1.16. Depletion of WIP in the presence of multiple lots and no lot streaming

This ratio gives a lower bound of 1/2, when $m \gg N$. This is identical to that obtained for the single-lot case.

1.6 A Brief Historical Perspective

1.6.1 *Evolution of Lot Streaming*

The lot streaming problem, although originally identified by Reiter [30] in 1966, was rediscovered in the late 1980s to early 1990s as a result of attempts to improve the effectiveness of a production system. Materials requirement planning (MRP), which basically provides a time-phased plan for the acquisition of raw materials, their processing on the basis of the bill of materials (BOM), and the procurement and lead times of the end products, was introduced in the 1970s. MRP serves as a central organizer that translates an overall production plan, developed based on demand forecast, into a series of specific steps to achieve the planned production. However, it suffers from the following three major drawbacks:

1. It implicitly assumes that production parameters such as lot sizes and lead times could be determined a priori, external to the system, and kept fixed. Hence, it is difficult for a MRP system to cope with the inherent uncertainties of production parameters. As a consequence, a MRP system is characterized by high work-in-process (WIP) and long lead times.

2. It ignores the finite capacity constraints and focuses on material flow, thereby, generating schedules that are not feasible.
3. A lot is treated as a single integral entity, i.e., all the operations of a lot are processed on a machine before transferring the lot to another machine.

To take care of (2), the MRP II methods took into account the limited capacity of relevant resources. The 1980s also saw an advent of the just-in-time (JIT) manufacturing philosophy that promoted the concept of minimizing waste, where waste is defined as anything that is not necessary for the manufacturing process or in excess of it, in the presence of dynamic demand changes in a growing "produce-to-order" environment. Implicitly, it took care of the drawbacks of an MRP-based system because of its following two inherent components. First, by using "Kanbans" to transfer lots, it limits WIP in between the machines, thereby, minimizing waste. Second, it supports overlapping the processing of a lot over multiple machines by requiring a one-item-at-a-time transfer between the machines (with setup time on each machine reduced to a minimum). In addition, the machines are required to be configured in accordance with the sequence of processing steps needed to produce a product, to simplify flow. Also, by using the concept of a "pull system" (i.e., a downstream machine withdrawing parts from an upstream machine based on local information), it takes into account availability of finite capacity. However, in spite of allowing the overlapping of operations, it fails to address the question of the optimality of using unit-size sublots given that these might be suboptimal in a majority of production environments where significant amounts of transfer times and setup times are incurred. Another technique known as the *optimized production technology* (OPT) appeared during the same time period. Like JIT, this technique also addresses the issue of reducing waste in a manufacturing system, but it is more focused in that it restricts its attention to the critical resources, i.e., the bottlenecks. It, first, separates the machines into bottleneck and non-bottleneck machines, and then, builds production plans such that the bottleneck is fully utilized. It does so by using large process batches (to eliminate setup costs) and small transfer batches (to reduce inventory carrying costs), thereby, reducing the overall cost incurred while also maximizing throughput. However, it relies on the following inherent assumptions:

1. The sublot sizes and the sequence in which to process the lots are known a priori.
2. Machines experience dissimilar workload and there exists a single bottleneck in the system.
3. The transfer batches are used only at the bottleneck machine and not at other machines.

The presence of long setup times, large production mixture (which affects the repetitiveness of a production system), process variability, and unbalanced workload are detrimental to the success of a JIT-type system implementation. By the late 1980s, a new technique, called constant work-in-process (CONWIP), was introduced to remedy the drawbacks of JIT. The basic idea behind a CONWIP system is to maintain a predetermined level of WIP in the entire system as opposed to maintaining a constant WIP at the machine level as it is proposed by JIT. In addition, a CONWIP system is more flexible than a JIT system since it allows for the simultaneous processing of different types of lots. In other words, the exit of a lot from the system triggers the entry of another lot, which may or may not be identical to the exiting lot. However, CONWIP does not address the sublot sizing or the lot sequencing issues.

1.6.2 *Research in Lot Streaming*

As alluded to above, overlapping the processing of a production lot over several machines (through the use of transfer lots) was promoted by both JIT and OPT in the 1980s. In the meantime, specific issues pertaining to lot streaming were addressed by Szendrovits [33] and Goyal [18]. Szendrovits [33] considered the m/1/E/NI/CV/{Lot Size} problem such that the sum of fixed costs per lot and the inventory holding costs of both the WIP and finished lots is minimized. Goyal [18] extended Szendrovits' work by developing a scheme for obtaining the optimal sublot sizes, which were assumed to be known in Szendrovits' model. Since this beginning, there has been an extensive amount of work reported in the literature that has explored specific issues pertaining to lot streaming and which constitutes its building blocks. This work considers the issue of lot streaming in a flow shop environment and addresses the basic question of how to split a lot into sublots and optimize a performance-related measure. The prominent measure of performance that has been considered is makespan. Our presentation of this work in this book is organized in accordance with the two-machine, three-machine, and the general m-machine flow shop environments.

1.6.2.1 Two-Machine Lot Streaming Problems

Clearly, a single-lot, lot streaming problem only involves determination of the number of sublots and their sizes, while a multiple lot, lot streaming problem involves the simultaneous determination of the number of sublots and sublot sizes for each lot, and the sequence in which to process the lots.

Trietsch and Baker [35] were the first to address the 2/1/C/NI/CV problem and show that the optimal sublot sizes are geometric in nature with a ratio of p_2/p_1. Although this result does not hold when sublot sizes are constrained to be integers, they developed an iterative algorithm to obtain the optimal integer sublot sizes. Sriskandarajah and Wagneur [32], then, addressed the 2/1/C/II/CV/{No-Wait Flow Shop, Lot-Detached Setups} problem and showed that the optimal sublot sizes are geometric in this case also. For the discrete version of the problem, they developed algorithms that utilize a solution to the continuous problem as a starting point. Chen and Steiner [9] have developed approximation algorithms, which generate a makespan value that is within $\min\{p_1, p_2\}$ of the optimal value, when the sublot sizes are integers.

Sen et al. [31] have addressed the $2/1/\{C,V\}/II/CV/\sum_{i=1}^{n} s_i C_{i2}$ problem, where the objective is to minimize the sum of the weighted sublot completion times. The continuous optimal sublot sizes are obtained by solving a quadratic program. Furthermore, it is shown that, under consistent sublot sizes, equal sublots are optimal, if $p_2/p_1 \leq 1$. For the case when $p_2/p_1 > 1$, an algorithm is developed, which gives geometric sublots (with ratio p_2/p_1) up to a sublot ν and same size sublots for the remaining. For the criterion under consideration, having variable-size sublots on the two machines might give a better solution since the completion time of each sublot contributes to the objective function value. In fact, the mean sublot completion time is minimized by equal sublots on both the machines if $p_2/p_1 \leq 1$, and by geometric sublots on the first machine and equal sublots on the second machine if $p_2/p_1 > 1$.

Bukchin et al. [5] addressed the $2/1/C/II/CV/\{\sum_{i=1}^{n} s_i C_{i2}$, Sublot-Attached Setup$\}$ problem. They developed a nonlinear formulation of the problem and a solution procedure based on the single machine bottleneck (SMB) property, which designates such a machine to be the one that has no idle time in between the processing of the sublots throughout the production process. The solution procedure first checks if the optimal solution has a unique bottleneck. If so, then it uses an algorithm to obtain the optimal solution. Otherwise, it uses another algorithm to find an approximate solution. Both of these algorithms are based on solving relaxed mathematical formulations of the problem under consideration. Their experimental results show that a solution, which minimizes the makespan, is not necessarily optimal for the criterion of flow time minimization.

Baker [2] and Vickson and Alfredsson [37] separately addressed the 2/N/C/II/DV problem. Clearly, in the absence of setup times, transfer and removal times, and intermingling amongst sublots of different lots, unit-size sublots are optimal. The optimal sequence in which to process the N lots is determined by using a modification of Johnson's algorithm [19]†, which relies on the concept of run-in and run-out times. Similarly, the 2/N/C/II/DV/{Lot-Detached/Attached Setup Times} problem with unit-size sublots [2,8] has been solved to optimality by using a modified version of Johnson's algorithm as well.

Vickson [36] has addressed the 2/N/C/II/{CV,DV}/{Lot-Detached Setups, Sublot Transfer Times} problem and has shown that the sublot sizing and lot sequencing problems are independent in this case. The sublot sizing problem is equivalent to the one without transfer times and its LP formulation has a special structure, which can be exploited to derive optimal solutions. The lot sequencing problem can be solved using Johnson's algorithm [19]. Centinkaya [7] has addressed the 2/N/C/II/CV/{Lot-Detached Setups and Removal Times} problem and has shown that the sublot sizing and lot sequencing problems are independent and the optimal sublot sizes are geometric. The optimal sequences of lots are determined using a modification of Johnson's algorithm [19].

† We use the terms 'Johnson's algorithm' and 'Johnson's rule' interchangeably in the sequel.

For the 2/N/C/II/{CV,DV}/{No-Wait, Lot-Detached Setup} problem, Sriskandarajah and Wagneur [32] have shown that the sublot sizing and lot sequencing problems are independent, the optimal continuous sublot sizes are geometric, and the optimal sequence for processing the lots can be obtained by using an algorithm proposed by Gilmore and Gomory [15] for a special type of the traveling salesman problem (TSP).

Kalir and Sarin [24] have addressed the 2/N/E/II/CV/{Sublot-Attached Setups} problem. When sublot sizes for all lots are the same, i.e., $s_i = s, \forall i = 1, \ldots, N$, then their algorithm iterates over all possible values of s. At each iteration, it uses the value of s on hand to compute the sublot sizes, which are then sequenced optimally using a modified Johnson's algorithm, with the processing time of a sublot being equal to the sum of the sublot setup and processing times. When the sublot sizes of different lots are not the same, the corresponding algorithm consists of a construction phase and improvement phase. The former optimally solves a two-machine single-lot problem, individually, for all the lots, and then, sequences them using a modified Johnson's algorithm. The improvement phase, then, reoptimizes the sublot sizes of each lot; however, the sequence is not changed.

A schematic presentation of the work reported on the two-machine lot streaming problems is given in Fig. 1.17.

1.6.2.2 Three-Machine Lot Streaming Problems

The three-machine lot streaming problem is of interest in its own right and also forms a stepping stone for the study of a more general m-machine lot streaming problem. A schematic presentation of the work reported on the three-machine lot streaming problem is given in Fig. 1.18. Trietsch and Baker [35] have addressed the 3/1/C/{NI,II}/CV problem and obtained optimal sublot sizes by solving a LP formulation, which minimizes the completion time of the last sublot on machine 3 or, in other words, the idle time appearing in between the processing of sublots on machine 3. They also present an algorithm for solving the discrete version of the problem. The no-idling and integer-size sublots can be obtained by making appropriate changes to the LP formulation. Glass et al. [16] have addressed the same problem but have obtained optimal sublot sizes in polynomial time by analyzing separately the structural properties of a network representation of the lot streaming problem when the following relationships among the processing times hold: $(p_2)^2 - p_1 p_3 > 0$, $(p_2)^2 - p_1 p_3 < 0$, and $(p_2)^2 = p_1 p_3$.

Trietsch and Baker [35] have also addressed the 3/1/V/{NI,II}/CV version of the problem and have shown that the optimal schedule must be no-wait. If $(p_2)^2 \leq p_1 p_3$, then the optimal continuous sublot sizes are geometric with a ratio of $(p_2 + p_3)/(p_1 + p_2)$, but this does not extend to the case of integer-size sublots. However, if $(p_2)^2 > p_1 p_3$, then the three-machine problem can be decomposed into two two-machine problems each of which can be solved to optimality by using the two-machine lot streaming procedure.

Chen and Steiner [11] have addressed the 3/1/C/II/CV/{Lot-Detached Setup} version of the problem and have developed closed-form expressions to determine

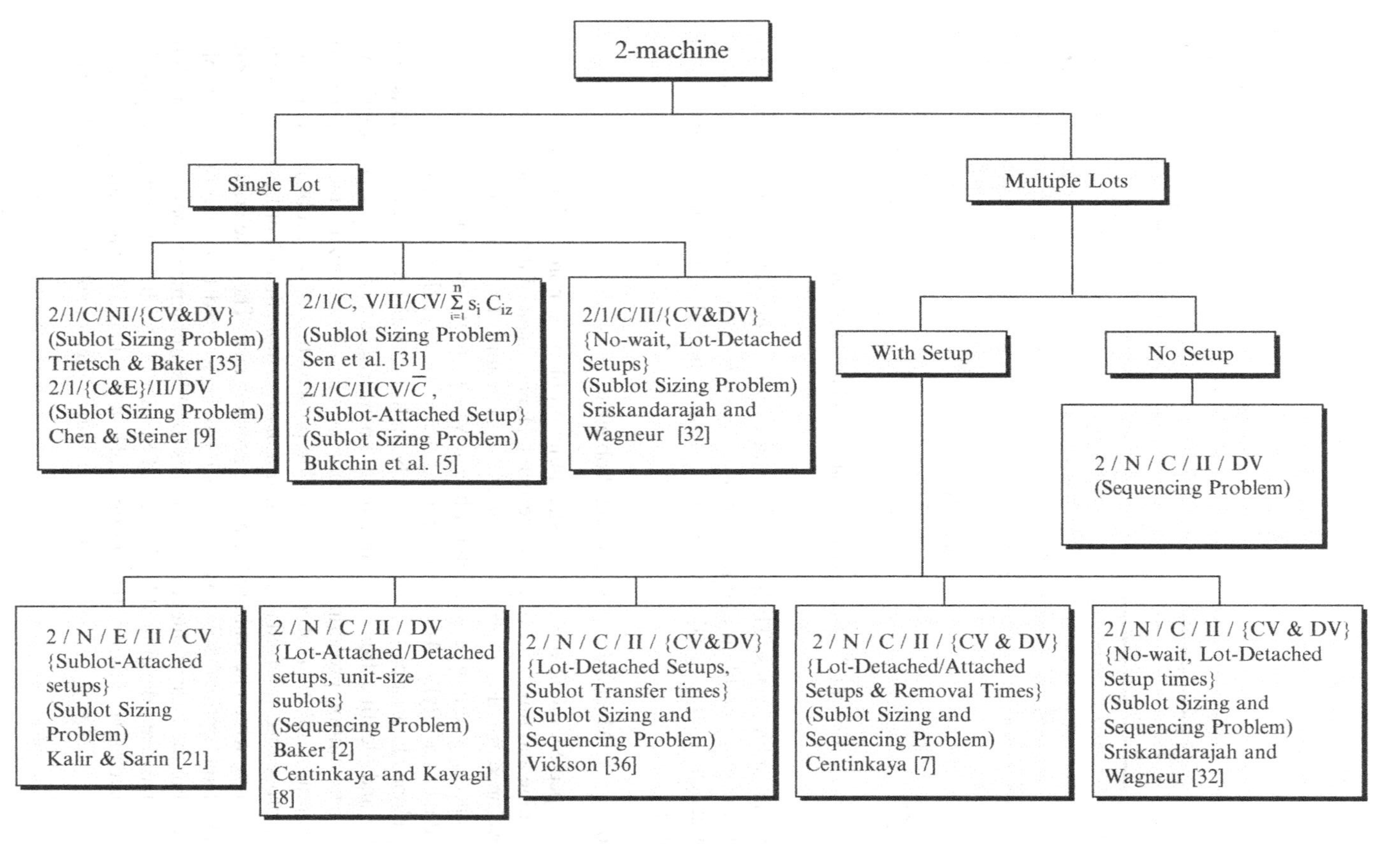

FIGURE 1.17. Overview of two-machine lot streaming problems

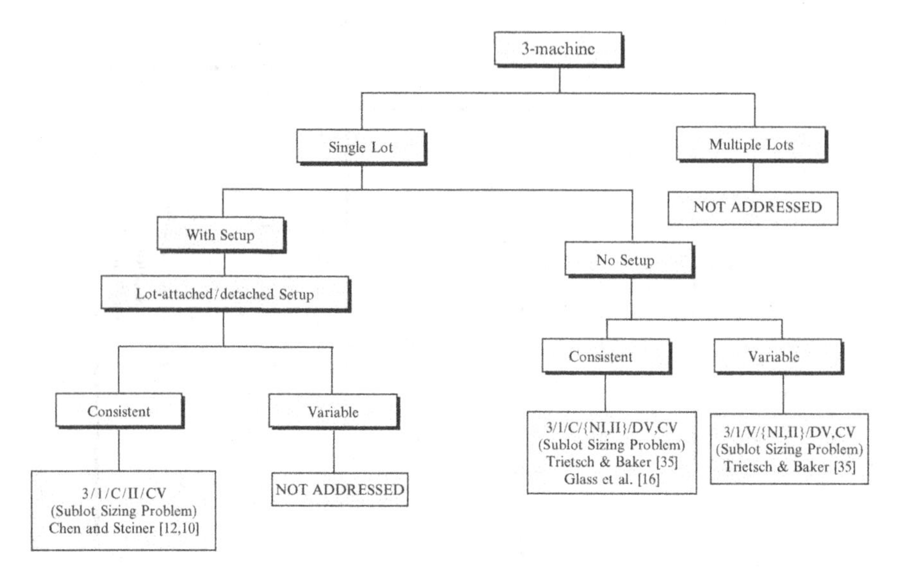

FIGURE 1.18. Overview of three-machine lot streaming problem

optimal sublot sizes. Their analysis relies on the network representation of the lot streaming problem like the one used by Glass et al. [16]. In the presence of lot-attached setups [10], however, they obtain optimal sublot sizes under special cases.

1.6.2.3 *m*-Machine Lot Streaming Problems

As expected, the m-machine lot streaming problem is more difficult than its two-machine and three-machine counterparts. Hence, most of the solution procedures developed for this problem are heuristic in nature. However, for some special instances, optimum seeking algorithms have also been developed.

Baker and Pyke [4] have addressed the m/1/C/{NI,II}/CV problem and have developed heuristic procedures that rely on the consideration of two sublots at-a-time and identification of a bottleneck machine. Glass and Potts [17] developed another solution procedure for the same problem by analyzing the properties of the optimal solution through a network representation of the problem. Their solution procedure, referred to as the *relaxation algorithm*, first identifies the dominant machines, which helps in reducing the problem size. Then, the critical path structure of the optimal solution is characterized and used to derive optimal sublot sizes. For the case of integer-size sublots for this problem, an approximation algorithm has been developed by Chen and Steiner [11]. Their algorithm assigns the fractional parts of the continuous optimal solution to the first u noninteger sublots, where $u = U - \sum_{j=1}^{n} \lfloor s_j \rfloor$, or over the entire range of sublots in a more balanced fashion.

Topaloglu et al. [34] have addressed the m-machine single-lot problem with consistent sublot sizes but with only two sublots on each machine, for the objective of minimizing the sum of the weighted sublot completion times. The feasible region of the problem under consideration is shown to be convex. Their first algorithm searches for the optimum solution by moving from one candidate point to an adjacent one. The second algorithm finds the optimal solution by performing a bisection search procedure on the candidate optimal points.

Kumar et al. [26] considered the m/1/C/II/{CV,DV}/{No-Wait, Lot-Detached Setup} problem. For the single-lot problem, they use a linear programming formulation to obtain continuous and optimal sublot sizes, and propose heuristic procedures to obtain integer-size sublots. Kalir and Sarin [23] have addressed the problem of finding the optimal equal-size sublots in the presence of sublot-attached setups. Their problem reduces to finding the optimal number of sublots n^*, since $s^* = (U/n^*)$. The algorithm that they present sequentially searches along the feasible region, which is convex and is defined by dominant makespan functions corresponding to the machines. Their algorithm finds an optimal solution in polynomial time.

Kalir et al. [25] have further addressed the m-machine single-lot problem with consistent and equal sublot sizes. The problem is to determine an optimal number of sublots that minimizes a weighted sum of the makespan, (sublot) mean flow time, average WIP, sublot-attached setup, and transfer time. The unified cost function so defined is a segmental strictly convex function and the corresponding algorithm searches for the minimizing solution over each segment of the function. It finds an optimal solution if it occurs at an intersection point, else it finds an approximate solution based on a closed-form expression for n^*.

The sequencing problem in the m-machine multiple-lot scenario has also been addressed by Kalir and Sarin [22]. They develop a sequencing heuristic, which attempts to maximize the time buffer prior to the bottleneck machine, thereby, minimizing potential bottleneck idleness, while also looking ahead to sequence lots with large remaining processing times early in the schedule.

Unlike the problems described above, where the objective is to find sublot sizes of lots such that the makespan is minimized, the problem of finding an optimal lot size (i.e., number of items in a lot so as to minimize the total cost for processing a single batch having a constant deterministic demand over an infinite time horizon) has been addressed by Ramasesh et al. [29]. The total cost is made up of sublot transfer times, waiting or delay times, which are independent of the lot size and setup time on each machine. The model considers equal/unequal production rates on different machines and assumes that sublot sizes are equal and the number of sublots is given.

Kumar et al. [26] have considered the m/N/C/II/{CV,DV}/{No-Wait, Lot-Detached Setup} problem and have shown that the lot sequencing and sublot sizing problems are not independent of each other, unlike the two-machine case. Their heuristic procedure for the sequencing of lots relies on the solution of a TSP, once the integer-size sublots for each lot have been obtained by the application of a heuristic developed for the single-lot case. They also develop a genetic algorithm

(GA) for the problem of simultaneously determining sublot sizes and the sequence in which to process the lots. Although the solution quality of GA is good, its computational requirements are reported to be high.

Yoon and Ventura [40] have addressed the problem of minimizing the mean absolute deviation of lot completion times from their due dates. For a given sequence of lots to be processed in a no-wait flow shop, they develop LP formulations to obtain the optimal sublot sizes for each lot when buffers between successive machines have infinite or finite capacity, sublots are equal or consistent, and when the flow shop is no-wait. The initial sequence is obtained by applying any of the following rules: EDD, smallest slack time on last machine, smallest overall slack time (OSL), or smallest overall weighted slack time. These sequences are improved upon by any of the following neighborhood search mechanisms: adjacent pairwise interchange of lots, nonadjacent pairwise interchange of lots (NAPI), selecting two lots and moving a lot in front/back of the other lot. Their experimental results show that the smallest overall slack time rule in conjunction with the nonadjacent lot interchange heuristic gives the best solutions. For the case of infinite buffer capacity and equal sublot sizes, they developed a hybrid genetic algorithm (HGA), which works in conjunction with OSL/NAPI to obtain a better sequence.

Dauzere-Peres and Lasserre [13] have developed an iterative heuristic consisting of a sublot sizing and lot sequencing procedure for the m-machine, n-lot, lot streaming problem but for a job shop environment and with no intermingling of sublots. The continuous optimal sublot sizes are obtained using a LP formulation, and these are, then, used to obtain integer-size sublots using a rounding procedure. The sequencing procedure treats each sublot as a separate job and sequences them so as to minimize the makespan using a modified shifting bottleneck procedure (SBP) presented in [14], which is an improved version of the original SBP of Adams et al. [1].

Baker and Jia [3] have conducted general experimental studies to determine the impact that various types of sublots may have in a flow shop environment. For the case of equal sublots, they show that $M^{E}/M^{C} < 1.53$ (where M^{E} and M^{C} are the makespan values for the case of equal and consistent sublot sizes, respectively). They also develop optimal lot streaming policies and improvement bounds, which are useful in determining appropriate number of sublots for the two-machine problem. Their experimentation has also shown that, for a three-machine problem, if machine 2 is dominant, the no-idling constraint has no impact since the optimal solution has no idle time, the consistent sublot size constraint has an impact of about 10% while the impact of the equal sublot size constraint is a little higher. Conversely, when machine 2 is dominated, the consistent sublot constraint has no impact whereas the no-idling constraint may have an impact of 50% or more.

Figure 1.19 gives a schematic presentation of the work reported on the m-machine lot streaming problem.

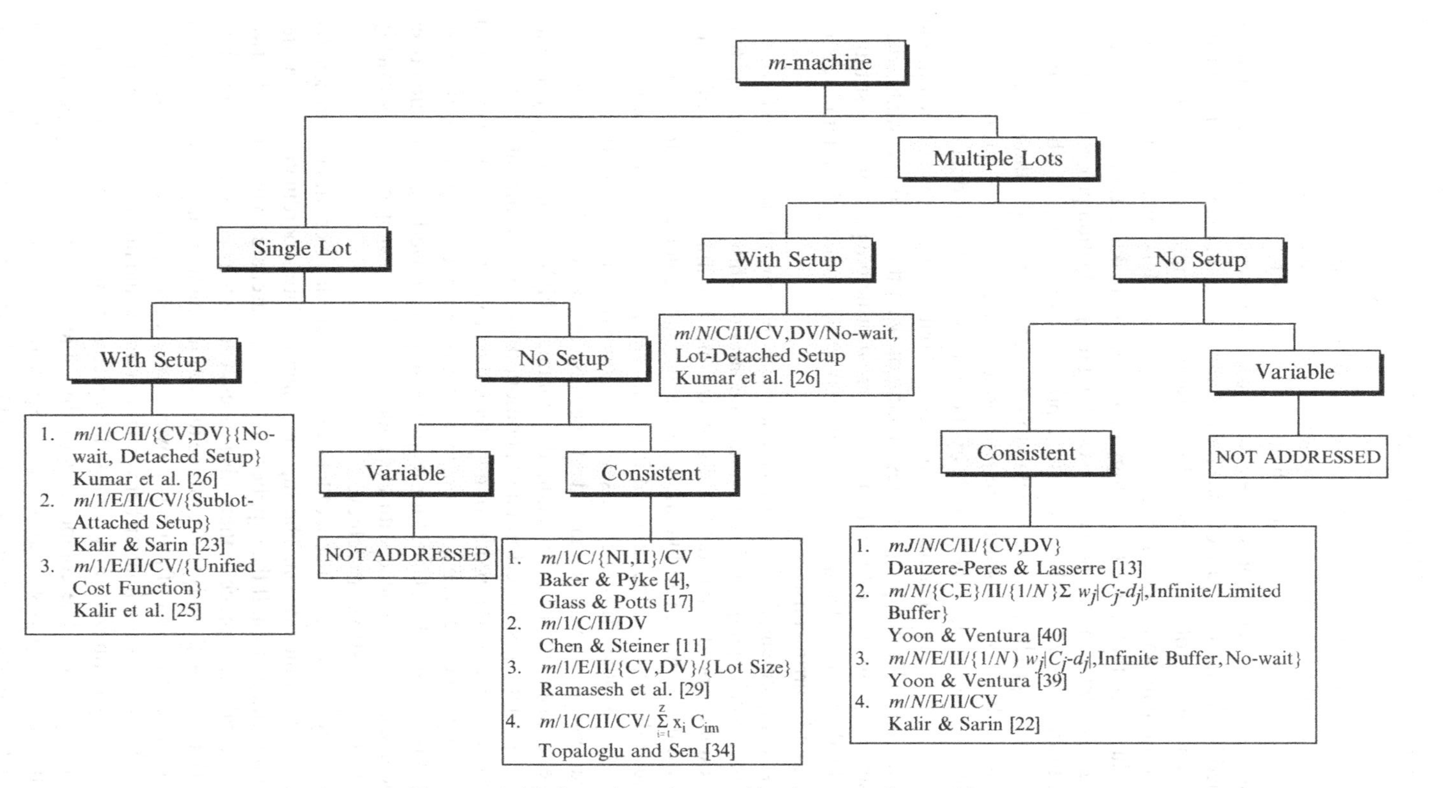

FIGURE 1.19. Overview of m-machine lot streaming problem

1.7 Applications of Lot Streaming

As we have noted in Sect. 1.6.1, the concept of lot streaming has evolved from the use of transfer batches in moving jobs from one machine to another. These transfer batches are smaller in size than a processing batch and help in reducing WIP and cycle time. This concept lies at the heart of one of the five lean principles identified by Womack and Jones [38] for reducing waste in manufacturing (or for the implementation of "lean thinking"). These five principles are value, the value stream, flow, pull, and perfection. The *value* principle refers to the utility or worth of a product (and that of the technology that goes into making it) to the customer. The *value stream* principle pertains to mapping the production process and determining those steps where the value of the product, as perceived by the customer, is added to the product; the idea being to eliminate, if possible, those steps that do not add value to the product. The *flow* principle refers to an unobstructive (or continuous) processing of the product by the system, i.e., the product is processed continuously from one machine to another without having to be batched at the machine. The *pull* principle promotes making of what the customer wants when the customer wants it. And finally, the *perfection* principle pertains to instituting the right values (those that the customer wants) in the product with minimal effort and without incurring mistakes.

The key implementation strategy behind the *flow* principle is to transform the traditional "batch and queue" system into a continuous flow system, the one that involves smaller production lots, ideally of size one. The implementation of such a system would require, in all likelihood, a drastic change in the machinery needed to produce the product, the one that incurs no setup for processing one product (or lot) after another. However, in several production (or fabrication) processes, the occurrence of a setup is unavoidable because of the downloading of requisite software, adjustments of tools required to maintain high product quality, installation of special tools to perform an operation, or because of the introduction of new products. For such a production system, the production lot size (also called *sublot*) need not be one. The question, then, is to determine the optimal sublot size (or sizes) in which to break a lot for its processing in an overlapping fashion over the machines, in accordance with the underlying *flow* principle. This is exactly what lot streaming strives to achieve.

By its very nature, the concept of lot streaming can play a central role in achieving supply chain integration. The supplier–manufacturer coordination lies at the heart of this integration. If the supplier and manufacturer are considered to be two steps of a production process, then this process can be viewed as a two-stage flow shop. The problem, then, is to determine the optimal number of sublots and sublot sizes for processing a lot of jobs in an overlapping fashion at each of these stages (facilities). A better coordination between the supplier and manufacturer can, then, be obtained by optimizing a system-wide objective function, which can be time-based or cost-based. The components of this objective function may pertain to production and setup as well as to the transportation of sublots from the supplier to the manufacturer, and then, from the manufacturer to the customer. The same concept can be extended for coordinating production along a series of facilities or to a network of facilities for a more general supply chain environment.

Mass customization constitutes another potential application area of lot streaming. Mass customization promotes the idea of postponing the final assembly step, which differentiates a product from another, until the very last stage of production. A successful implementation of this concept requires an appropriate redesign of the product, now composing of separate and standard modules. These modules, once produced in central facilities, are shipped to regional distribution/assembly facilities where they are assembled in response to customer demand. The key motivations behind mass customization are (a) a quick response to customer demand, where the product specifications vary with customers and (b) reduction in system cost (due to production, inventory, and distribution). It also leads to the designing of standard modules, which help in achieving customization of products. Application of the lot streaming concept to the production of these modules can help in their timely availabilities for assembly at the final stage. Also, the final stage is an assembly process dedicated to assembling different batches of products with a batch containing customer-specified modules. This assembly process is suitable for the application of lot streaming in its own right.

In a high-tech environment, especially pertaining to semiconductor manufacturing and related areas, a constant shrinkage of device sizes has led to the management of microscopic dimensions with high precision. It is, therefore, imperative to continuously monitor the production process to determine whether it is in control. Lot streaming is particularly helpful in this regard since a smaller-size sublot can be used to quickly detect an out-of-control machine, which, thus, helps in avoiding processing of the remaining lot, thereby, improving process yield and reducing rework.

1.8 Chapter Summary

In the prevailing competitive manufacturing environment, it is essential to have production cycle times that are as small as possible. This requirement arises not only because of the desire to get the product to the market as quickly as possible, but also due to the need of holding minimum levels of WIP to reduce tied-up capital. In this chapter, we have introduced a technique called *lot streaming*, which can help in accomplishing this objective. Basically, lot streaming accelerates the flow of a product through a production process.

As a motivation for the forthcoming chapters, we have provided a brief historical perspective of lot streaming and an overview of the work that has been reported in the literature regarding the flow shop lot streaming problems. First, the terminology and concepts necessary for a better understanding of this subject matter are presented. This is followed by an introduction of assumptions that are relevant for the discussion presented in subsequent chapters. Also, the notation that is used in the sequel is described. Next, a scheme for classifying the lot streaming problems is introduced that helps in organizing the work to be presented. A dominance relationship among various lot streaming problems is also presented. To emphasize the importance of lot streaming, analytical expressions of its potential benefits are developed for various problem scenarios. This is followed by an overview of the work on lot streaming for the two-machine, three-machine, and m-machine flow shop environments. Finally, some application areas of lot streaming are described.

2 Generic Mathematical Models for the Flow Shop Lot Streaming Problem

2.1 Introduction

To comprehend the intricacies of a problem situation, it is best to represent it, if possible, as a mathematical model. A mathematical model can also help in possibly identifying some inherent structural properties of the problem and in devising an appropriate algorithm for its solution. Chapter 1 contains a brief review of work on the flow shop lot streaming problems. This work has focused on addressing two-machine, three-machine, and the general m-machine scenarios. In this chapter, we develop some generic mathematical models for the lot streaming problem that encompass all of these scenarios and also that address various features pertinent to lot streaming. We present these models in Sect. 2.2 and also give their illustrations using simple examples. The key features of the models that are presented in this section are summarized in Table 2.1. In Sect. 2.3, we introduce mathematical models for some special cases of the flow shop lot streaming problem that have been presented in the literature.

2.2 Some Generic Mathematical Models for the Flow Shop Lot Streaming Problem

2.2.1 *Notation*

We define the following notation in addition to that presented in Sect. 1.3.2.

Parameters:

RT_{jk}	Removal time of lot j on machine k
FT_j	Fixed transfer time for lot j
VT_j	Variable transfer time per unit for lot j
τ_{jk}	Sublot-attached setup time for a sublot of lot j on machine k
G	A large positive number used to make a constraint redundant

TABLE 2.1. Key features of the mathematical models presented in Sect. 2.2

Section	Number of machine	Number of lots	Setup	Removal time	Transfer time	Intermingling	Wait–No Wait	Sublot type	Intermediate/ No Intermediate Idle Time	Continuous or Discrete Sublots
2.2.2	m	N	Lot–Attached	Yes	Yes	No	Both	C, E	Both	Both
2.2.3	m	N	Lot–Detached	Yes	Yes	No	Both	C, E	Both	Both
2.2.4	m	N	Sublot–Attached	Yes	Yes	Yes	Both	C, E	Both	Both
2.2.5	m	N	Sublot–Detached	Yes	Yes	Yes	Both	C, E	Both	Both
2.2.6	This section addresses the issue of handling variable sublot sizes in the models presented in the above sections									

Variables:

s_{ijk} Sublot size of the ith sublot of lot j on machine k; a generalization of the definition of sublot sizes presented in Sect. 1.3.2

C_{ijk} Completion time of the ith sublot of lot j on machine k; the subscript j is omitted for problems involving a single lot

$$y_{ij} = \begin{cases} 1, \text{ if lot } i \text{ precedes lot } j \\ 0, \text{ otherwise.} \end{cases}$$

2.2.2 *m/N/{C,E,V}/{II,NI}/{CV,DV}/{Lot-Attached Setup and Removal Times, Sublot Transfer Times, No Intermingling}*

The lot streaming problem involving multiple lots deals with the issue of finding the sublot sizes for each lot and the sequence in which to process the lots in order to optimize a performance measure. Here, we consider the objective of minimizing the makespan. However, other performance measures can be conveniently included in the formulations that we present. We make the following assumptions.

1. The sublot transfer times are variable and comprise of two parts, a fixed component, which remains the same for all the sublots of a particular lot and a variable component, which depends on the size of the sublot and is given by $VT_j \cdot s_{ijk}$.
2. The removal times are attached to the last sublot of each lot and are independent of the sequence in which the lots are processed.
3. The number of sublots for all lots is known in advance.

Generic Model 1 (GM1):

Minimize: $C_{\max}$

Subject to:

1. Makespan Constraint:

$$C_{\max} \geq C_{n_j jm} + RT_{jm}, \forall n_j, j = 1, \ldots, N.$$

This constraint captures the makespan $C_{\max}$, which is the largest among the completion times of the last sublots of all the lots on the last machine (m).

2. Item Allocation Constraint:

$$\sum_{u=1}^{n_j} s_{ujk} = U_j, \quad \forall j = 1, \ldots, N, k = 1, \ldots, m.$$

This constraint ensures that the sum of the items in the sublots of a lot, $j (j = 1, \ldots, N)$ that is processed on machine $k (k = 1, \ldots, N)$ must be equal to the total number of items in that lot.

The next two constraints capture the type of sublots involved. Constraint (3) can be used in case we have consistent sublots while Constraints (3) and (4) together, capture the requirement of equal sublot sizes. We consider the case of variable sublot sizes later.

3. **Consistent Sublot Constraint:**

$$s_{ijk} = s_{ij(k+1)}, \quad \forall i = 1, \ldots, n_j, j = 1, \ldots, N, k = 1, \ldots, (m-1).$$

4. **Equal Sublot Constraint:**

$$s_{ijk} = s_{(i+1)jk}, \quad \forall j = 1, \ldots, N, k = 1, \ldots, m.$$

5. **Lot-attached Setup Constraint:**

$$s_{1jk} \geq \Psi, \quad \forall j = 1, \ldots, N, k = 1, \ldots, m,$$

where ψ is the minimum number of items required to perform a setup on any machine. In the presence of lot-attached setups, the setup time is associated with the first sublot of every lot. However, there might be technological constraints on the minimum number of items required to perform a setup. The constraint above ensures that the size of the first sublot of all the lots is greater than ψ, thus ensuring that a setup can always be performed once the first sublot has been transferred to machine k from machine $(k-1)$.

6. **Sublot Size Constraint:**

$$s_{ijk} \geq 0, \quad \forall i = 2, \ldots, n_j, j = 1, \ldots, N, k = 1, \ldots, m.$$

This constraint ensures nonnegative sublot sizes. These may also be restricted to take integer or real (continuous) values.

7. **Sequential Processing Constraint:**

(a) **First sublot:**

$$C_{1j(k+1)} - p_{j(k+1)} s_{1j(k+1)} \geq C_{1jk} + t_{j(k+1)} + FT_j + VT_j s_{1jk},$$
$$\forall j = 1, \ldots, N, k = 1, \ldots, (m-1).$$

This constraint ensures that the first sublot begins processing on machine $(k+1)$ only after it has completed processing on machine k, has been transferred to machine $(k+1)$ and the setup on machine $(k+1)$ has been completed.

(b) **For sublots 2,...,n_j:**

$$C_{ij(k+1)} - p_{j(k+1)} s_{ij(k+1)} \geq C_{ijk} + FT_j + VT_j s_{ijk},$$
$$\forall i = 2, \ldots, n_i, j = 1, \ldots, N, k = 1, \ldots, (m-1).$$

This constraint ensures that all the sublots, excluding the first one, begin processing on the $(k+1)$ th machine only after they have finished processing on the kth machine and have been transferred to machine $(k+1)$.

By replacing the above inequalities with equalities, the formulation can be adapted to the no-wait flow shop.

8. No-Intermingling Constraint for Machines 1,. . ., m:

(a) (i, j) **precedes** (i', j')

$$\left(C_{i'j'k} - p_{j'k}s_{i'j'k}\right) - \left(C_{ijk} - p_{jk}s_{ijk}\right) + G\left(1 - y_{jj'}\right) \geq \left(U_j - \sum_{u=1}^{i-1} s_{ujk}\right) p_{jk} + RT_{jk} + t_{j'k} + p_{j'k} \sum_{u=1}^{i'-1} s_{uj'k},$$
$$\forall (i, j) \text{ and } (i'j') : j \neq j', i = 1, \ldots, n_j, j = 1, \ldots, N, i' = 1, \ldots, n_{j'}, j' = 1, \ldots, N, k = 1, \ldots, m.$$

(b) (i', j') **precedes** (i, j)

$$\left(C_{ijk} - p_{jk}s_{ijk}\right) - \left(C_{i'j'k} - p_{j'k}s_{i'j'k}\right) + Gy_{jj'} \geq \left(U_{j'} - \sum_{u=1}^{i'-1} s_{uj'k}\right) p_{j'k} + RT_{j'k} + t_{jk} + p_{jk} \sum_{u=1}^{i-1} s_{ujk},$$
$$\forall (i, j) \text{ and } (i', j') : j \neq j', i = 1, \ldots, n_j, j = 1, \ldots, N, i' = 1, \ldots, n_{j'}, j' = 1, \ldots, N, k = 1, \ldots, m.$$

For any two lots j and $j'(j \neq j')$, we have two possibilities, namely, j precedes j' or j' precedes j. Since, either one must hold, these are referred to as *disjunctive constraints*. To model these into the formulation, we define a binary variable $y_{jj'}$ which takes a value of 1 if j precedes j', and 0, otherwise. If it takes a value 1, then (8a) holds true since $G(1 - y_{jj'}) = 0$ and (8b) becomes redundant. On the other hand, if $y_{jj'}$ takes a value of zero, then (8b) is enforced and (8a) becomes redundant. For any pair of sublots (i, j) and $(i', j') : j \neq j'$, the terms on the right hand side of (8a) ensure that the difference between the start times of sublots i and i' is atleast equal to the sum of the processing times of the sublots i to n_j of lot j and 1 to $(i' - 1)$ of lot j', the removal time for lot j and setup time for lot j'. These constraints are enforced for all pairs of sublots belonging to lots j and j', and on all the machines.

By replacing the above inequalities with equalities, the formulation can be adapted to the scenario when no intermittent idling is permitted.

9. Station Capacity Constraint:

(a) **First sublot of any lot on Machine 1:**

$$c_{1j1} - p_{j1}s_{1j1} \geq t_{j1}, \quad \forall j = 1, \ldots, N.$$

This constraint ensures that the processing of the first sublot, of any lot appearing first in the sequence, begins after its setup has been completed.

(b) **Sublots 2, . . ., n_j of any lot on Machine 1:**

$$C_{(i+1)j1} - p_{j1}s_{(i+1)j1} = C_{ij1}, \quad \forall i = 1, \ldots, (n_j - 1), j = 1, \ldots, N.$$

This constraint captures the fact that the $(i+1)$th sublot of lot j should begin processing on machine 1 only after the completion of its ith sublot.

(c) **All sublots on machines $k = 2, \ldots, m$:**

$$C_{(i+1)jk} - p_{jk}s_{(i+1)jk} \geq C_{ijk}, \quad \forall i = 1, \ldots, (n_j - 1), j = 1, \ldots, N, k = 2, \ldots, m.$$

This constraint ensures that for all the lots processed on machines $k = 2, \ldots, m$, the $(i + 1)$th sublot of lot j begins processing on machine k only after the completion of $i$ts ith sublot on that machine.

Example 2.1 To illustrate the above model, consider a two-machine, three-lot flow shop with the data shown in Tables 2.2 and 2.3. The sublot sizes are consistent, restricted to take integer values and intermittent idling is permitted. Recall, G is a large positive number; $G = 5,000$ was used in this and subsequent problems.

In lieu of the above data, model **GM1** can be written as follows.
Minimize: $C_{\max}$
Subject to:

Makespan Constraint:

$$C_{\max} \geq C_{n_j j2} + RT_{j2}, \quad \forall n_j, j = 1, 2, 3.$$

Item Allocation Constraint:

$$\sum_{u=1}^{n_j} s_{ujk} = U_j, \quad \forall j = 1, 2, 3, k = 1, 2.$$

TABLE 2.2. Data for the Illustration of lot-attached setup model

	Processing time		Setup time		Removal time	
	M/C 1	M/C 2	M/C 1	M/C 2	M/C 1	M/C 2
Lot 1	2	1	1	2	2	1
Lot 2	2	3	2	1	2	2
Lot 3	1	2	2	2	1	2

TABLE 2.3. Data for the Illustration of lot-attached setup model

	n_j	U_j	r_j	FT_j	VT_j
Lot 1	2	4	0	1	1
Lot 2	4	6	0	2	1
Lot 3	3	5	0	1	1

Consistent Sublot Constraint:

$$s_{ij1} = s_{ij2}, \quad \forall i = 1, \ldots, n_j, j = 1, 2, 3.$$

Attached-Setup Constraint:

$$s_{1jk} \geq 1, \quad \forall j = 1, 2, 3, k = 1, 2.$$

Sublot Size Constraint:

$$s_{ijk} \geq 0, \quad \text{integer}, \quad \forall i = 2, \ldots, n_j, j = 1, 2, 3, k = 1, 2.$$

Sequential Processing Constraint:

(a) **First sublot:**

$$C_{1j2} - p_{j2}s_{1j2} + FT_j + VT_j s_{1j1} + t_{j2}, \quad \forall j = 1, 2, 3.$$

(b) **For sublots** $2, \ldots, n_j$**:**

$$C_{ij2} - p_{j2}s_{ij2} \geq C_{ij1} + FT_j + VT_j \cdot s_{ij1}, \quad \forall i = 2, \ldots, n_j, j = 1, 2, 3.$$

No-Intermingling Constraint for Machines 1 & 2:

(a) (i, j) **precedes** (i', j')

$$\begin{aligned} &\left(C_{i'j'k} - p_{j'k}s_{i'j'k}\right) - \left(C_{ijk} - p_{jk}s_{ijk}\right) + G(1 - y_{jj'}) \\ &\geq \left(U_j - \sum_{u=1}^{i=1} s_{ujk}\right) p_{jk} + RT_{jk} + t_{j'k} + p_{j'k} \sum_{u=1}^{i'-1} s_{uj'k}, \\ &\forall (i, j) \text{ and } (i', j') : j \neq j', i = 1, \ldots, n_j, j = 1, 2, 3, i' = 1, \ldots, n_{j'}, \\ &j' = 1, 2, 3, k = 1, 2. \end{aligned}$$

(b) (i', j') **precedes** (i, j)

$$\begin{aligned} &\left(C_{ijk} - p_{jk}s_{ijk}\right) - \left(C_{i'j'k} - p_{j'k}s_{i'j'k}\right) + Gy_{ij'} \\ &\geq \left(U_{j'} - \sum_{u=1}^{i'-1} s_{uj'k}\right) p_{j'k} + RT_{j'k} + t_{jk} + p_{jk} \sum_{u=1}^{i=1} s_{ujk}, \\ &\forall (i, j) \text{ and } (i', j') : j \neq j', i = 1, \ldots, n_j, j = 1, 2, 3, i' = 1, \ldots, n_{j'}, \\ &j' = 1, 2, 3, k = 1, 2. \end{aligned}$$

Station Capacity Constraint:

(a) **First sublot of any lot on Machine 1:**

$$C_{1j1} - p_{j1}s_{1j1} \geq t_{j1}, \quad \forall j = 1, 2, 3.$$

TABLE 2.4. Solution for the illustrative Example 2.1

	Lot 1		Lot 2				Lot 3		
Consistent sublot sizes	s_1	s_2	s_1	s_2	s_3	s_4	s_1	s_2	s_3
	2	2	1	1	1	3	1	1	3
Start time on machine 1	33	37	11	13	15	17	2	3	4
Start time on machine 2	42	44	20	23	26	29	7	9	11
Optimal sequence of lots					3–2–1				
Optimal makespan					47				

(b) **Sublots 2, ..., n_j of any lot on Machine 1:**

$$C_{(i+1)j1} - p_{j1} \cdot s_{(i+1)j1} = C_{ij1}, \quad \forall i = 2, \ldots, (n_j - 1), j = 1, 2, 3.$$

(c) **All sublots on machines 2:**

$$C_{(i+1)j2} - p_{j2} \cdot s_{(i+1)j2} \geq C_{ij2}, \forall i = 1, \ldots, (n_j - 1), \quad j = 1, 2, 3.$$

The above model was coded using AMPL and was solved using the CPLEX optimization software. The optimal sublot sizes and the sequence in which to process the lots are shown in Table 2.4.

2.2.3 *m/N/{C,E,V}/{II,NI}/{CV,DV}/{Lot-Detached Setup and Removal Times, Sublot Transfer Times, No Intermingling}*

The generic formulation above (**GM1**) can easily be adapted to the case of detached setup (designated as model **GM2**) by making the following changes.

1. The lot-attached setup constraint (5) can be relaxed since the setups are detached.
2. The Sequential Processing constraints (7a) and (7b) can be combined to give a single constraint as follows:

$$C_{ij(k+1)} - p_{j(k+1)}s_{ij(k+1)} \geq C_{ijk} + FT_j + VT_j s_{ijk},$$
$$\forall i = 1, \ldots, n_j, j = 1, \ldots, N, k = 1, \ldots, (m-1).$$

3. The Station Capacity constraint (9a) now becomes

$$C_{1jk} - p_{jk}s_{1jk} \geq t_{jk}, \quad \forall j = 1, \ldots, N, k = 1, \ldots, m.$$

This constraint ensures that the first sublot of any lot starts after the setup has been completed. It needs to be enforced for machines 2, ..., m explicitly and is not implied by the Sequential Processing constraint since the setups are detached.

TABLE 2.5. Data for the illustrative lot-detached setup problem

	Processing time		Setup time		Removal time	
	M/C 1	M/C 2	M/C 1	M/C 2	M/C 1	M/C 2
Lot 1	2	1	1	2	2	1
Lot 2	3	2	2	1	2	2
Lot 3	2	3	2	4	1	2

TABLE 2.6. Data for the illustrative lot-detached setup problem

	n_j	U_j	r_j	FT_j	VT_j
Lot 1	2	4	0	1	1
Lot 2	4	6	0	2	1
Lot 3	3	5	0	1	1

Example 2.2 To illustrate the above model, consider a two-machine, three-lot system with the data shown in Tables 2.5 and 2.6. The sublot sizes are consistent, restricted to take integer values and intermittent idling is permitted.

In lieu of the above data, model **GM2** can be written as follows.
Minimize: $C_{\max}$
Subject to:

Makespan Constraint:

$$C_{\max} \geq C_{n_j j2} + RT_{j2}, \quad \forall n_j, j = 1, 2, 3.$$

Item Allocation Constraint:

$$\sum_{u=1}^{n_j} s_{ujk} = U_j, \quad \forall j = 1, 2, 3, k = 1, 2.$$

Consistent Sublot Constraint:

$$s_{ij1} = s_{ij2}, \quad \forall i = 1, \ldots n_j, j = 1, 2, 3.$$

Sublot Size Constraint:

$$s_{ijk} \geq 0, \quad \text{integer}, \quad \forall i = 2, \ldots, n_j, j = 1, 2, 3, k = 1, 2.$$

Release Time Constraint:

$$\sum_{t=0}^{200} tX_{1j1t} \geq 0, \quad \forall j = 1, 2, 3.$$

Sequential Processing Constraint:

$$C_{ij2} - p_{j2}s_{ij2} \geq C_{ij2} + FT_j + VT_j s_{ij1}, \quad \forall i = 1, \ldots, n_j, j = 1, 2, 3.$$

TABLE 2.7. Solution for the illustrative Example 2.2

	Lot 1		Lot 2				Lot 3		
Consistent sublot sizes	s_1	s_2	s_1	s_2	s_3	s_4	s_1	s_2	s_3
	2	2	1	1	2	2	1	1	3
Start time on machine 1	36	40	15	18	21	27	2	4	6
Start time on machine 2	45	47	28	31	33	37	7	10	16
Optimal sequence of lots					3–2–1				
Optimal makespan					50				

No-Intermingling Constraint for Machines 1 and 2:

(a) (i, j) **precedes** (i', j')

$$\left(C_{i'j'k} - p_{j'k}s_{i'j'k}\right) - \left(C_{ijk} - p_{jk}s_{ijk}\right) + G\left(1 - y_{jj'}\right) \geq \left(U_j - \sum_{u=1}^{i-1} s_{ujk}\right) p_{jk} + RT_{jk} + t_{j'k} + p_{j'k}\sum_{u=1}^{i'-1} s_{uj'k}$$

$\forall(i, j)$ and $(i', j') : j \neq j', i = 1, \ldots, n_j, j = 1, 2, 3, i' = 1, \ldots n_j,$
$j' = 1, 2, 3, k = 1, 2.$

(b) (i', j') **precedes** (i, j)

$$\left(C_{ijk} - p_{jk}s_{ijk}\right) - \left(C_{i'j'k} - p_{j'k}s_{i'j'k}\right) + Gy_{jj'} \geq \left(U_{j'} - \sum_{u=1}^{i'-1} s_{uj'k}\right) p_{j'k} + RT_{j'k} + t_{jk} + p_{jk}\sum_{u=1}^{i-1} s_{ujk}$$

$\forall(i, j)$ and $(i', j') : j \neq j', i = 1, \ldots, n_j, j = 1, 2, 3, i' = 1, \ldots, n_{j'},$
$j' = 1, 2, 3, k = 1, 2.$

Station Capacity Constraint:

(a) $C_{1j1} - p_{j1}s_{1j1} \geq t_{j1}, \quad \forall j = 1, 2, 3, \forall k = 1, 2.$
(b) $C_{(i+1)j1} - p_{j1}s_{(i+1)j1} = C_{1j1}, \quad \forall i = 1, \ldots, n_{j-1}, \forall j = 1, 2, 3.$
(c) $C_{(i+1)j2} - p_{j2}s_{(i+1)j2} \geq C_{ij2}, \quad \forall i = 1, \ldots, n_{j-1}, \forall j = 1, 2, 3.$

The optimal sublot sizes and the sequence in which to process the lots are shown in Table 2.7.

2.2.4 *m/N/{C,E,V}/{II,NI}/{CV,DV}/{Sublot-Attached Setup and Removal Times, Sublot Transfer Times, Intermingling}*

This problem is identical to the one discussed in Sect. 2.2.2 except for the fact that the setup involved is sublot attached instead of lot attached considered earlier, and also, we permit intermingling of the sublots. The formulation for this problem (designated as Generic Model **GM3**) follows from that presented in Sect. 2.2.2. The constraints (1), (2), (3), (4), and (6) are the same for this problem as well. Constraint (5) is no longer relevant as we now have sublot-attached setups.

The Sequential Processing Constraint for this case is as follows:

$$C_{ij(k+1)} - p_{j(k+1)}s_{ij(k+1)} \geq C_{ijk} + \tau_{j(k+1)} + FT_j + VT_j s_{ijk}$$
$$\forall i = 1, \ldots, n_j, j = 1, \ldots, N, k = 1, \ldots, (m-1).$$

This constraint ensures that any sublot i begins processing on machine $(k+1)$ only after it has completed processing on machine k has been transferred to machine $(k+1)$ and the setup on machine $(k+1)$ for sublot i has been completed.

By replacing the above inequalities with equalities, the formulation can be adapted to the no-wait flow shop scenario.

The intermingling constraint for machines $1, \ldots, m$ can be expressed as follows:

(a) (i, j) precedes (i', j')

$$C_{i'j'k} - p_{j'k}s_{i'j'k} - C_{ijk} + G(1 - y_{iji'j'}) \geq RT_{jk} + \tau_{j'k},$$
$$\forall (i, j) \text{ and } (i', j'), i = 1, \ldots, n_j, j = 1, \ldots, N, i' = 1, \ldots, n_{j'}$$
$$j' = 1, \ldots, N, k = 1, \ldots, m: \text{ if} j = j', \text{then } i \neq i'.$$

(b) (i', j') precedes (i, j)

$$\left(C_{ijk} - p_{jk}s_{ijk}\right) - C_{i'j'k} + Gy_{iji'j'} \geq RT_{j'k} + \tau_{jk}$$
$$\forall (i, j) \text{ and } (i', j'), i = 1, \ldots, n_j, j = 1, \ldots, N, i' = 1, \ldots, n_{j'},$$
$$j' = 1, \ldots, N, k = 1, \ldots, m: \text{ if } j = j', \text{then } i \neq i'.$$

These disjunctive constraints are identical to those presented in Sect. 2.1.2, except that now, since intermingling is allowed, we define a new binary variable $y_{iji'j'}$, which takes a value of 1 if sublot (i, j) precedes (i', j'), and a value of 0 if (i', j') precedes (i, j). For any pair of sublots (i, j) and (i', j') if $j = j'$ then $i \neq i'$ The terms on the right hand side in (a) above ensure that the difference between the start times of sublots (i, j) and (i', j') is at least equal to the sum of processing times of sublot (i, j), the removal time for sublot (i, j), and setup time for(i', j'). These constraints are enforced for all pairs of sublots scheduled on any machine k.

By replacing the inequalities with equalities in the above expressions, the formulation can be adapted to the case of no-intermittent idling.

The station capacity constraint for this case is as follows:

$$C_{ij1} - p_{jk}s_{ij1} \geq \tau_{j1}, \quad \forall i = 1, \ldots n_j, j = 1, \ldots, N.$$

This constraint ensures that any sublot i of any job j begins processing on machine 1 only after its setup has been completed.

Example 2.3 To illustrate the above model, consider a two-machine, two-lot system with data as shown in Tables 2.8 and 2.9. The sublot sizes are consistent, restricted to take integer values and intermittent idling is permitted.

TABLE 2.8. Data for the illustrative sublot-attached setup problem

	Processing time		Setup time		Removal time	
	M/C 1	M/C 2	M/C 1	M/C 2	M/C 1	M/C 2
Lot 1	2	1	1	1	1	1
Lot 2	2	1	1	1	1	1

TABLE 2.9. Data for the illustrative sublot-attached setup problem

	n_j	U_j	r_j	FT_j	VT_j
Lot 1	2	4	0	1	1
Lot 2	3	5	0	1	1

In lieu of the above data, model **GM3** can be written as follows.
Minimize: $C_{\max}$
Subject to:

Makespan Constraint:

$$C_{\max} \geq C_{n_j j2} + RT_{j2}, \quad \forall i = 1, \ldots n_j, j = 1, 2.$$

Item Allocation Constraint:

$$\sum_{u=1}^{n_j} s_{ujk} = U_j, \quad \forall j = 1, 2, k = 1, 2.$$

Consistent Sublot Constraint:

$$s_{ij1} = s_{ij2}, \quad \forall i = 1, \ldots, n_j, j = 1, 2.$$

Sublot Size Constraint:

$$s_{ijk} \geq 0, \text{ integer}, \quad \forall i = 2, \ldots, n_j, j = 1, 2, k = 1, 2.$$

Sequential Processing Constraint:

$$C_{ij2} - p_{j2}s_{ij2} \geq c_{ij1} + \tau_{j2} + FT_j + VT_j s_{ij1}, \quad \forall i = 1, \ldots, n_j, j = 1, 2.$$

Intermingling Constraint for Machines 1 and 2:

(a) (i, j) **precedes** (i', j')

$$\left(C_{i'j'k} - p_{j'k}s_{i'j'k}\right) - \left(C_{ijk} - p_{jk}s_{ijk}\right) + G(1 - y_{iji'j'}) \geq RT_{jk} + \tau_{j'k},$$
$$\forall (i, j) \text{ and } (i', j'), i = 1, \ldots, n_j, j = 1, 2, i' = 1, \ldots, n_{j'},$$
$$j' = 1, 2, k = 1, 2 : \text{if } j = j', \text{then } i \neq i'.$$

TABLE 2.10. Solution for the illustrative Example 2.3

	Lot 1		Lot 2		
Consistent sublot sizes	s_1	s_2	s_1	s_2	s_3
	1	3	1	2	2
Start time on machine 1	21	1	25	15	9
Start time on machine 2	27	12	30	23	17
Optimal sequence of lots (s_{ij})			21–32–22–11–12		
Optimal makespan			32		

(b) (i', j') **precedes** (i, j)

$$\begin{aligned}&\left(C_{ijk} - p_{jk}s_{ijk}\right) - \left(C_{i'j'k} - p_{j'k}s_{i'j'k}\right) + Gy_{iji'j'}\\&\geq RT_{j'k} + \tau_{jk}, \forall (i, j) \text{ and } (i', j'), i = 1, \ldots, n_j, j = 1, 2,\\&i' = 1, \ldots, n_{j'} j' = 1, 2, k = 1, 2 : \text{if} j = j', \text{then } i \neq i'.\end{aligned}$$

Station Capacity Constraint:

$$C_{ij1} - p_{j1}s_{ij1} \geq \tau_{j1}, \quad \forall i = 1, \ldots, n_j, j = 1, 2.$$

The optimal solution is shown in Table 2.10. Note that in this solution, the sublots of lot 1 are not processed continuously. The third and the second sublots of lot 2 are processed in between the second and the first sublots of lot 1. Note that the numbering of the sublots of a lot is arbitrary.

2.2.5 *m/N/{C,E,V}/{II,NI}/{CV,DV}/{Sublot-Detached Setup and Sublot-Attached Removal Times, Sublot Transfer Times, Intermingling}*

The generic formulation of the sublot-attached setup (**GM3**) problem can be adapted to the case when detached setups are present. We designate the resulting model as **GM4**. The changes that need to be incorporated are as follows.

1. The sequential processing constraint can be replaced with the following constraint.

$$\begin{aligned}&C_{ij(k+1)} - p_{(k+1)}s_{ij(k+1)} \geq C_{ijk} + FT_j + VT_j s_{ijk},\\&\forall i = 1, \ldots, n_j, j = 1, \ldots, N, k = 1, \ldots, (m-1).\end{aligned}$$

   This constraint is similar to that for the case of sublot-attached setups, except that the setup time for sublot i on machine $(k + 1)$ is not considered since the setup is detached.
2. The station capacity constraint is modified as follows.

$$C_{ijk} - p_{jk}s_{ijk} \geq \tau_{jk}, \quad \forall i = 1, \ldots, n_j, j = 1, \ldots, N, k = 1, \ldots, m.$$

This constraint ensures that if any sublot i of lot j is scheduled first on machine k, then it can begin processing only after the setup has been completed.

Example 2.4 If the setup in Example 2.3 were detached, then the model **GM4** can be written as follows.

Minimize: $C_{\max}$
Subject to:

Makespan Constraint:

$$C_{\max} \geq C_{n_j j2} + RT_{j2}, \quad \forall n_j, j = 1, 2.$$

Item Allocation Constraint:

$$\sum_{u=1}^{n_j} s_{ujk} = U_j, \quad \forall j = 1, 2, k = 1, 2.$$

Consistent Sublot Constraint:

$$s_{ij1} = s_{ij2}, \quad \forall i = 1, \ldots, n_j, j = 1, 2.$$

Sublot Size Constraint:

$$s_{ijk} \geq 0, \text{ integer}, \quad \forall i = 2, \ldots, n_j, j = 1, 2, k = 1, 2.$$

Sequential Processing Constraint:

$$C_{ij2} - p_{j2}s_{ij2} \geq C_{ij2} + FT_j + VT_j s_{ij1}, \quad \forall i = 1, \ldots, n_j, j = 1, 2.$$

Intermingling Constraint for Machines 1 and 2:

(a) (i, j) precedes (i', j')

$$\begin{aligned}&\left(C_{i'j'k} - p_{j'k}s_{i'j'k}\right) - \left(C_{ijk} - p_{jk}s_{ijk}\right) + G(1 - y_{iji'j'})\\ &\quad \geq RT_{jk} + \tau_{j'k}, \forall (i, j) \text{ and } (i', j'), i = 1, \ldots, n_j, j = 1, 2,\\ &\quad i' = 1, \ldots, n_{j'}, j = 1, 2, i' = 1, \ldots, n_{j'}, j' = 1, 2,\\ &\quad k = 1, 2 : \text{if} j = j', \text{then } i \neq i'.\end{aligned}$$

(b) (i', j') precedes (i, j)

$$\begin{aligned}&\left(C_{ijk} - p_{jk}s_{ijk}\right) - \left(C_{i'j'k} - p_{j'k}s_{i'j'k}\right) + Gy_{iji'j'}\\ &\quad \geq RT_{j'k} + \tau_{jk} \forall (i, j) \text{ and } (i', j'), i = 1, \ldots, n_j, j = 1, 2,\\ &\quad i' = 1, \ldots, n_{j'}, j' = 1, 2, k = 1, 2 : \text{if} j = j', \text{then } i \neq i'.\end{aligned}$$

Station Capacity Constraint:

$$C_{ijk} - p_{jk}s_{ijk} \geq \tau_{jk}, \quad \forall i = 1, \ldots, n_j, j = 1, 2, k = 1, 2.$$

The optimal solution is shown in Table 2.11. Note that in the optimal solution, the sublots of lot 1 are not processed continuously. The second sublot of lot 2 is processed in between the first and second sublots of lot 1.

TABLE 2.11. Solution for the illustrative sublot-detached setup problem

	Lot 1		Lot 2		
Consistent sublot sizes	s_1	s_2	s_1	s_2	s_3
	2	2	1	3	1
Start time on machine 1	1	15	25	7	21
Start time on machine 2	10	22	29	17	26
Optimal sequence of lots (s_{ij})			11–22–21–32–12		
Optimal makespan			31		

2.2.6 *The Case of Variable Sublots*

Next, we consider the case of variable sublot sizes as a lot moves from one machine to another. There are the following two ways in which a new sublot can be configured for processing on a machine, after the items constituting that sublot have been processed on the preceding machine.

Case (1). The items constituting a new sublot can be reconfigured to form this sublot only after the completion of the entire sublots to which they belong.

Case (2). The items constituting a new sublot can be reconfigured to form this sublot without the completion of the entire sublots to which they belong.

Consider Case (1) and the scenario of the generic model **GM1**. Constraints (3) and (4) are no longer valid for this case. Also, the sequential processing constraints, are impacted as follows:

(a) **First sublot:**

$$C_{1jk} - p_{jk}s_{1jk} - C_{i'j(k-1)} - FT_j - VT_j s_{i'j(k-1)} - t_{jk} \\ + G(1 - x_{i'1jk}) \geq 0, \forall i' = 1, \ldots, n_j, j = 1, \ldots, N, k = 2, \ldots, m.$$

(b) **For sublot 2, . . ., n_j:**

$$C_{ijk} - p_{jk}s_{ijk} - C_{i'j(k-1)} - FT_j - VT_j s_{i'j(k-1)} + G(1 - x_{i'ijk}) \geq 0, \\ \forall i = 2, \ldots, n_j, i' = 1, \ldots, n_j, j = 1, \ldots, N, k = 2, \ldots, m.$$

Also, we need to add a new constraint, termed the variable sublot constraint, as follows:

Variable Sublot Constraint:

$$\sum_{h=1}^{i'-1} s_{hj(k-1)} - \sum_{h=1}^{i} s_{hjk} + Gx_{i'ijk} \geq 0,$$
$$\forall i = 1, \ldots, n_j, i' = 2, \ldots, n_j, j = 1, \ldots, N, k = 2, \ldots, m$$
$$x_{1ijk} = 1, \forall i = 1, \ldots, n_j, j = 1, \ldots, N, k = 2, \ldots, m,$$

where $x_{i'ijk} = 1$, if sublot i of lot j on machine k is started no earlier than the completion time of the sublot i' of the same lot on machine $k-1$, and $= 0$, otherwise. Thus, in accordance with Constraint (a) above, if the first sublot of a lot j on machine k starts no earlier than the completion time of sublot i' on machine $k-1$, then, the appropriate relationship between the starting time of this sublot to the completion time of sublot i' and the requisite transfer and setup times must be maintained; otherwise it is relaxed. In a similar manner, Constraint (b) captures this relationship for any other sublot, other than the first sublot. However, if a sublot i on machine k starts earlier than the completion time of a sublot i' on machine $k-1$, then the sum of all the sublots until sublot i on machine k must not exceed the sum of the sublots until sublot $i'-1$ on machine $k-1$. This is captured by the variable sublot constraint.

The above development is applicable for the other generic models as well except that in the case of sublot-attached setup, we need to include a setup time, τ_{jk}, for every sublot rather than just for first sublot. The corresponding constraint is as follows:

$$C_{ijk} - p_{jk}s_{ijk} - C_{i'j(k-1)} - FT_j - VT_j s_{i'j(k-1)} - \tau_{jk}$$
$$+ G \cdot (1 - x_{i'ijk}) \geq 0 \quad \forall i, i = 1, \ldots, n_j, j = 1, \ldots, N, k = 2, \ldots, m.$$

Next, consider Case (2). The sequential processing constraints for this case under the scenario of generic model **GM1** are as follows:

(a) First sublot:

$$C_{1jk} - p_{jk}s_{1jk} - C_{i'j(k-1)} + p_{j(k-1)}s_{i'j(k-1)} - FT_j - VT_j s_{1jk} - t_{jk}$$
$$+ G(1 - x_{i'1jk}) \geq \max\left\{ p_{j(k-1)} \left(s_{1jk} - \sum_{h=1}^{i'-1} s_{hj(k-1)} \right), 0 \right\},$$
$$\forall i' = 1, \ldots, n_j, j = 1, \ldots, N, k = 2, \ldots, m.$$

(b) For sublot $2, \ldots, n_j$:

$$C_{ijk} - p_{jk}s_{ijk} - C_{i'j(k-1)} + p_{j(k-1)}s_{i'j(k-1)} - FT_j - VT_j s_{ijk}$$
$$+ G(1 - x_{i'ijk}) \geq \max\left\{ p_{j(k-1)} \left(\sum_{h=1}^{i} s_{hjk} - \sum_{h=1}^{i'-1} s_{hj(k-1)} \right), 0 \right\},$$
$$\forall i = 2, \ldots, n_j, i' = 1, \ldots, n_j, j = 1, \ldots, N, k = 2, \ldots, m.$$

Variable Sublot Constraint:

$$\sum_{h=1}^{i'-1} s_{hj(k-1)} - \sum_{h=1}^{i} s_{hjk} + Gx_{i'ijk} \geq 0,$$

$$\forall i = 1, \ldots, n_j, i' = 2, \ldots, n_j, j = 1, \ldots, N, k = 2, \ldots, m$$

$$x_{1ijk} = 1, \forall i = 1, \ldots n_j, j = 1, \ldots, N, k = 2, \ldots, m.$$

Note that, in this case, the definition of x is different from that for Case (1). In particular, $x_{i'ijk} = 1$, if sublot i of lot j on machine k is started no earlier than the starting time of sublot i' of the same lot on machine $k - 1$, and $=0$, otherwise. Accordingly, if the first sublot of lot j on machine k starts no earlier than the starting time of sublot i' of the same sublot on machine $k - 1$, then the starting time of sublot i on machine k should be no earlier than the starting time of sublot i' on machine $(k - 1)$ plus the processing time of the jobs from sublot i' to be contained in sublot i on machine k, i.e., $\left(\sum_{h=1}^{i} s_{hjk} - \sum_{h=1}^{i'-1} s_{hj(k-1)}\right) p_{i'j(k-1)}$, along with the transfer and setup times. Note that the maximum operator is needed here since $\sum_{h=1}^{i'-1} s_{hj(k-1)}$ could be larger than $\sum_{h=1}^{i} s_{hjk}$. The corresponding constraints for the first sublot are shown in (a), and for other sublots in (b) above. However, in case the sublot i of lot j on machine k starts earlier than the starting time of i' on machine $k-1$, then the sum of all the sublots until sublot i on machine k must not exceed the sum of the sublots until sublot $i'-1$ on machine $k-1$. This is captured by the variable sublot constraint.

As alluded to earlier for Case (1), the above development is applicable for the other generic models as well except that, in the case of sublot-attached setup, we need to include setup time for all sublots as follows:

$$C_{ijk} - p_{jk}s_{ijk} - C_{i'j(k-1)} + p_{j(k-1)}s_{i'j(k-1)} - FT_j - VT_j s_{ijk} - \tau_{jk}$$

$$+ G(1 - x_{i'ijk}) \geq \max\left\{ p_{j(k-1)}\left(\sum_{h=1}^{i} s_{hjk} - \sum_{h=1}^{i'-1} s_{hj(k-1)}\right), 0\right\},$$

$$\forall i = 1, \ldots, n_j, i' = 1, \ldots, n_j, j = 1, \ldots, N, k = 2, \ldots, m.$$

The above models are illustrated using a three-lot, three-machine problem. The data is given in Table 2.12. The results are depicted in Table 2.13. For the sake of comparison, we have also given results for the case of consistent sublot sizes. As expected, the makespan value obtained for Case (2) of the variable sublots is the smallest, namely, 203, while that for Case (1) of the variable sublots is 208. For the consistent sublots, the makespan value obtained is 213.

TABLE 2.12. Data for the illustration of lot-attached setup

	Processing time			Setup time			Removal time		
	M/C 1	M/C 2	M/C 3	M/C 1	M/C 2	M/C 3	M/C 1	M/C 2	M/C 3
Lot 1	2	1	2	1	2	2	2	1	2
Lot 2	2	4	1	2	1	3	2	2	4
Lot 3	4	2	2	2	2	1	1	2	1

	n_j	U_j	r_j	FT_j	VT_j
Lot 1	5	14	0	4	5
Lot 2	4	16	0	5	4
Lot 3	3	15	0	8	5

Above, we have presented fairly general mathematical models of the m-machine, N-lot streaming problems. There are some mathematical models presented in the literature that are suitable for the special cases of the lot streaming problem that they consider. We present these next.

2.3 Mathematical Models for Special Cases

This section presents mathematical formulations for some special cases of the lot streaming problem, each of which is further analyzed in the following chapters. The key features of these models are summarized in Table 2.14.

2.3.1 *2/1/C/{II,NI}/{CV,DV}/{Lot-Detached Setup, No-Wait}*

This problem addresses the issue of finding the continuous optimal sublot sizes for a single batch in a no-wait flow shop, in the presence of detached setup times [32]. In a no-wait flow shop, idle time can appear before the processing of any sublot i on machine 1 or machine 2. The expression for the makespan in terms of Δ_i (see Fig. 1.10), the idle time on machine 2 immediately preceding the ith sublot, is given as

$$C_{\max} = t_2 + p_2 \cdot U + \Delta_1 + \sum_{i=2}^{n} \Delta_i,$$

where t_1 is the setup time on machine 1, t_2 is the setup time on machine 2, $\Delta_1 = \max\{0, t_1 + p_1 s_1 - t_2\}$, and $\Delta_i = \max\{0, p_1 s_i - p_2 s_{i-1}\}$.

TABLE 2.13. Solutions for the consistent and variable sublot cases

1. Consistent sublot case

		Lot 1					Lot 2				Lot 3		
Consistent sublot sizes		s_1	s_2	s_3	s_4	s_5	s_1	s_2	s_3	s_4	s_1	s_2	s_3
	Sublot size	1	5	2	4	2	3	4	5	4	6	5	4
Machine 1	Start time	64	66	76	80	88	96	102	110	120	2	26	46
	Finish time	66	76	80	88	92	102	110	120	128	26	46	62
	Sublot size	1	5	2	4	2	3	4	5	4	6	5	4
Machine 2	Start time	104	105	110	112	116	120	132	148	168	68	80	90
	Finish time	105	110	112	116	118	132	148	168	184	80	90	98
	Sublot size	1	5	2	4	2	3	4	5	4	6	5	4
Machine 3	Start time	160	162	172	176	184	193	196	200	205	119	131	141
	Finish time	162	172	176	184	188	196	200	205	209	131	141	149
Optimal sequence of lots							3–1–2						
Optimal makespan							213						

2. Sublot availability case

		Lot 1					Lot 2				Lot 3		
Variable sublot sizes		s_1	s_2	s_3	s_4	s_5	s_1	s_2	s_3	s_4	s_1	s_2	s_3
	Sublot size	3	4	0	4	3	3	2	5	6	6	5	4
Machine 1	Start time	64	70	78	78	86	96	102	106	116	2	26	46
	Finish time	70	78	78	86	92	102	106	116	128	26	46	62
	Sublot size	7	0	0	0	7	5	5	4	2	5	6	4
Machine 2	Start time	104	111	111	111	111	121	141	161	177	69	79	91
	Finish time	111	111	111	111	118	141	161	177	185	79	91	99
Machine 3	Sublot size	1	0	0	0	13	5	5	6	0	5	0	10
	Start time	155	157	157	157	157	188	193	198	204	119	129	129
	Finish time	157	157	157	157	183	193	198	204	204	129	129	149
Optimal sequence of lots							3–1–2						
Optimal makespan							208						

3. Item availability case

		Lot 1					Lot 2				Lot 3		
Variable sublot sizes		s_1	s_2	s_3	s_4	s_5	s_1	s_2	s_3	s_4	s_1	s_2	s_3
	Sublot size	1	0	0	0	13	1	0	0	15	1	0	14
Machine 1	Start time	64	66	66	66	66	96	98	98	98	2	6	6
	Finish time	66	66	66	66	92	98	98	98	128	6	6	62
	Sublot size	4	4	3	0	3	3	3	5	5	6	5	4
Machine 2	Start time	102	106	110	113	113	120	132	144	164	68	80	90
	Finish time	106	110	113	113	116	132	144	164	184	80	90	98
	Sublot size	1	0	0	5	8	6	5	3	2	6	5	5
Machine 3	Start time	148	150	150	150	160	183	189	194	197	112	122	132
	Finish time	150	150	150	160	176	189	194	197	199	122	132	142
Optimal sequence of lots							3–1–2						
Optimal makespan							203						

TABLE 2.14. Key features of the mathematical models presented in Sect. 2.3

Section	Number of machines	Number of lots	Sublot type	Inter./No Inter. Idle Time	Continuous/Discrete Sublot Sizes	Setup	Removal time	Transfer time	Interminging	Wait/no wait	Objective function
2.3.1	2	1	C	Both	Both	Lot–Detached	No	No	N/A	No-wait	Makespan
2.3.2	2	N	C	Both	Both	Lot–Attached and lot Detached	No	Yes	No	Wait	Makespan
2.3.3	2	1	C	II	CV	None	No	No	N/A	Wait	Total Sublot Completion Times
2.3.4	2	1	C	II	CV	Sublot–Attached	No	No	N/A	Wait	Total Sublot Completion Times
2.3.5	3	1	C	Both	Both	None	No	No	N/A	Wait	Makespan
2.3.6	m	1	C (two sublots)	II	CS	None	No	No	N/A	Wait	Total Sublot Completion Times
2.3.7	m	1	E	II	CV	Sublot–Attached	No	Yes	N/A	Wait	Unified Cost Function

Let I represent the total idle time on machine 2. In order to minimize the makespan $C_{\max}$, it is sufficient to minimize the total idle time I on machine 2. This problem can be formulated as a linear program.

$$\text{Minimize}: \quad I = \sum_{i=1}^{n} \Delta_i.$$

Subject to :

$$\Delta_1 \geq t_1 + p_1 s_1 - t_2$$

$$\Delta_i \geq p_1 s_i - p_2 s_{i-1}, \quad \forall i, i = 2, \ldots, n$$

$$\sum_{i=1}^{n} s_i = U$$

$$\Delta_i \geq 0, \quad \forall i, i = 1, \ldots, n$$

$$s_i \geq 0, \quad \forall i, i = 1, \ldots, n.$$

A solution to the above model will give the desired sublot sizes and the order of their processing on the machines.

2.3.2 *2/N/C/{II,NI}/{CV,DV}/{Lot-Attached/Detached Setup, Sublot Transfer Times}*

This problem addresses the issue of finding the continuous, optimal sublot sizes and the sequence in which to process the lots in the presence of lot-attached/detached setup times and variable sublot transfer times [36]. These transfer times are made up of a fixed component FT_j and a variable component VT_j, which depends on the size of a sublot.

For ease of understanding, the situation on hand is depicted in Fig. 2.1 for $N = 1$. In this figure, F and V represent fixed and variable transfer times; and t_1 and t_2 are lot-detached setup times on machines 1 and 2, respectively. Note that Δ_1, the idle time on machine 2 before the start of sublot 1 on that machine can be

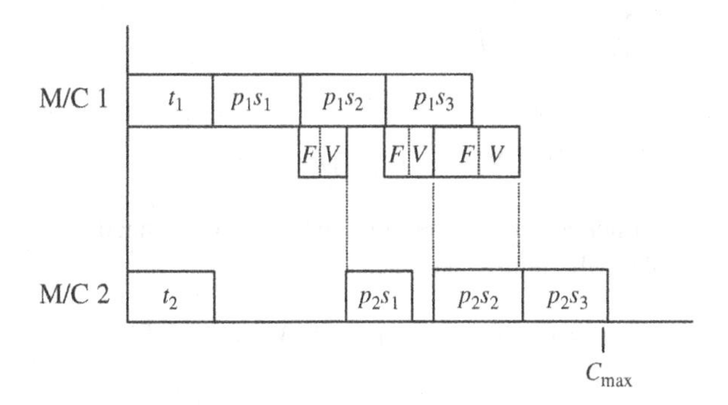

FIGURE 2.1. Graphical depiction of sublot-attached transfer times

expressed as follows:

$$\Delta_1 = \max\{0, t_1 + p_1 \cdot s_1 + VT \cdot s_1 + FT - t_2\}.$$

If we let $t_1' = t_1 + FT, \quad p_1' = p_1 + VT$, and $t_2' = t_2$, then we have,

$$\Delta_1 = \max\left\{0, t_1' + p_1' - t_2'\right\}.$$

Similarly, Δ_i, the idle time on machine 2 before the start of sublot i, can be given as follows:

$$\begin{aligned} \Delta_i &= \max\left\{0, t_1 + p_1 \sum_{u=1}^{i-1} s_u + p_1 s_i + VT s_i + FT - t_2 - p_2 \sum_{u=1}^{i-1} s_u - \sum_{u=1}^{i-1} \Delta_u\right\} \\ &= \max\left\{0, t_1' + p_1 \sum_{u=1}^{i} s_u - t_2' - p_2' \sum_{u=1}^{i-1} s_u - \sum_{u=1}^{i-1} \Delta_u\right\}, \quad \forall i = 2, \ldots, n_j, \end{aligned}$$

where $p_2' = p_2' + VT$.

Now, if we designate by I_j^{DS} the total idle time on machine 2 under sublot-detached setup for lot j, then a formulation for the problem of determining optimal sublot sizes for lot j that minimizes the makespan (or, equivalently I_j^{DS}), can be given as follows.

$$\begin{aligned} &\text{Minimize :} \quad I_j^{\text{DS}}. \\ &\text{Subject to :} \\ &I_j^{\text{DS}} \geq t_{j1}' - t_{j2}' + p_{j1}' s_{1j} \\ &I_j^{\text{DS}} \geq t_{j1}' - t_{j2}' + p_{j1}'(s_{1j} + s_{2j}) - p_{j2}' s_1 j \\ &\vdots \\ &I_j^{\text{DS}} \geq t_{j1}' - t_{j2}' + p_{j1}' \sum_{u=1}^{n_j} s_{uj} - p_{j2}' \sum_{u=1}^{n_j - 1} s_{uj} \\ &\sum_{u=1}^{n_j} s_{uj} = U_j \\ &I_j^{\text{DS}} \geq 0 \\ &s_{ij} \geq 0, \quad \forall i, i = 1, \ldots, n_j. \end{aligned}$$

In the case of lot-attached setups, the only change that we need to make is in the determination of Δ_1, which now becomes,

$$\Delta_1 = \max\{0, t_1 + p_1 \cdot s_1 + VT s_1 + FT\}.$$

Accordingly, the formulation for the lot-attached setup is as follows:

$$\text{Minimize}: \quad I_j^{\text{AS}}.$$

Subject to :

$$I_j^{\text{AS}} \geq t'_{j1} + p'_{j1}s_{1j}$$

$$\vdots$$

$$I_j^{\text{AS}} \geq t'_{j1} - t'_{j2} + p'_{j1}\sum_{u=1}^{n_j} s_{uj} - p'_{j2}\sum_{u=1}^{n_j=1} s_{uj}$$

$$\sum_{u=1}^{n_j} s_{uj} = U_j$$

$$I_j^{\text{AS}} \geq 0$$

$$s_{1j} \geq 1$$

$$s_{ij} \geq 0, \quad \forall i = 1, \ldots, n_j,$$

where I_j^{AS} is the total idle time on machine 2 under sublot-attached setup for lot j. Once the sublot sizes have been obtained for each lot for either the lot-detached or lot-attached setup case, the lots are sequenced in accordance with the Johnson's rule [19]. This is further explained in Chap. 3.

A slightly different version of the above formulation is presented in [8] for the detached setup case, which includes removal time for each lot, and is based on the concept of run-in and run-out times.

2.3.3 *2/1/C/II/CV/*$\sum_{i=1}^{n} s_i C_{i2}$

This problem can be described as follows: Given a two-machine flow shop with a single lot, determine the continuous and consistent sublot sizes such that the total weighted sublot completion time, i.e., $\sum_{i=1}^{n} s_i C_{i2}$, is minimized [31]. This is essentially a sublot sizing problem, and can be formulated as a linear program as follows.

$$\text{Minimize}: \quad F(s, C) \equiv \sum_{i=1}^{n} s_i C_{i2}.$$

Subject to :

$$C_{ik} \geq C_{i-1k} + p_2 s_i, \quad \forall i = 2, \ldots, n, k = 1, 2, \tag{2.1}$$

$$C_{i2} \geq C_{i1} + p_2 s_i, \quad \forall i = 2, \ldots, n, \tag{2.2}$$

$$C_{11} \geq s_1 p_1, \tag{2.3}$$

$$\sum_{i=1}^{n} s_i = U, \tag{2.4}$$

$$s_i \geq 0, \quad \forall i = 1, \ldots, n, \tag{2.5}$$

$$C_{i,k} \geq 0, \quad \forall i = 2, \ldots, n, k = 1, 2. \tag{2.6}$$

As mentioned above, the objective function minimizes the total weighted sublot completion time. Constraint (2.1) ensures that sublots on any machine are processed only after the preceding sublot has finished processing. Constraint (2.2) captures the fact that machine 2 processes sublots only after it has finished processing on machine 1. Constraint (2.3) ensures that the completion time of the first sublot is greater than or equal to its processing time. Constraint (2.4) imposes the requirement that the sublot sizes add up to the lot size. Constraint (2.5) and (2.6) represent the nonnegativity of the sublot sizes and the completion times.

2.3.4 *2/1/C/II/CV/*$\sum_{i=1}^{\bar{n}} s_i C_{i2}$*, Sublot-Attached Setup*

This problem is like the one in Sect. 2.3.3 except that, now, the sublot-attached setups are present and also the number of sublots is not known *a priori* [5]. Let $\bar{n}$ be an upper bound on the number of sublots. A mathematical model of this problem is as follows.

$$\text{Minimize}: F(\text{s}, \text{C}) \equiv \sum_{i=1}^{\bar{n}} s_i C_{i2} \tag{2.7}$$

Subject to :

$$C_{i2} = i \cdot t_2 + p_2 \sum_{j=1}^{i} s_j + I_i, \quad \forall i = 1, \dots, \bar{n}, \tag{2.8}$$

$$\begin{aligned} I_1 &= t_1 + p_1 s_1 \\ I_i &\geq I_{i-1}, \quad \forall i = 2, \dots, \bar{n}, \end{aligned} \tag{2.9}$$

$$I_i \geq \left(i t_1 + p_1 \sum_{j=1}^{i} s_j \right) - \left((i-1) t_2 + p_2 \sum_{j=1}^{i-1} s_j \right), \quad \forall i = 2, \dots, \bar{n}, \tag{2.10}$$

$$\sum_{i=1}^{M} s_i = U, \tag{2.11}$$

$$s_i \geq 0, \quad \forall i = 1, \dots, \bar{n}. \tag{2.12}$$

The objective function $F(s, C)$ seeks to minimize the total weighted sublot completion time of all the $\bar{n}$ possible positive sublots. Constraint (2.8) defines the completion time of any sublot i on machine 2 as the sum of

(i) Setup times of all previous sublots including sublot i on machine 2
(ii) Processing times of all previous sublots including sublot i on machine 2
(iii) Cumulative idle time appearing before sublot i on machine 2

Constraint (2.9) defines the idle time appearing before sublot 1 on machine 2 as the sum of its setup and processing time on machine 1. Constraint (2.10) defines the cumulative idle time on machine 2 for sublots $2, \ldots, \bar{n}$. The following two cases are possible:

(i) The cumulative idle time remains the same i.e., $I_i = I_{i-1}$, implying that sublot $(i-1)$ finishes processing on machine 2 later than the completion of sublot i on machine 1

(ii) The cumulative idle time increases implying that sublot $(i-1)$ finishes processing on machine 2 before the completion of sublot i on machine 1

Constraint (2.11) ensures that the sum of the sublot sizes does not exceed the given lot size. The last constraint restricts the sublot sizes to be nonnegative.

2.3.5 *3/1/C/{NI,II}/{CV,DV}/{No Setup}*

This problem addresses the sublot sizing problem for a three-machine flow shop by minimizing the completion time of the last sublot on machine 3 when the sublot sizes are consistent [35]. Let C_{ik} denote the completion time of the ith sublot on machine k. Then, we have

$$\text{Minimize} : C_{3n}$$

$$\text{Subject to} :$$

$$C_{11} \geq s_1 p_1 \tag{2.13}$$

$$C_{ik} \geq C_{i,(k-1)} + p_k s_i, \quad \forall i = 1, 2, \ldots, n, k = 2, 3, \tag{2.14}$$

$$C_{ik} \geq C_{(i-1),k} + p_k s_i, \quad \forall i = 2, \ldots, n, k = 1, 2, 3, \tag{2.15}$$

$$\sum_{i=1}^{n} s_i = U, \tag{2.16}$$

$$s_i \geq 0, \quad \forall i = 1, 2, \ldots, n.$$

Constraints (2.14) and (2.15) ensure that any sublot i begins processing on machine k after its completion on the previous machine or the processing of the $(i-1)$th sublot on machine k, whichever is maximum. Constraint (2.16) imposes that the total number of items in all sublots equals U. The no-idling and discrete version can be obtained by replacing the inequalities with equalities and by restricting the sublot sizes to take integer values, respectively.

2.3.6 *m/1/C/II/CV/*$\sum_{i=1}^{2} x_i C_{im}$

We, now, consider the problem of minimizing the total weighted sublot completion time in an m-machine flow shop consisting of a single lot [34]. The number of sublots is restricted to two on each machine, the sublots sizes are consistent and can take real values. Let x_1 and $x_2 = (1 - x_1)$ be the proportion of work allocated to the first and second sublots, respectively. Let $C_{i,k}$ denote the completion time of the ith sublot on machine k and p_k be the processing time per item on machine k.

The mathematical formulation for this problem is as follows.

$$
\begin{aligned}
\text{Minimize :} \quad & F(x_1, x_2) = (x_1 C_{1m} + x_2 C_{2m}). \\
\text{Subject to :} \quad & C_{11} \geq x_1 p_1, \\
& C_{2k} \geq C_{1k} + x_2 p_k, \quad \forall k = 1, \ldots, m, \\
& C_{ik+1} \geq C_{ik} + x_i p_{k+1}, \quad \forall i = 1, 2; k = 1, \ldots, (m-1), \\
& x_1 + x_2 = 1, \\
& C_{ik} \geq 0, \quad \forall i = 1, 2, \ k = 1, \ldots, m \text{ and } x_1, x_2 > 0.
\end{aligned}
$$

The completion time of the sublots can be written as

$$
\begin{aligned}
C_{1m} &= x_1 \sum_{k=1}^{m} p_k \text{ and} \\
C_{2m} &= \max_{1 \leq k \leq m} \left\{ x_1 \sum_{l=1}^{k} p_l + x_2 \sum_{l=k}^{m} p_l \right\}.
\end{aligned}
$$

Making the above substitutions along with $x_2 = 1 - x_1$, in the expression for flowtime, we have

$$
F(x_1) = x_1^2 \sum_{k=1}^{m} p_k + (1 - x_1) \max_{1 \leq k \leq m} \left\{ x_1 \sum_{l=1}^{k} p_l + (1 - x_1) \sum_{l=k}^{m} p_l \right\}.
$$

Simplification of the above expression gives

$$
F(x_1) = \max_{1 \leq k \leq m} \left\{ x_1^2 \left(\left(2 \sum_{l=k}^{m} p_l \right) - p_k \right) + x_1 \left(\sum_{l=1}^{k} p_l - 2 \sum_{l=k}^{m} p_l \right) + \sum_{l=k}^{m} p_l \right\}.
$$

Let

$$
a_k = \left(2 \cdot \sum_{l=k}^{m} p_l \right) - p_k \; b_k = \left(\sum_{l=1}^{k} p_l - 2 \sum_{l=k}^{m} p_l \right) \quad \text{and} \quad c_k = \sum_{l=k}^{m} p_l.
$$

Therefore,

$$
F(x_1) = \max_{1 \leq k \leq m} \left\{ a_k x_1^2 + b_k x_1 + c_k \right\}.
$$

Hence, an equivalent formulation can be written as,

$$
\begin{aligned}
& \text{Minimize : } F(x_1) \\
& \text{Subject to :} \\
& F(x_1) \geq a_k x_1^2 + b_k x_1 + c_k, \quad k = 1, \ldots, m
\end{aligned}
$$

where

$$a_k = \left(2\sum_{l=k}^{m} p_l\right) - p_k,$$

$$b_k = \left(\sum_{l=1}^{k} p_l - 2\sum_{l=k}^{m} p_l\right), \quad \text{and}$$

$$c_k = \sum_{l=k}^{m} p_l.$$

2.3.7 *m/1/E/II/CV/Sublot-Attached Setup, Transfer Times/Unified Cost Function*

We now consider a hybrid objective function consisting of a weighted sum of the makespan ($C_{\max}$), (sublot) mean flow time (MFT), average work-in-process (WIP), sublot-attached setup (SAS), and transfer time (TT), in an m-machine flow shop with a single lot and continuous and equal sublot sizes [25].

The problem is to determine an optimal number of sublots (n) so as to minimize the above hybrid cost function. This problem can be formulated as an integer program as follows.

Minimize : $\quad Z(n) \equiv c_1 C_{\max}(n) + c_2\mathrm{MFT}(n) + c_3\mathrm{WIP}(n) + c_4 t_k(n) + c_5\mathrm{TT}(n).$

Subject to:

$$C_{\max}(n) = \left\{\frac{U}{n}\sum_{k=1}^{m} p_k + \sum_{k=1}^{m} t_k\right\} + (n-1)\max_{1\le k\le m}\left\{\frac{U}{n}p_k + t_k\right\},$$

$$\mathrm{MFT}(n) = \frac{U}{n}\sum_{k=1}^{m} p_k + \sum_{k=1}^{m} t_k + \frac{n-1}{2}\max_{1\le k\le m}\left\{\frac{U}{n}p_k + t_k\right\},$$

$$\mathrm{WIP}(n) = U\left\{\frac{\frac{U}{n}\sum_{k=1}^{m} p_k + \sum_{k=1}^{m} t_k + \frac{n-1}{2}\max_{1\le k\le m}\left\{\frac{U}{n}p_k + t_k\right\}}{\frac{U}{n}\sum_{k=1}^{m} p_k + \sum_{k=1}^{m} t_k + (n-1)\max_{1\le k\le m}\left\{\frac{U}{n}p_k + t_k\right\}}\right\},$$

$$\mathrm{SAS}(n) = n\sum_{k=1}^{m} t_k,$$

$$\mathrm{TT}(n) = n(m-1)\mathrm{TT},$$

$$1 \le n \le U \quad \text{and} \quad \text{integer}.$$

2.4 Chapter Summary

In this chapter, we have presented some generic mathematical models for the flow shop lot streaming problem. These generic models capture the various important features that may be encountered in practice. These include lot-attached (detached) setup, sublot-attached (detached) setup, lot removal time, and sublot transfer time. The removal time of a lot is assumed to be attached to the last sublot of a lot and is independent of the sequence in which the lots are processed or the size of the last sublot. The sublot transfer time, on the other hand, is assumed to be comprised of two components, one being fixed and identical for all the sublots of a lot while the other depends on the sublot size. The transfer time and removal time differ in that, during the occurrence of the former, the machine from where the transfer occurs is free to process another sublot, while, when the latter is encountered, the machine is occupied and cannot process the next lot. We also consider the situations of equal, consistent, and variable sublot sizes. In the case of variable sublots, as a lot moves from one machine to another, a new sublot can be formed in two ways. According to one of these ways, the jobs constituting a new sublot can be reconfigured to form this sublot only after the completion of the entire sublots from the previous machine to which they belong. The other way is for the jobs constituting a new sublot to be reconfigured to form this sublot without having completed the entire sublots to which they belong. We present models for both of these situations. We also consider situations in which the sublots belonging to different lots may or may not be intermingled.

We have provided illustrations for the use of several of the models that we have developed, which depict optimal sublot sizes and the sequence in which to process the lots to achieve minimum makespan values. These models are integer programs due to the presence of disjunctive constraints (for determining the sequence in which to process the lots) and the requirement of integer sublot sizes. They are solved using the CPLEX solver.

Mathematical models for some special cases of the flow shop lot streaming problem have been discussed in the literature. We have also presented these models in this chapter.

3
Two-Machine Lot Streaming Models

3.1 A Brief Overview

The two-machine lot streaming problem has been investigated extensively and encompasses both the single-lot and multiple-lot scenarios. In the presence of a single lot and a given number of sublots in which to process this lot, the problem reduces to finding sublot sizes in order to optimize a criterion under consideration. For an ordinary flow shop and under the objective of minimizing the makespan, the optimal continuous sublot sizes are geometric in nature with a ratio of p_2/p_1, where p_j is the unit processing time on machine j. However, this result does not hold when sublot sizes are constrained to be integers. This result also extends to the case of no-wait flow shops with lot-detached setups.

When multiple lots are present, an additional problem of sequencing the lots also arises besides the problem of determining optimal sublot sizes for each lot. However, for a given number of sublots for each lot, the problem of determining optimal sublot sizes and that of optimally sequencing the lots are independent of each other. This remains true when lot-attached or lot-detached setups are present as well as when transfer and removal times are encountered. As before, this result also extends to the case of no-wait flow shops.

Depending upon the scenario addressed, closed-form expressions can be developed to determine optimal sublot sizes or they can be determined by solving an underlying linear programming formulation of the problem. The lot sequencing problem is solved using a variation of Johnson's algorithm [19] except for the no-wait flow shops in which case the optimal sequence is determined by using the algorithm proposed by Gilmore and Gomory [15] for a special type of traveling salesman problem. When the number of sublots is not given, then the problem of determining sublot sizes and the lot sequencing problem need not be independent.

A schematic representation of work on the two-machine lot streaming models is shown in Fig. 3.1. Table 3.1 depicts status of the various two-machine lot streaming problems involving single and multiple lots. Note that $\Im^L(s)$ is used to represent a lot sizing problem, $\Im^L(S)$ to represent a sequencing problem, and $\Im^L(s, S)$ to represent a problem that involves both lot sizing and lot sequencing. Also, the

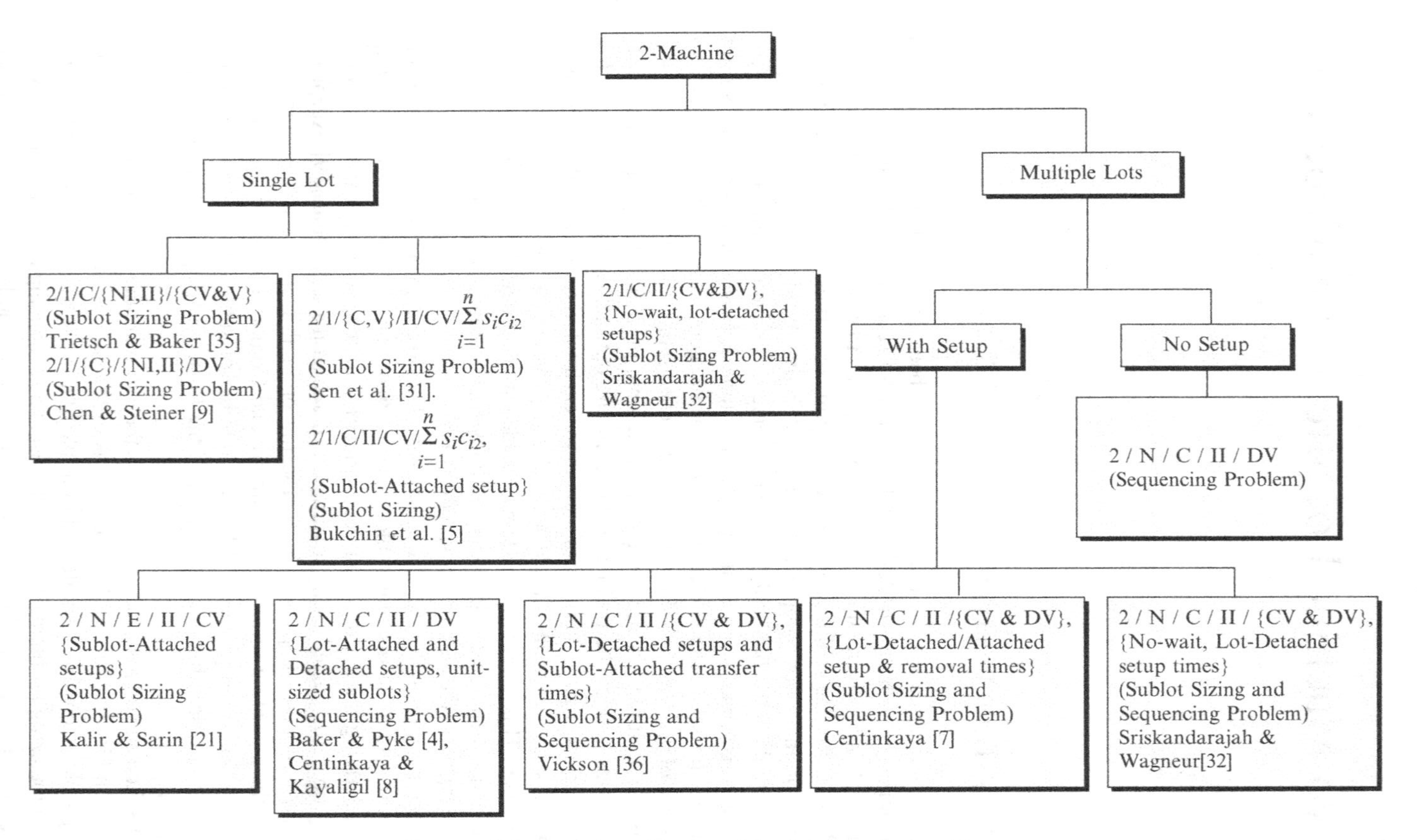

FIGURE 3.1. Overview of two-machine lot streaming problems

TABLE 3.1. Problem status of two-machine lot streaming problems

No.	Problem	Problem type	Status
A. Makespan objective function			
A1. Single lot			
1.	2/1/C/{NI,II}/{CV,DV}	$\Im^L(s)$	Sects. 3.2 and 3.3
2.	2/1/C/II/{CV,DV}/{Lot-Detached Setup, No-Wait}	$\Im^L(s)$	Sects. 3.4 and 3.5
3.	2/1/C/II/{CV,DV}/{Release Times}	$\Im^L(s)$	Open
A2. Multiple lots			
1.	2/N/C/II/DV/{No Setup, Unit-Size Sublots}	$\Im^L(s)$	Sect. 3.8
2.	2/N/C/II/DV/{Lot-Detached/Attached Setup Times, Unit-Size Sublots}	$\Im^L(s, S)$	Sects. 3.9 and 3.10
3.	2/N/C/II/{CV,DV}/{Lot-Detached/Attached Setup and Lot-Attached Removal Times}	$\Im^L(s, S)$	Sect. 3.11
4.	2/N/C/II/{CV,DV}/{Lot-Detached/Attached Setup and Sublot-Attached Transfer Times}	$\Im^L(s, S)$	Sect. 3.12
5.	2/N/C/II/{CV,DV}/{Lot-Detached Setup Times, No-Wait}	$\Im^L(s, S)$	Sect. 3.13
6.	2/N/E/II/DV/{Sublot-Attached Setups}	$\Im^L(s, S)$	Sect. 3.14
7.	2/N/C/II/{CV,DV}/{Release Times, Sublot Intermingling}	$\Im^L(s, S)$	Open
B. Other objective functions			
1.	2/1/{C,V}/II/CV/$\sum s_i C_{i2}$	$\Im^L(s)$	Sect. 3.6
2.	2/1/C/II/CV/$\sum s_i C_{i2}$, Sublot-Attached Setups	$\Im^L(s)$	Sect. 3.7
3.	2/1/{C,V}/II/DV/$\{\sum s_i C_{i2}, \sum C_{i2}\}$	$\Im^L(s)$	Open
4.	2/1/{C,V}/II/DV/$\sum s_i C_{i2}$, Sublot-Attached Setups	$\Im^L(s)$	Open
5.	2/{1,N}/{C,V}/II/{CV,DV}/$\{\sum C_{i2}, \sum s_i C_{i2}\}$, Release Times, Sublot-Detached Setup, Sublot Transfer Times, and Removal Times, Sublot Intermingling	$\Im^L(s, S)$	Open

status column contains section numbers of the chapter in which the corresponding problems are studied. An "open" designation indicates that the corresponding problem is not discussed in the literature. Table 3.2 gives a brief summary of both the major results and solution methodologies available for the two-machine lot streaming problem.

We now discuss each of these problems. We present important results that are pertinent for the model being discussed, develop expressions for optimal sublot sizes, and present algorithms for the sequencing of lots, wherever relevant.

3.2 2/1/C/{NI,II}/CV

The analyses and algorithms of diverse lot streaming problems rely on various results from the classical scheduling theory. To begin the discussion, we invoke one such result that is applicable to the lot streaming problem on-hand.

TABLE 3.2. Summary of major results/solution methodology for the two-machine lot streaming problems

No.	Problem	Major result/solution methodology	Complexity	
			$\Im^L(s)$	$\Im^L(S)$
1.	2/1/C/{NI,II}/CV (Sect. 3.2) [35]	(a) Optimal sublot sizes are geometric with ratio of p_2/p_1. (b) The resulting schedule has a block structure which is compact.	Polynomial	–
2.	2/1/C/{NI,II}/DV (Sect. 3.3) [9, 35]	Optimal sublot sizes can be obtained by using an algorithm which utilizes the continuous solution as a staring solution and iteratively increments the makespan by rounding up the sublot size closest to an integer. Another method: (a) For equal sublot sizes, when $p_1 < p_2$ and U/n is not an integer, we round up U/n till sublot $(U - n\lfloor U/n \rfloor)$ and round down the remaining sublots. When $p_1 = p_2$, then any solution satisfying $s_i \leq \lceil U/n \rceil$, $\forall i$, is optimal. When $p_1 > p_2$, then we interchange the processing times and follow the procedure for $p_1 < p_2$. (b) For consistent sublot sizes, when $p_1 < p_2$, we check if an optimal solution with $s_1 = 1$ and s_1 critical exists by solving a set of equations recursively. If the sum of the sublot sizes equals the lot size, then this solution is optimal. Else, we use a trial value for the makespan and compute the sublot sizes by solving another set of equations recursively. If they satisfy certain conditions, they are optimal. When this is not the case, the optimal sublot sizes can be obtained by iteratively transferring items from the critical sublots to the last sublot until no further improvement in the makespan is obtained. The procedure for obtaining consistent sublot sizes when $p_1 = p_2$ and $p_1 > p_2$ remains identical to that described above for equal sublot sizes.	Polynomial	–
3.	2/1/C/II/CV/{Lot-Detached Setup Times, No-Wait} (Sect. 3.4) [32]	Optimal sublot sizes are geometric as in (1) and the resulting schedule has a compact block structure, but idle time may be present before the first sublot on machine 1 or machine 2 depending on the values of the setup times on machine 1, machine 2, and the size of the first sublot.	Polynomial	–

4.	2/1/C/II/DV/ {Lot-Detached Setup Times, No-Wait} (Sect. 3.5) [32]	(a) Unequal sublot sizes can be obtained by using an algorithm which first rounds up the sublot sizes obtained from the continuous solution. The additional items resulting due to the rounding up are then allocated iteratively to the sublot which is closest to an integer when rounded down. (b) Almost equal sublot sizes can be obtained by first calculating the minimum equal sublot size as $\lfloor U/n \rfloor$ and then allocating the remaining items to the sublots in sequential order.	$O(n)$	–
5.	2/1/{C, V}/II/CV/ $\sum s_i C_{i2}$ (Sect. 3.6) [31]	(a) When $q = p_2/p_1 \leq 1$, the optimal consistent sublot sizes are equal and the variable sublot sizes might also be equal (if an associated conjecture is true). (b) When $q > 1$, then the optimal consistent sublot sizes can be obtained by using an algorithm, which searches for the index v of a sublot before which the sublot sizes are geometric and equal thereafter. The variable sublot sizes might be geometric on machine 1 and equal on machine 2, if the associated conjecture is true.	NA	–
6.	2/1/C/II/CV/ $\sum s_i C_{i2}$, Sublot-Attached Setup Times (Sect. 3.7) [5]	The solution procedure first checks if the optimal solution has a unique bottleneck. If so, then it uses an algorithm to find the same; else, it uses another algorithm to find an approximate solution. Both of these algorithms are based on solving relaxed mathematical formulations of the problem under consideration.	NA	–
7.	2/N/C/II/DV/ {No Setup, Unit-Size Sublots} (Sect. 3.8)	Unit-size sublots are optimal. Lots can be sequenced optimally by using a modified Johnson's algorithm for the run-in or run-out times or transfer lags.	–	Polynomial
8.	2/N/C/II/DV/ {Lot-Attached/ Detached Setup, Unit-Sized Sublots} (Sects. 3.9 and 3.10) [4, 8]	Unit-size sublots are assumed in order to make the problem tractable for analysis. $\Im^L(S)$ is solved by using the run-in or run-out time version of Johnson's algorithm, where the run-in and run-out times are calculated by taking into account the presence of lot-detached/attached setups.	–	Polynomial
9.	2/N/C/II/CV/ {Lot-Detached/ Attached Setup, Lot-Attached Removal Times} (Sect. 3.11) [7]	$\Im^L(s)$ and $\Im^L(S)$ can be solved independently. The continuous sizes can be obtained by analyzing the LP formulation of $\Im^L(s)$, which minimizes the idle time on machine 2 appearing in between the sublots. $\Im^L(S)$ can be solved by using Johnson's algorithm with modified processing times.	Exponential	Polynomial

TABLE 3.2. continued

No.	Problem	Major result/solution methodology	Complexity	
			$\Im^L(s)$	$\Im^L(S)$
10.	2/N/C/II/DV/ {Lot-Detached/ Attached Setup, Lot-Attached Removal Times} (Sect. 3.11) [7]	The algorithm for calculating the integer-size sublots is based on the concept of iteratively increasing the maximum idle time Z appearing on machine 2 in between the sublots in the absence of setup times. Similar to (9), $\Im^L(S)$ can be solved by using Johnson's algorithm with modified processing times.	NA	Polynomial
11.	2/N/C/II/CV/ {Lot-Detached/ Attached Setup and Sublot-Attached Transfer Times} (Sect. 3.12) [36]	$\Im^L(s)$ and $\Im^L(S)$ are independent of each other and the optimal sublot sizes are geometric as in (1). $\Im^L(S)$ can be solved by using the run-in or run-out version of Johnson's algorithm where the run-in and run-out times are calculated by taking into account the lot-detached setup times and lot-attached removal times	Polynomial	Polynomial
12.	2/N/C/II/DV/ {Lot-Detached/ Attached Setup and Sublot-Attached Transfer Times} (Sect. 3.12) [36]	The algorithm to obtain the integer sublot sizes iteratively increments the maximum optimal idle time appearing on machine 2 in between the sublots obtained by solving the corresponding LP formulation. $\Im^L(S)$ can be solved as in (11).	NA	Polynomial
13.	2/N/C/II/CV/ {Lot-Detached Setup Times, No-Wait} (Sect. 3.13) [32]	$\Im^L(s)$ and $\Im^L(S)$ are independent of each other. The optimal sublot sizes for 2/1/C/II/CV/Lot-Detached Setup, No-Wait flow shop are optimal for this problem also. $\Im^L(S)$ can be solved by the algorithm of Gilmore and Gomory [15].	NA	Polynomial
14.	2/N/C/II/DV/ {Lot-Detached Setup Times, No-Wait} (Sect. 3.13) [32]	Integer-size sublots can be obtained by applying the algorithm for 2/1/C/II/DV/Lot-Detached Setup, No-Wait flow shop to each lot separately. The lots can be sequenced using the algorithm of Gilmore and Gomory [15].	NA	$O(N \log N)$
15.	2/N/E/II/DV/ {Sublot-Attached Setup Times} (Sect. 3.14) [21]	When sublot sizes of all the lots are the same, i.e., $s_i = s, \forall i = 1, \ldots, N$, then the algorithm iterates over all possible values of s. At each iteration, it uses the value of s to compute the sublot sizes, which are then sequenced optimally using a modified Johnson's algorithm [19], in which the processing times are given by the sum of the sublot setup and processing times. When the sublot sizes are not the same, the corresponding algorithm consists of a construction phase and improvement phase. The former optimally solves the two-machine single-lot problem for all lots, and then, sequences them using a modified Johnson's algorithm. The improvement phase, then, reoptimizes the sublot sizes; however, the sequence is not changed.	$O(N \log(N) U_{\max})$ and $O(N^2 \log(N) U)$, respectively	

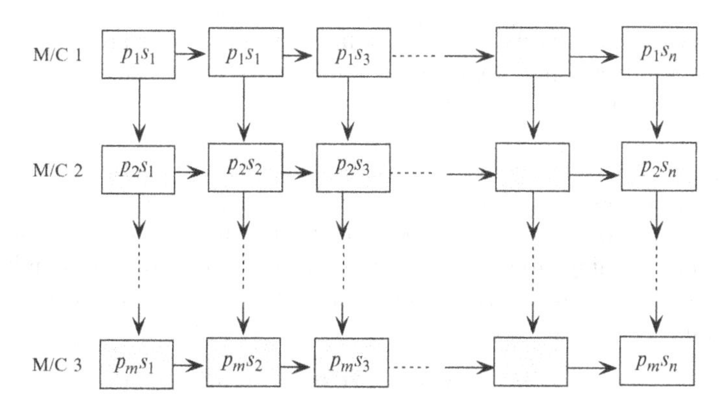

FIGURE 3.2. Processing of n sublots on m machines

Lemma 3.1 *The reversibility property holds true for the two-machine, lot streaming problem for the makespan criterion.*

This follows by the fact that, once the sublot sizes are known, the forward and reverse schedules will give the same makespan value.

In general, this property holds for the m-machine lot streaming problem as well so long as the objective is to minimize makespan. To see this, the processing of the n sublots (with known sizes $s_1, s_2, \ldots, s_n$) on m machines is depicted in Fig. 3.2. A node (i, k) represents the processing of the ith sublot on machine k and carries a weight of $p_k S_i$, which is its processing time. A directed arc from node (i, k) to $(i, k + 1)$ captures the requirement that sublot i cannot begin processing on machine $k + 1$ unless it has finished processing on machine k. Similarly, a directed arc from (i, k) to $(i + 1, k)$ indicates that machine k cannot start processing on sublot $i + 1$ until it has finished processing sublot i. The length of a path is the sum of the weights of the nodes that lie on that path. Clearly, the underlying network representation remains the same when the arrows are reversed, thereby maintaining the reversibility property for the m-machine lot streaming problem.

From the classical scheduling theory, we know that, for the makespan criterion, the optimal sequence in which to process jobs on the first two machines of a flow shop is the same, and also, the same is true for the jobs processed on the last two machines. For the lot streaming problem, we can show the following.

Theorem 3.1 (Potts and Baker [28]) *For a given number of sublots, there exists an optimal schedule for the makespan criterion in which* $s_{i1} = s_{i2}$ *and* $s_{im} = s_{im-1}$, *where* s_{ij} *is the size of sublot* i *on machine* j.

An immediate consequence of this result is that, for $m = 2$ and $m = 3$, consistent sublot sizes are optimal.

Lemma 3.2 *The no-idling restriction does not impact the optimal makespan value of the two-machine lot streaming problem for both continuous and discrete sublot sizes.*

This result follows by the fact that the first machine always processes sublots without intermittent idling. By the reversibility property, the same is true for the second machine as well.

However, this result need not hold true for $m > 2$. An example to this fact is shown in Fig. 3.3. Note that the schedule in (a) that permits intermittent idle time between the sublots on machine 2 gives a better makespan value.

The notion of critical sublots is important for determining the optimal makespan value of lot streaming problems. A sublot is called *critical* if a change in its size (whether continuous or discrete) impacts the optimal makespan value. For the two-machine problem on-hand

$$M^* \geq M(s_i), \quad \forall i = 1, \ldots, n,$$

where M^* is the optimal makespan value and $M(s_i) = p_1 \sum_{k=1}^{i} s_k + p_2 \sum_{k=i}^{n} s_k$. For a critical sublot, c,

$$M^* = M(s_c) = p_1 \sum_{k=1}^{c} s_k + p_2 \sum_{k=c}^{n} s_k.$$

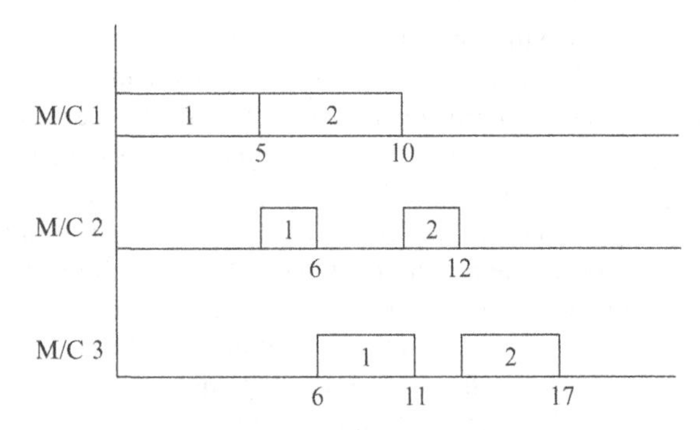

(a) Idle times in between the processing of sublots on machines 2 and 3.

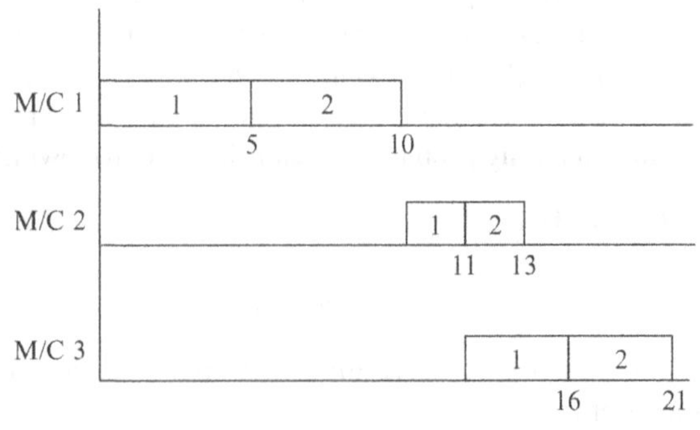

(b) No-intermittent idle times.

FIGURE 3.3. A schedule with idle time between the processing of sublots (see (**a**)) having a better makespan value than the case of no-intermittent idle time (see (**b**)) for $m > 2$

Theorem 3.2 *In an optimal solution for the 2/1/C/{NI,II}/CV problem, all the sublots of the lot are critical.*

We can show this result by contradiction. Suppose that there exist optimal sublot sizes $s_1, s_2, \ldots, s_n$, in which a sublot j is noncritical. We designate this set of sublot sizes by S. Then, by the definition of a critical sublot, we have

$$M^*(S) - p_1(s_1 + \cdots + s_j) + p_2(s_j + \cdots + s_n) = \Delta, \quad \text{for some } \Delta > 0.$$

Now, construct another set of sublot sizes $s'_1, \ldots, s'_n$ designated by S' as follows:

$$\begin{aligned} s'_j &= s_j(1-\theta) + U\theta, \quad 0 \le \theta \le 1, \\ s'_k &= s_k(1-\theta) \quad \text{for all} \quad k \ne j. \end{aligned}$$

We have

$$M^*(S') = p_1\left(s'_1 + \cdots + s'_{c'}\right) + p_2\left(s'_{c'} + \cdots + s'_n\right).$$

If $c' \ne j$,

$$M^*(S') = \begin{cases} (1-\theta)\left[p_1(s_1 + \cdots + s_{c'}) + p_2(s_{c'} + \cdots + s_n)\right] + p_2\theta U, & \text{if } c' < j, \\ (1-\theta)\left[p_1(s_1 + \cdots + s_{c'}) + p_2(s_{c'} + \cdots + s_n)\right] + p_1\theta U, & \text{if } c' > j. \end{cases}$$

In other words,

$$M^*(S') \le (1-\theta)\left[p_1(s_1 + \cdots + s_{c'}) + p_2(s_{c'} + \cdots + s_n)\right] + \theta \max\{p_1 U, p_2 U\}.$$

Since $p_1(s_1 + \cdots + s_{c'}) + p_2(s_{c'} + \cdots + s_n) \le M^*(S)$ and $\max\{p_1 U, p_2 U\} < M^*(S)$, we have

$$M^*(S') < (1-\theta)M^*(S) + \theta M^*(S) < M^*(S),$$

which contradicts the fact that $M^*(S)$ is optimal.

If $c' = j$, then

$$\begin{aligned} M^*(S') &= (1-\theta)\left[p_1\left(s_1 + \cdots + s_j\right) + p_2\left(s_j + \cdots + s_n\right)\right] + \theta\left(p_1 U + p_2 U\right) \\ &= (1-\theta)\left[M^*(S) - \Delta\right] + \theta\left(p_1 U + p_2 U\right) \\ &= (1-\theta)M^*(S) - (1-\theta)\Delta + \theta\left(p_1 U + p_2 U\right) + \theta M^*(S) - \theta M^*(S) \\ &< (1-\theta)M^*(S) + \theta M^*(S) \\ &< M^*(S), \end{aligned}$$

$$\text{for} \quad 0 < \theta < \frac{\Delta}{(p_1 + p_2)U + \Delta - M^*(S)} \le 1,$$

which contradicts the fact that $M^*(S)$ is optimal. Thus, our supposition is wrong and sublot j must be critical.

Thus, if the sublot sizes are continuous, all of the sublots are critical. In addition, by Lemma 3.2, there exists no intermediate idle time in between the sublots. Therefore, we have

$$s_i p_1 = s_{i-1} p_2, \quad \forall i = 2, \ldots, n.$$

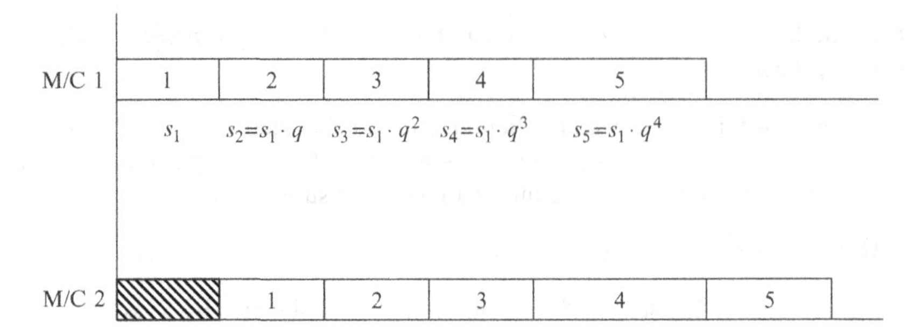

FIGURE 3.4. An example of geometric sublot sizes

If we let $q = p_2/p_1$, we have

$$s_i = qs_{i-1}, \quad \forall i = 2, \ldots, n, \tag{3.1}$$

or in general,

$$s_i = q^{i-1}s_1, \quad \forall i = 1, \ldots, n.$$

Thus, the sublot sizes are geometric in nature (see Fig. 3.4). In case $p_2 = p_1$, all sublots are equal in size ($= U/n$).

Let,

$$x_i = \frac{s_i}{U}, \quad \forall i = 1, \ldots, n. \tag{3.2a}$$

where $U = \sum_{i=1}^{n} s_i$. Then, the above expression that links s_i and s_1 can also be written in terms of x_i as follows:

$$x_i = q^{i-1}x_1, \quad \forall i = 1, \ldots, n.$$

Since

$$x_1 = \frac{s_1}{U} = \frac{s_1}{\sum_{k=1}^{n} s_k},$$

we have

$$x_i = \frac{q^{i-1}s_1}{\sum_{k=1}^{n} s_k}, \quad \forall i = 1, \ldots, n,$$

or,

$$x_i = \frac{q^{i-1}}{1 + q + q^2 + \cdots + q^{n-1}}, \quad \forall i = 1, \ldots, n. \tag{3.2b}$$

Hence, the proportions of sublot sizes, $x_i, \forall i = 1, \ldots, n$, are given by

$$x_i = \left\{ \begin{array}{ll} \frac{q^{i-1}-q^i}{1-q^n}, & \text{when } p_1 \neq p_2 \\ \frac{1}{n}, & \text{when } p_1 = p_2 \end{array} \right\}. \tag{3.3}$$

The sublot sizes, s_i, can be obtained by using (3.2a). The optimal makespan value, M, can be expressed in terms of s_1 or s_n, as follows:

$$M_1 = p_1 s_1 + p_2 U \quad \text{or} \quad M_2 = p_1 U + p_2 s_n. \tag{3.4}$$

Note that, since there is no inserted idle time among the sublots on machine 1 and those on machine 2, the sublots 2, ..., n on machine 1 and sublots 1, ..., $(n-1)$ on machine 2 form a "block structure," which is compact (see Fig. 3.5). Also, if $p_1 > p_2$, then the optimal sublots are decreasing in size. This holds true since the completion time of the $(j+1)$th sublot on machine 1 and that of the jth sublot on machine 2 should be equal (from (3.1)). Since machine 2 processes items faster than machine 1, the sublot sizes must be decreasing for (3.1) to hold. By the same logic, when $p_1 < p_2$, the sublot sizes should be increasing in size.

3.3 2/1/C/{NI,II}/DV

As alluded to earlier, the intermittent idle time and non-intermittent idle time solutions are identical in this case as the former can be right shifted on the second machine to obtain the latter (Fig. 3.6). However, there is no guarantee of this solution resulting in a compact block structure of the type shown in Fig. 3.5.

The latest time by which the ith sublot must begin processing on machine 2, in order to achieve makespan M, is given by $\{M - p_2(U - S_{i-1})\}$, where $S_i = \sum_{j=1}^{i} s_j$ represents the total number of items up to sublot i. For the schedule to be feasible, the items in the first i sublots must be completed no later than $p_1 S_i$ on machine 1 since no-idling is permitted. Hence,

$$\begin{aligned} p_1 S_i &\leq M - p_2\,(U - S_{i-1})\,, \\ S_i &\leq \min\{(M - p_2\,(U - S_{i-1}))\,/p_1, U\}. \end{aligned} \tag{3.5}$$

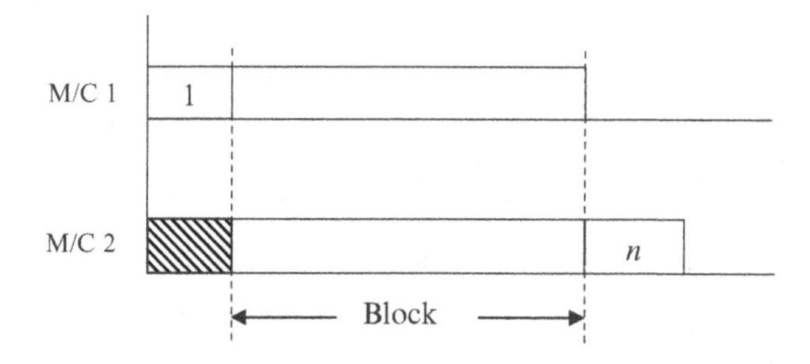

FIGURE 3.5. Compact block structure

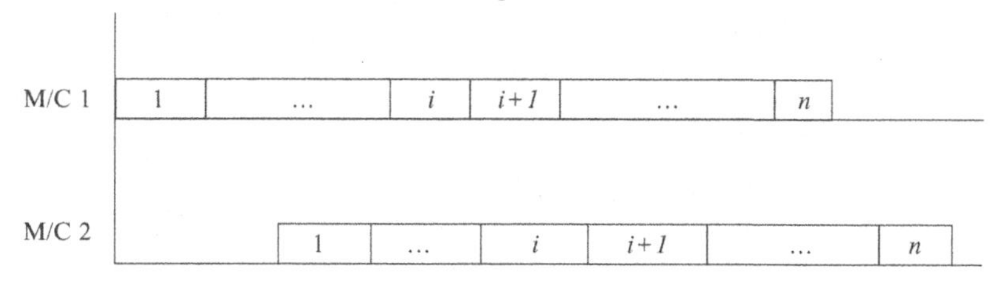

FIGURE 3.6. Illustration for the 2/1/C/{NI,II}/DV problem

We initialize the procedure by using $S_0 = 0$ and $i = 1$. The initial value of M is taken as the optimal solution value obtained from the continuous version of the problem (see Sect. 3.2). The S_k values are calculated at every stage using (3.5) and are the largest integer values permitted by the inequality. In the end, if $S_n = U$, then the solution is optimal. However, if $S_n < U$, then the current value of M cannot accommodate all the items and must be incremented. The new value of M, $\hat{M}$, is given by

$$\hat{M} = M + p_1 \min_{1 \le i \le n} (1 - e_i),$$

where e_i represents the fractional difference between the continuous and discrete sublot sizes, i.e.,

$$e_i = \min \left\{ \frac{M - p_2(U - S_{i-1})}{p_1}, U \right\} - S_i.$$

The underlying idea is to compensate for the error introduced due to the rounding down of the continuous values. Toward this end, the incrementing procedure selects amongst all S_i, the one closest to a rounded-up integer and, by doing so, it increments M by the minimum amount. In other words, it increments the total number of items at the stage which is closest to an integer (rounded-up value), thus incrementing it by the minimum amount. This process is repeated until $S_n = U$. This algorithm can be implemented in polynomial time of $O(n^2)$. The corresponding algorithm to determine the optimal sublot sizes is given in Fig. 3.7.

Next, we present an example to illustrate implementation of this algorithm.

<u>Step 1:</u> Let $S_0 = 0$

<u>Step 2:</u> For $i = 1, 2, \ldots, n$

Calculate $S_i = \lfloor \min\{(M - p_2(U - s_{i-1}))/p_1, U\} \rfloor$

where S_i = Cumulative number of items in the first i sublots

M = Optimal makespan from the continuous solution

U = Number of items in the batch

<u>Step 3:</u> If $S_n = U$

STOP

ELSE

Find new value of M as

$\hat{M} = M + p_1 \min_{1 \le i \le n} (1 - e_i)$

where $e_i = \min\{M - p_2(U - S_{i-1}))/p_1, U\} - S_i$

Go to Step 2

FIGURE 3.7. Algorithm for the 2/1/C/{NI,II}/DV problem

TABLE 3.4. Data for an illustrative example of the 2/1/C/{NI,II}/DV problem

Lot size (U)	80
Number of sublots (n)	5
Processing time on machine 1 (p_1)	4
Processing time of machine 2 (p_2)	3

Example 3.1 Consider a problem with the data presented in Table 3.4.

Note that for this data, $q = 3/4 = 0.75$.

First, calculate the continuous sublot sizes and the makespan value using (3.3) and (3.4), respectively. These are shown below along with the corresponding schedule.

	Continuous sublot sizes
s_1	26.22
s_2	19.67
s_3	14.75
s_4	11.06
s_5	8.30
Makespan $= \max\{344.8896, 344.8904\} = 344.8904$	

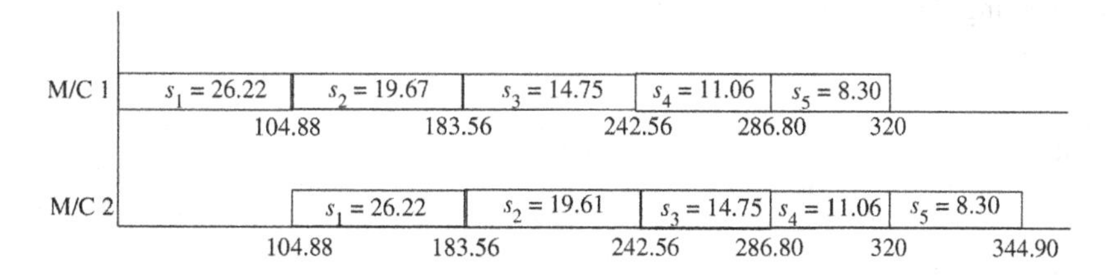

Using Step 2, we compute cumulative integer sublot sizes, $s_i, i = 1, \ldots, 5$. These are shown in the table below. Since $s_5 = 78 < u$, we increment the value of M using Step 3, as shown in the table.

Cumulative sublot sizes	e_i	$(1 - e_i)$	Incremented makespan value
$S_1 = \lfloor \min(26.2226, 80) \rfloor = 26$	0.2226	0.7774	
$S_2 = \lfloor \min(45.7226, 80) \rfloor = 45$	0.7226	0.2774	
$S_3 = \lfloor \min(59.9726, 80) \rfloor = 59$	0.9726	0.0274	$344.8904 + 4 \times (0.0274) = 345$
$S_4 = \lfloor \min(70.4726, 80) \rfloor = 70$	0.4726	0.5274	
$S_5 = \lfloor \min(78.7226, 80) \rfloor = 78$	0.7226	0.2774	

This incremented makespan value is used in Step 2 to determine integer sublot sizes once again. The new value of S_5 is 79, which is still less than U. We further increment the value of M using Step 3, which now becomes 346 (see the table below).

Cumulative sublot sizes	e_i	$(1-e_i)$	Incremented makespan value
$S_1 = \lfloor \min(26.25, 80) \rfloor = 26$	0.25	0.75	
$S_2 = \lfloor \min(45.75, 80) \rfloor = 45$	0.75	0.25	$345 + 4 \times (0.25) = 346$
$S_3 = \lfloor \min(60, 80) \rfloor = 60$	0	1	
$S_4 = \lfloor \min(71.25, 80) \rfloor = 71$	0.25	0.75	
$S_5 = \lfloor \min(79.5, 80) \rfloor = 79$	0.5	0.5	

Upon recalculating the integer sublot sizes using Step 2 based on this incremented value of the makespan, we obtain:

Cumulative sublot sizes
$S_1 = \lfloor \min(26.5, 80) \rfloor = 26$
$S_2 = \lfloor \min(46, 80) \rfloor = 46$
$S_3 = \lfloor \min(61, 80) \rfloor = 61$
$S_4 = \lfloor \min(72.25, 80) \rfloor = 72$
$S_5 = \lfloor \min(80.5, 80) \rfloor = 80$

Since $S_n = U$, the solution is optimal, and we stop. The final optimal integer sublot sizes are 26, 20, 15, 11, and 8, and the optimal makespan is 346. The corresponding schedule is as follows:

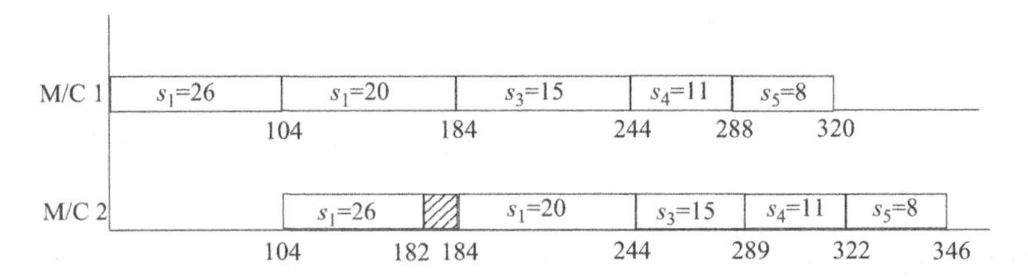

Note that, contrary to the schedule for continuous sublot sizes, the above schedule for the discrete case results in an intermittent idle time on machine 2. However, the start times of the appropriate sublots on machine 2 can be delayed to obtain the NI solution without any impact on the makespan.

The above approach for the 2/1/C/{NI,II}/DV problem determines the optimal integer sublot sizes through an attempt to generate a solution with no-intermittent idle times among the sublots on machine 2 by sequentially updating the makespan value, starting from the optimal makespan value obtained for continuous sublot sizes. However, optimal discrete sublot sizes can also be obtained using the concept of critical path [10]. A critical path for a given set of sublot sizes is constituted by those sublots on machines 1 and 2 that give the makespan value. A network representation for the processing of sublots over the two machines is shown in Fig. 3.8.

FIGURE 3.8. Network representation of the 2/1/C/{NI,II}/DV problem

The longest path from (1, 1) to $(n, 2)$ is known as the *critical path* and defines the makespan, i.e.,

$$M(\mathrm{s}) = \max_{1 \le i \le n} \{M(s_i)\},$$

where

$$M(s_i) = p_1 \sum_{j=1}^{i} s_j + p_2 \sum_{j=i}^{n} s_j.$$

When $p_1 \le p_2$, machine 2 determines the longest path, and we have

$$M(\mathrm{s}) \ge p_1 s_1 + p_2 \sum_{j=i}^{n} s_j = p_1 s_1 + p_2 U \ge p_1 + p_2 U,$$

where i is the first positive-size sublot. Similarly, if $p_1 > p_2$, machine 1 determines the longest path, and $p_2 + p_1 U$ is a lower bound on the makespan value. Hence, a lower bound for the problem under consideration can be written as

$$M(\mathrm{s}) \ge \min\{p_1, p_2\} + U \max\{p_1, p_2\}. \tag{3.5a}$$

Furthermore, if $U \le n$, then the optimal sublot sizes are given as

$$s_i = 1, \quad \forall 1 \le i \le U \quad \text{and} \quad s_i = 0, \quad \forall U < i \le n.$$

The above result must hold since, when $p_1 \le p_2$, the path (1, 1)–(1, 2)–$\cdots$–$(n, 2)$ is critical with length

$$M(\mathrm{s}) = p_1 + p_2 \sum_{j=1}^{n} s_j = p_1 + p_2 U,$$

which equals the lower bound given above, and hence, must be optimal. When $p_1 > p_2$, then the path (1, 1)–(2, 1)–$\cdots$–$(n, 1)$–$(n, 2)$ must be critical with length

$$M(\mathrm{s}) = p_2 + p_1 \sum_{j=1}^{n} s_j = p_2 + p_1 U,$$

which, once again, equals the lower bound given above, and therefore, must be optimal. Hence, we assume that $U > n$. We consider three cases depending on $p_1 < p_2$, $p_1 = p_2$, or $p_1 > p_2$.

Case 1a. $p_1 < p_2$

When the optimal makespan is equal to the lower bound given above (by (3.5a)), then the first sublot has only one item and is critical, i.e., it is processed in a no-wait fashion and there is no idle time in between the processing of sublots on machine 2. We refer to this situation by saying that the *extreme solution* is optimal. Hence, if $\boldsymbol{s}$ is an extreme solution which is not optimal, then it implies that the first sublot is not critical or it is critical but $s_1 \neq 1$. Similar to the 2/1/C/{NI,II}/CV problem, where every sublot must have a positive size in the optimal solution and the makespan is a strictly decreasing function of the number of sublots, the 2/1/C/{NI,II}/DV problem has the same property, but only if the extreme solution is not optimal. This is formally stated below.

Theorem 3.3 (Chen and Steiner [9]) *If $p_1 < p_2$, and the extreme solution is not optimal, then all sublot sizes are positive, i.e., $s_i > 0, \forall i$, in every optimal solution $\boldsymbol{s}$. Furthermore, the optimal makespan $M(\boldsymbol{s})$ is a strictly decreasing function of n.*

Recall that the optimal solution to the 2/1/C/{NI,II}/CV problem was defined to be compact, i.e., all the sublots are processed in a no-wait fashion and any path $(1,1)-\cdots-(i,1)-(i,2)-\cdots-(n,2), \forall i$, has the same length, i.e., all complete paths are critical. For the integer version of the problem, this is almost true, as stated in the theorem below, unless we are dealing with the extreme solution case.

Theorem 3.4 (Chen and Steiner [9]) *If $p_1 < p_2$, and the extreme solution is not optimal, then there exists an optimal solution $\boldsymbol{s}'$ for which every complete path, except possibly the last one, has a length which is within p_1 of the longest path, i.e.,*

$$M(s') < M_i(s') + p_1, \quad \forall i = 1, \ldots, (n-1).$$

Theorem 3.4 states that there is an optimal solution in which every path except for one of the complete paths has a length close to the length of the critical path, i.e., the network is balanced in this sense.

The solution procedures for the determination of sublot sizes when an extreme solution is optimal as well as when an extreme solution is not optimal are as follows. First, consider the situation when there exists an extreme optimal solution, i.e., there exists a set of sublot sizes $\boldsymbol{s} = (s_1, s_2, \ldots, s_n)$ which satisfy the constraints:

1. $s_1 = 1$.
2. s_1 is critical.

This implies that segment $(1,1) - \cdots - (i,1) - (i,2)$ should be no longer than segment $(1,1) - (1,2) - \cdots - (i,2)$ for $1 < i \leq n$, i.e.,

$$p_1 s_1 + p_2 U \geq p_1 s_1 + p_1 \sum_{j=2}^{i} s_j + p_2 \sum_{j=i}^{n} s_j$$

or

$$p_2 \sum_{j=1}^{i-1} s_j \geq p_1 \sum_{j=2}^{i} s_j, \quad \forall 1 < i \leq n.$$

Let $s_i (1 < i \leq n)$ be the largest nonnegative integer satisfying the above equation, subject to the additional constraint $\sum_{j=1}^{i} s_j \leq U$. This, in conjunction with $s_1 = 1$, can be used recursively to determine the sublot sizes s_i, $i = 2, \ldots, n$, which requires $O(n)$ time. If $\sum_{j=1}^{n} s_j = U$, then $\boldsymbol{s}$ is an optimal solution since $M(s) = M_1(s) = p_1 + p_2 U$ achieves the lower bound given above. If $\sum_{j=1}^{n} s_j < U$, then it can be shown by contradiction that there does not exist an optimal solution with s_1 critical and $s_1 = 1$. Hence, we can determine in $O(n)$ time whether there exists an extreme optimal solution.

Next, we consider the case when there does not exist an extreme optimal solution. Let M_0 denote a trial value for the length of the critical path in an optimal network. If M_0 is the optimal value for the makespan, then there is also an optimal solution $s^{\mathbf{0}} = \left(s_1^0, s_2^0, \ldots, s_n^0\right)$ which satisfies the conditions of Theorem 3.4, i.e., all complete paths, except for the last one, are within p_1 of the optimal makespan, i.e.,

$$M_1(s^{\mathbf{0}}) > M_0 - p_1,$$

where

$$M_1(s^{\mathbf{0}}) = p_1 s_1^0 + p_2 \sum_{i=1}^{n} s_i^0 = p_1 s_1^0 + p_2 U.$$

Hence,

$$M_0 \geq p_1 s_1^0 + p_2 U > M_0 - p_1,$$

where the first inequality holds from the optimality of M_0. This implies that s_1^0 must satisfy the following:

$$\frac{M_0 - p_2 U}{p_1} \geq s_1^0 > \frac{M_0 - p_2 U}{p_1} - 1.$$

Note that there is a unique value of $3.0^{s_1^0}$, which satisfies the above relationship, since the end points differ by 1. Similarly, by Theorem 3.4 we have the following relationship for sublots 2 to n.

$$M_0 \geq M_i(s^{\mathbf{0}}) > M_0 - p_1, \quad \forall 1 < i < n,$$

where

$$M_i(s^{\mathbf{0}}) = p_1 \sum_{j=i}^{i} s_j^0 + p_2 \sum_{j=i}^{n} s_j^0.$$

The above relationship is equivalent to

$$\frac{M_0 - p_1 \sum_{j=1}^{i-1} s_j^0 - p_2 \left(U - \sum_{j=1}^{i-1} s_j^0\right)}{p_1} \geq s_i^0 > \frac{M_0 - p_1 \sum_{j=1}^{i-1} s_j^0 - p_2 \left(U - \sum_{j=1}^{i-1} s_j^0\right)}{p_1} - 1,$$

which, once again, uniquely determines the sublot sizes 2 to $n-1$. Finally, these values can be used to determine the size of the last sublot as $s_n = U - \sum_{j=1}^{n-1} s_j^0$. Hence, for a given M_0, we can obtain the optimal sublot sizes $\boldsymbol{s}^{\mathbf{0}}$ in $O(n)$ time.

It can be shown [9] that there exists a rounded integer solution s' which is within $\min\{p_1, p_2\}$ of the optimal continuous makespan, i.e., it satisfies the following relationship

$$M(\mathrm{s}^*) \leq M(\mathrm{s}') < M(\mathrm{s}^{\mathrm{C}}) + \min\{p_1, p_2\},$$

where $M(s^{\mathrm{C}})$ is the optimal continuous makespan corresponding to $s^{\mathrm{C}} = \left(s_1^{\mathrm{C}}, s_2^{\mathrm{C}}, \ldots, s_n^{\mathrm{C}}\right)$. Since, the optimal makespan M^* satisfies $M^* \in [M(\mathrm{s}^{\mathrm{C}}), M(\mathrm{s}^{\mathrm{C}}) + p_1)$, we can use $M_0 = M(\mathrm{s}^{\mathrm{C}}) + p_1$ to obtain a feasible and balanced integer solution. Further, it can be shown that the resulting sublot sizes $\boldsymbol{s}^{\mathbf{0}}$ are either optimal for the discrete problem or some simple adjustments can be made to them in order to derive an optimal integer solution. These adjustments utilize the following facts:

1. $s_j^0 - s_j^* = 1$, where j is the smallest index for which $s_i^0 \neq s_i^*$.
2. $s_n^0 + 1 \leq s_n^*$.

Note that the transfer of a single unit from s_j^0 to the last sublot does not change $M_i(\boldsymbol{s}^{\mathbf{0}}), \forall i < j$, reduces $M_j(\boldsymbol{s}^{\mathbf{0}})$ by p_1, and increases every $M_i(\boldsymbol{s}^{\mathbf{0}}), \forall j < i < n$, by $p_2 - p_1$. Thus, in order to achieve a reduction in the makespan, we must reduce the length of every critical path in the network representation corresponding to $\boldsymbol{s}^{\mathbf{0}}$. Hence, the adjustment procedure iteratively transfers a single item from each critical sublot obtained in the above solution to the last sublot, until no further reduction in the makespan is possible. For an improved balanced optimal solution to exist, all of the following conditions must be satisfied.

1. $M_n(\boldsymbol{s}^{\mathbf{0}}) - rp_2 < M(\boldsymbol{s}^{\mathbf{0}})$, where r is the number of critical sublots.
2. $M_{i_j}(\boldsymbol{s}^{\mathbf{0}}) - p_1 + (j-1)(p_2 - p_1) < M(\boldsymbol{s}^{\mathbf{0}}), j = 1, \ldots, r$, where $i_1, i_2, \ldots, i_r$ are the critical sublots in the network representation using $\boldsymbol{s}^{\mathbf{0}} = \left(s_1^0, s_2^0, \ldots, s_n^0\right)$.
3. $M_t(\boldsymbol{s}^{\mathbf{0}}) + j(p_2 - p_1) < M(\boldsymbol{s}^{\mathbf{0}}), t \in (i_j, i_{j+1}), j = 1, \ldots, r$ (assuming that $i_{r+1} = n$).

Case 1b. $p_1 = p_2$

This case is similar to Case 1a, discussed above.

Case 1c. $p_1 > p_2$

The problem under consideration is equivalent to its inverse. Hence, we can simply interchange the processing times, i.e., p_2 is the processing time per item on machine 1 and p_1 is the processing time per item on machine 2. The resulting situation is the same as in Case 1a discussed above.

The entire solution procedure for obtaining consistent sublot sizes is given below and is self-explanatory.

3.3.1 *Sublot Sizing Algorithm for 2/1/C/{NI,II}/DV*

Case 1. $p_1 < p_2$

Step 1. Check if there exists an optimal integer solution with $s_1 = 1$ and s_1 is critical by solving the following equations recursively to obtain $s_2, \ldots, s_n$

$$p_2 \sum_{j=1}^{i-1} s_j \geq p_1 \sum_{j=2}^{i} s_j, \quad \forall i = 2, \cdots, n,$$

$$\sum_{j=2}^{i} s_j \leq U, \quad \text{and} \quad s_1 = 1.$$

Step 2. If $\sum_{j=1}^{n} s_j = U$, then $\boldsymbol{s} = (s_1, s_2, \ldots, s_n)$ is optimal and the corresponding makespan, $M(\boldsymbol{s}) = p_1 + p_2 U$, and Stop; otherwise, continue.

Step 3. If $\sum_{j=1}^{n} s_j < U$, then there does not exist an optimal solution with $s_1 = 1$ and s_1 critical. Go to Step 4.

Step 4. Solve the continuous version of the problem and obtain the optimal continuous makespan $M(\boldsymbol{s}^{\mathrm{C}})$. Set $M_0 = M(\boldsymbol{s}^{\mathrm{C}}) + p_1$.

Step 5. Determine $\boldsymbol{s}^{\mathbf{0}}$, the set of integer sublot sizes as below:

$$\frac{M_0 - p_2 U}{p_1} \geq s_1^0 > \frac{M_0 - p_2 U}{p_1} - 1,$$

$$\frac{M_0 - p_1 \sum_{j=1}^{i-1} s_j^0 - p_2 \left(U - \sum_{j=1}^{i-1} s_j^0\right)}{p_1}$$

$$\geq s_i^0 > \frac{M_0 - p_1 \sum_{j=1}^{i-1} s_j^0 - p_2 \left(U - \sum_{j=1}^{i-1} s_j^0\right)}{p_1} - 1, \quad \forall i = 2, \ldots, n,$$

$$s_n = U - \sum_{j=1}^{n-1} s_j^0.$$

Step 6. Determine the critical sublots in the above solution and denote them as $i_1, i_2, \ldots, i_r$. If all of the following conditions are satisfied, then there exists an improved solution, and we go to Step 7a; else, $\boldsymbol{s}^{\mathbf{0}}$ is optimal and we Stop:

1. $M_n(\boldsymbol{s}^{\mathbf{0}}) < M(\boldsymbol{s}^{\mathbf{0}}) - r p_2$.
2. $M_{i_j}(\boldsymbol{s}^{\mathbf{0}}) - p_1 + (j-1)(p_2 - p_1) < M(\boldsymbol{s}^{\mathbf{0}}), \forall j = 1, \ldots, r$, where $i_1, i_2, \ldots, i_r$ are the critical sublots in the network representation using $\boldsymbol{s}^{\mathbf{0}} = \left(s_1^0, s_2^0, \ldots, s_n^0\right)$.
3. $M_t(\boldsymbol{s}^{\mathbf{0}}) + j(p_2 - p_1) < M(\boldsymbol{s}^{\mathbf{0}}), t \in (i_j, i_{j+1}), j = 1, \ldots, r$ (assuming that $i_{r+1} = n$).

Step 7a. Reduce $s_{i_1}^0, s_{i_2}^0, \ldots, s_{i_r}^0$, i.e., the critical sublots by one item each and increase the size of the last sublot by r, i.e.,

$$s_{i_j}^0 = s_{i_j}^0 - 1, \quad \forall j = 1, \ldots, r \quad \text{and} \quad s_n^0 = s_n^0 + r.$$

Step 7b. Calculate the makespan $M'(\mathbf{s^0})$ with the new sublot sizes.
Step 7c. If $M'(\mathbf{s^0}) < M(\mathbf{s^0})$, then set $M(\mathbf{s^0}) = M'(\mathbf{s^0})$ and go to Step 7a; else, the current set of sublot sizes is optimal, and Stop.

Case 2. $p_1 = p_2$

Any solution satisfying $s_i \leq \lceil U/n \rceil\ \forall i = 1, \cdots, n$, is optimal.

Case 3. $p_1 > p_2$

Let $p_1' = p_2$, $p_2' = p_1$, and apply the procedure for Case 1: $p_1' < p_2'$, above.

Next, we address the special case of discrete but equal sublot sizes. We consider the following three cases depending on the processing time values.

Case 1a. $p_1 < p_2$

When the lot size U is divisible by n (the number of sublots), the equal sublot sizes are given as

$$s_i^{\mathrm{EQ}} = \frac{U}{n}, \forall i = 1, \ldots, n.$$

When this is not the case, we find a sublot v for which the total increment by rounding up sublots 1 to v is equal to the total decrement resulting from rounding down of the sublots $v+1$ to n. Such a v is given by $U - n\lfloor U/n \rfloor$. Hence, the sublots sizes are given as

$$\hat{s}_i^{\mathrm{EQ}} = \left\lceil \frac{U}{n} \right\rceil, \quad \forall i = 1, \cdots, U - n\lfloor U/n \rfloor,$$

$$\hat{s}_i^{\mathrm{EQ}} = \left\lfloor \frac{U}{n} \right\rfloor, \quad \forall i = U - n\lfloor U/n \rfloor + 1, \cdots, n.$$

Note that,

$$\frac{M\left(\hat{s}^{\mathrm{EQ}}\right) - M\left(s^{\mathrm{EQ}}\right)}{M\left(s^{\mathrm{EQ}}\right)} = \frac{\left\{\left\lceil \frac{U}{n} \right\rceil - \frac{U}{n}\right\} p_1}{\left(\frac{U}{n}\right) p_1 + np_2} < \frac{\left\lceil \frac{U}{n} \right\rceil - \frac{U}{n}}{\left(\frac{U}{n}\right) + n} \text{ (since } p_1 < p_2)$$

$$< \frac{1}{n}\left(\text{since } \left\lceil \frac{U}{n} \right\rceil - \frac{U}{n} < 1 \quad \text{and} \quad \frac{U}{n} > 1, \text{ by assumption}\right).$$

Thus, the relative error due to this approximation is less than $1/n$.

Case 1b. $p_1 = p_2$

In this case, since $p_1 = p_2$, we have

$$M(s) = p_1 \max_{1 \leq i \leq n} \{s_i + U\}.$$

Since the sublot sizes are restricted to take integer values, we must have $s_i = \lceil \forall U/n \rceil$ for some i. Therefore,

$$M(s) \geq p_1 (\lceil U/n \rceil + U).$$

Hence, since any solution satisfying $s_i \leq \lceil U/n \rceil, \forall i = 1, \cdots, n$, with at least one sublot of size $\lceil U/n \rceil$ achieves a makespan equal to this lower bound, it is optimal.

Case 1c. $p_1 > p_2$

The problem under consideration is equivalent to its inverse in which the roles of machines 1 and 2 are reversed, and the schedule is read from right to left. Hence, we can simply interchange the processing times, i.e., p_2 is the processing time per item on machine 1 and p_1 is the processing time per item on machine 2. The resulting situation is the same as in Case 1a discussed above.

The solution procedure for this case is summarized in Fig. 3.9.

3.4 2/1/C/II/CV/{Lot-Detached Setup Times, No-Wait}

Next, we consider a no-wait flow shop. The problem is to determine continuous optimal sizes of n (given) sublots of a single lot in the presence of detached setups. Let Φ_i and Δ_i denote the idle times immediately preceding the ith sublot on machines 1 and 2, respectively (see Fig. 1.10). The makespan can be expressed in terms of Δ_i and Φ_i as follows

$$M = t_2 + p_2 U + \Delta_1 + \sum_{i=2}^{n} \Delta_i \quad \text{(in terms of } \Delta_i\text{)},$$

where

$$\Delta_1 = \max\{0, t_1 + p_1 s_1 - t_2\}$$

and

$$\Delta_i = \max\{0, p_1 s_i - p_2 s_{i-1}\}, \quad \forall i = 2, \ldots, n,$$

and

$$M = t_1 + p_1 U + \Phi_1 + \sum_{i=2}^{n} \Phi_i + p_2 s_n \quad \text{(in terms of } \Phi_i\text{)},$$

where

$$\Phi_1 = \max\{0, t_2 - (t_1 + p_1 s_1)\}$$

and

$$\Phi_i = \max\{0, p_2 s_{i-1} - p_1 s_i\}, \quad \forall i = 2, \ldots, n.$$

Let $\boldsymbol{I}$ represent the total idle time on machine 2, i.e., $\mathbf{I} = \sum_{i=1}^{n} \Delta_i$. In order to minimize the makespan M, it is sufficient to minimize the total idle time $\boldsymbol{I}$ on machine 2. A linear programming formulation to achieve this is as follows:

$$\text{Minimize: } \sum_{i=1}^{n} \Delta_i$$

Subject to:

$$\begin{aligned}
&\Delta_1 \geq t_1 + p_1 s_1 - t_2, \\
&\Delta_i \geq p_1 s_i - p_2 s_{i-1}, \quad \forall i = 2, 3, \ldots, n, \\
&\sum_{i=1}^{n} s_i = U, \\
&\Delta_i \geq 0, \quad \forall i = 1, \ldots, n, \\
&s_i \geq 0, \quad \forall i = 1, \ldots, n.
\end{aligned}$$

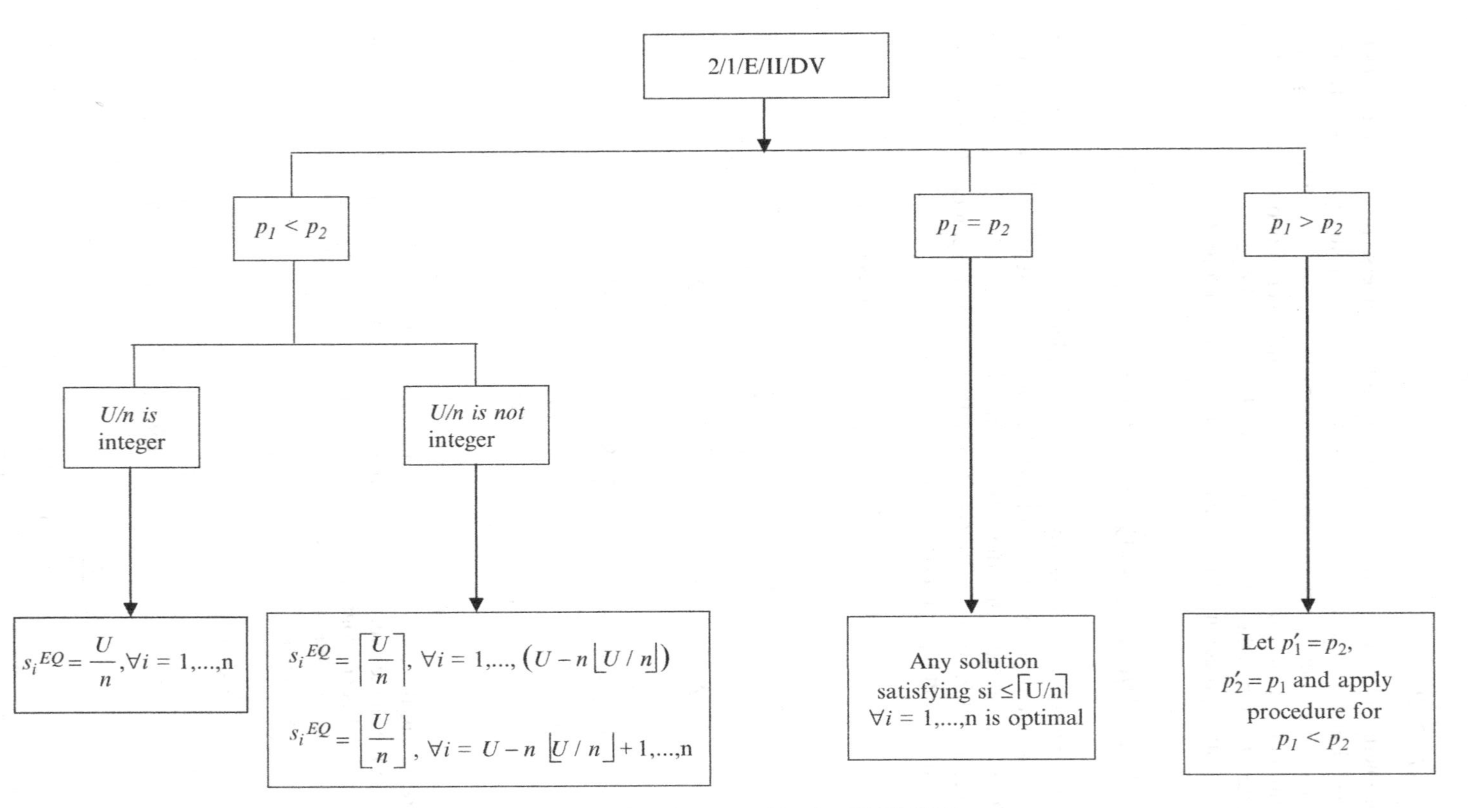

FIGURE 3.9. Solution procedure for 2/1/E/II/DV

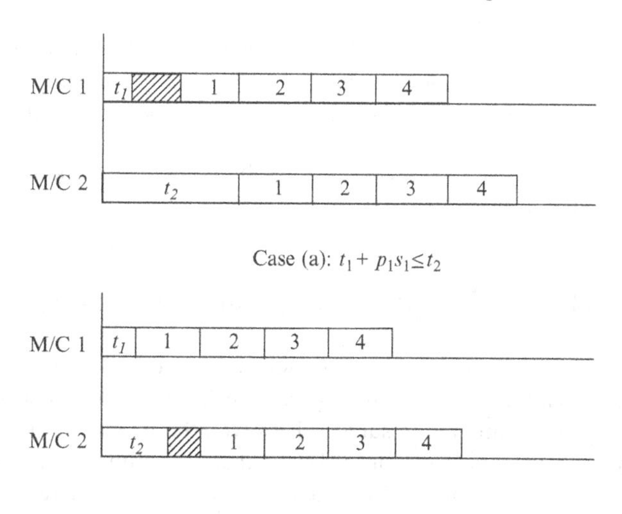

FIGURE 3.10. Implementation of geometric sublot sizes for the solution of the 2/1/C/II/CV/Lot-Detached Setup Times, No-Wait Problem

Note that the first two constraints determine the idle times present in front of the sublots on machine 2 while the third constraint ensures that the sublot sizes add up to U. The optimal sublot sizes can be obtained directly by solving this model.

Consider the geometric sublot sizes given by (3.3). In case $t_1 + p_1 s_1 \leq t_2$ (see Fig. 3.10, Case (a)), no idle time occurs on machine 2, and therefore, the geometric sublot sizes are optimal. In case $t_1 + p_1 s_1 > t_2$ (see Fig. 3.10, Case (b)), an idle time is incurred on machine 2 before the processing of sublot 1. However, any reduction in size of the first sublot (in order to reduce this idle time on machine 2) will not result in the decrement of makespan since (1) the number of items by which the size of the first sublot is reduced will need to be processed eventually and (2) the schedule of the sublots on machine 2 is tight. Hence, we can state the following properties of the optimal solution for the problem on-hand.

Lemma 3.3 *The minimum value of total idle time $\boldsymbol{I}^*$ is such that*

$$\Delta_1^* = \max\{0, (t_1 + s_1 p_1) - t_2\} \geq 0, \quad \text{while } \Delta_i^* = 0,\ i = 2, \ldots, n,$$

where

$$s_1 = \frac{U}{\sum_{k=0}^{n-1} q^k}.$$

Theorem 3.5 *The optimal sublot sizes are geometric in q and are given by*

$$s_i^* = \frac{Uq^{i-1}}{\sum\limits_{k=0}^{n-1} q^k}, \forall i = 1, 2, 3, \ldots, n.$$

Corollary 1 *The optimal sublot sizes form a compact block structure as shown in Fig. 3.5; however, the idle time may be present before the first sublot on machine 1 or before the first sublot on machine 2, depending on the values of* t_1, t_2, *and* s_1.

As in the previous case (see Sect. 3.2), the sublots are increasing in size if $p_1 < p_2$, and decreasing in size if $p_1 > p_2$.

3.5 2/1/C/II/DV/{Lot-Detached Setup Times, No-Wait}

This is the integer version of the problem discussed in Sect. 3.4. Due to the requirement of discrete sublot sizes, there is no guarantee that a solution exists that would result in no idle time among the sublots. However, the intent is to find one, if it exists. Sriskandarajah and Wagneur [32] show that the optimal sublot sizes, s_i^{INT}, for all i, $i = 1, \ldots, n$, for this problem are either $\lfloor s_i \rfloor$ or $\lceil s_i \rceil$, where

$$s_i = \frac{U - \sum_{k=1}^{i-1} s_k^{\mathrm{INT}}}{\sum_{k=1}^{n-i+1} q^{k-1}}, \qquad \forall i = 1, \ldots, n.$$

This expression corresponds to iteratively sizing the ith sublot at the ith stage, as if it were the first sublot of a lot containing $U - \sum_{i'=1}^{i-1} s_{i'}^{\mathrm{INT}}$ items. However, the exploitation of this structure of the optimal integer sublot sizes would require the enumeration of all 2^n possible solutions, thereby, leading to a nonpolynomial algorithm. A mathematical programming formulation of this problem follows from the formulation for the continuous version of the problem given in Sect. 3.4.

$$\text{Minimize:} \quad \sum_{i=1}^{n} \Delta_i^+$$

Subject to:

$$\Delta_1^+ - \Delta_1^- = t_1 + p_1 s_1 - t_2,$$

$$\Delta_i^+ - \Delta_i^- = p_1 s_i - p_2 s_{i-1} \quad \forall i = 2, 3, \ldots, n,$$

$$\sum_{i=1}^{n} s_i = U,$$

$$\Delta_i^+, \Delta_i^- \geq 0, \quad \forall i = 1, \ldots, n,$$

$$s_i \geq 0, \quad \forall i = 1, \ldots, n,$$

$$s_i \text{ integer}, \quad \forall i = 1, \ldots, n.$$

A heuristic solution procedure for the solution of this problem is as follows.

3.5.1 *A Heuristic Solution Procedure for Unequal Integer Sublot Sizes* $\left(H_{UE}^{INT}\right)$

Step 1. Initialize the procedure by setting $U_0 = 0$, $U_1 = U$, $\hat{S} = 0$. Calculate the continuous sublot sizes s_i using Expressions (3.2a) and (3.3).

Step 2. Find integer sublot sizes by rounding down and adding 1 to the optimal continuous sublot sizes. Maintain a count of the total number of items allocated, i.e.,

For $i = 1$ to n

$$s_i^{\mathrm{INT}} = \lfloor s_i \rfloor + 1$$
$$U_0 = U_0 + s_i^{\mathrm{INT}}.$$

Step 3. Find the number of additional items allocated as follows:

$$U_0 = U_0 - U_1.$$

Step 4. Identify all sublots, which contain more than one item, and include them in set $\hat{S}$, i.e.,

$$\hat{S} = \left\{ i : s_i^{\mathrm{INT}} > 1, i = 1, \ldots, n \right\}.$$

Step 5. WHILE the number of additional items, U_0, is greater than zero:

(a) Calculate the difference between integer and continuous sublot sizes for all sublots in set $\hat{S}$ as follows:

$$d_i = s_i^{\mathrm{INT}} - s_i, \quad \forall s_i^{\mathrm{INT}} \in \hat{S}.$$

(b) Identify the sublot with the maximum difference, i.e.,

$$d_m = \max_{s_i^{\mathrm{INT}} \in \hat{S}} \{d_i\}.$$

(c) Decrease the number of items in this sublot by 1, i.e.,

$$s_m^{\mathrm{INT}} = s_m^{\mathrm{INT}} - 1.$$

(d) If this sublot has only one item remaining, then remove it from set $\hat{S}$, to ensure positive sublot sizes, i.e.,

$$\text{If } s_m^{\mathrm{INT}} = 1, \quad \hat{S} = \hat{S} - \left\{ s_m^{\mathrm{INT}} \right\}.$$

(e) Decrease the number of additional items by 1:

$$U_0 = U_0 - 1.$$

END

TABLE 3.5. Data for the illustration of $H_{\text{UE}}^{\text{INT}}$

Lot size (U)	30
Number of sublots (n)	5
Processing time on machine 1 (p_1)	2
Processing time of machine 2 (p_2)	3

Example 3.2 Consider a problem with the data shown in Table 3.5.

Step 1. Set $U_0 = 0$ and $U_1 = 30$.
Step 2. Calculation of integer sublot sizes.

	Continuous sublot size	Integer sublot size	Total items allocated
s_1	2.3	$\lfloor 2.3 \rfloor + 1 = 3$	3
s_2	3.4	$\lfloor 3.4 \rfloor + 1 = 4$	7
s_3	5.1	$\lfloor 5.1 \rfloor + 1 = 6$	13
s_4	7.7	$\lfloor 7.7 \rfloor + 1 = 8$	21
s_5	11.5	$\lfloor 11.5 \rfloor + 1 = 12$	33

Step 3. $U_0 = 33 - 30 = 3$.
Step 4. $\hat{S} = \{1, 2, 3, 4, 5\}$.
Step 5.

Iteration 1:

d_i	$d_{\max}$	s_i^{INT}	Total items allocated
$d_1 = 0.7$		3	3
$d_2 = 0.6$		4	7
$d_3 = 0.9$	0.9	5	12
$d_4 = 0.3$		8	20
$d_5 = 0.5$		12	32

Iteration 2:

d_i	$d_{\max}$	s_i^{INT}	Total items allocated
$d_1 = 0.7$	0.7	2	2
$d_2 = 0.6$		4	6
$d_3 = -0.1$		5	11
$d_4 = 0.3$		8	19
$d_5 = 0.5$		12	31

Iteration 3:

d_i	d_{max}	s_i^{INT}	Total items allocated
$d_1 = -0.3$		2	2
$d_2 = 0.6$	0.6	3	5
$d_3 = -0.1$		5	10
$d_4 = 0.3$		8	18
$d_5 = 0.5$		12	30

Since the total number of allocated items is 30, the algorithm stops. The integer sublot sizes are 2, 3, 5, 8, and 12.

Consider a variation of the above heuristic to obtain almost equal sublot sizes.

3.5.2 A Heuristic Solution Procedure for Equal Integer Sublot Sizes $\left(H_{EQ}^{INT}\right)$

Step 1. Initialize U_0 by setting $U_0 = U$.

Step 2. If U/n is an integer, the optimal integer sublot sizes are given by U/n; otherwise, find the integer-size sublots as $\lfloor U/n \rfloor$. Find the number of items remaining to be allocated as $U_0 - n\lfloor U/n \rfloor$, i.e.,

IF $U/n = \lfloor U/n \rfloor$,
STOP;
ELSE
$s^{INT} = \lfloor U/n \rfloor$,
$U_0 = U_0 - ns^{INT}$.

Step 3. Starting with the first sublot, increment the number of items in each sublot until the number of remaining items becomes zero.

IF $U_0 > 0$
WHILE $\sum_{i=1}^{v} \hat{s}_i^{INT} + \sum_{i=v+1}^{n} s^{INT} \neq U$,
$\hat{s}_i^{INT} = s^{INT} + 1$.

This results in v sublots ($< n$) of size $s^{INT} + 1$ and the remaining sublots of size s^{INT}.
END

Example 3.3 Consider a problem with the following data (Table 3.6).

Step 1. $U_0 = 30$.
Step 2. $s^{INT} = \lfloor 30/4 \rfloor = \lfloor 7.5 \rfloor = 7$, $U_0 = 30 - 4 \times 7 = 2$.
Step 3. Because, $U_0 > 0$,

$$\hat{s}_1^{INT} = 7 + 1 = 8 \quad \text{and} \quad \hat{s}_2^{INT} = 7 + 1 = 8.$$

TABLE 3.6. Data for the illustration of H_{EQ}^{INT}

Lot size (U)	30
Number of sublots (n)	4
Processing time on machine 1 (p_1)	2
Processing time of machine 2 (p_2)	3

Hence, the sublot sizes are 8, 8, 7, and 7. Apparently, the above heuristic gives optimal sublot sizes if U/n is an integer. Otherwise, it generates sublots of nearly identical integer values.

3.6 $2/1/\{C,V\}/II/CV/\sum_{i=1}^{n} s_i C_{i2}$

Next, we consider the situation involving a different objective function, namely, that of minimizing the sum of the weighted completion times of sublots, where the weight of a sublot is defined as the number of jobs in that sublot, i.e., $\sum_{i=1}^{n} s_i C_{i2}$. The sublot sizes are consistent and continuous. In order to develop a solution methodology, we make use of the following mathematical model for this problem.

$$\text{Minimize:} \quad F(\text{s}, \text{C}) \equiv \sum_{i=1}^{n} s_i C_{i2}$$

Subject to:

$$C_{ik} \geq C_{i-1k} + p_k s_i, \quad \forall i = 2, \ldots, n, k = 1, 2. \tag{3.6}$$

$$C_{i2} \geq C_{i1} + p_2 s_i, \quad \forall i = 2, \ldots, n. \tag{3.7}$$

$$C_{11} \geq s_1 p_1. \tag{3.8}$$

$$\sum_{i=1}^{n} s_i = U. \tag{3.9}$$

$$s_i \geq 0, \quad \forall i = 1, \ldots, n. \tag{3.10}$$

$$C_{ik} \geq 0, \quad \forall i = 2, \ldots, n, k = 1, 2. \tag{3.11}$$

The objective function minimizes the sum of the weighted completion times of all the sublots. Constraint (3.6) ensures that a sublot on any machine is processed only after its preceding sublot has finished processing on that machine. The fact that machine 2 processes a sublot only after that sublot has finished processing on machine 1 is captured in constraint (3.7). The completion time of the first sublot must be greater than or equal to its processing time and this is represented by constraint (3.8). Constraint (3.9) ensures that the sublot sizes sum up to the lot size. Constraints (3.10) and (3.11) are the nonnegativity requirements on the sublots sizes and the completion times.

In order to develop a solution procedure for the above model, we consider two cases as below:

Case 1. $q \leq 1$

For this case, it can be shown [31] that the optimal sublots sizes are nondecreasing, i.e., $s_i \leq s_{i+1}, \forall i = 1, \ldots, n$. As a result, the completion times of the sublots on machine 2 can be written explicitly, and consequently, the above model formulation becomes,

$$\text{Minimize:} \quad F(\text{s}) \equiv \sum_{i=1}^{n} s_i \left(p_1 \sum_{j=1}^{i} s_j + p_2 s_i \right)$$

$$\text{Subject to:}$$

$$\sum_{i=1}^{n} s_i = U.$$

Associating a Lagrange multiplier, λ, with the constraint, we have,

$$L(\text{s}, \lambda) \equiv \sum_{i=1}^{n} s_i \left(p_1 \sum_{j=1}^{i} s_j + p_2 s_i \right) + \lambda \left(\sum_{i=1}^{n} s_i - U \right),$$

or,

$$L(\text{s}, \lambda) \equiv p_1 \sum_{i=1}^{n} s_i \sum_{j=1}^{i} s_j + p_2 \sum_{i=1}^{n} s_i^2 + \lambda \left(\sum_{i=1}^{n} s_i - U \right).$$

Differentiating $L(s, \lambda)$ with respect to $s_1, \ldots, s_n$ and λ and setting them equal to zero, we have

$$\frac{\partial}{\partial s_i} L(\text{s}, \lambda) = 0 \Rightarrow p_1 \left(\sum_{j=1}^{n} s_j + s_i \right) + 2 p_2 s_i + \lambda = 0, \quad \forall i = 1, \ldots, n.$$

$$\frac{\partial}{\partial \lambda} L(\text{s}, \lambda) = 0 \Rightarrow \sum_{i=1}^{n} s_i = U.$$

$$\frac{\partial}{\partial s_{i+1}} L(\text{s}, \lambda) - \frac{\partial}{\partial s_i} L(\text{s}, \lambda) = 0 \Rightarrow (p_1 + 2 p_2)(s_{i+1} - s_i) = 0.$$

Since $(p_1 + 2p_2) > 0$, we must have

$$s_{i+1} = s_i, \quad \forall i = 1, \ldots, (n-1).$$

And, since $\sum_{i=1}^{n} s_i = U$, this results in $s_i = U/n, \forall i = 1, \ldots, n$.

Furthermore, we can show that $F(s)$ is a convex function. To that end, consider the partial derivatives of $F(s)$.

$$\frac{\partial}{\partial s_i} F(\mathrm{s}) = p_1\left(\sum_{i=1}^{n} s_i + s_i\right) + 2p_2 s_i, \quad \forall i = 1, \ldots, n,$$

$$\frac{\partial^2}{\partial s_i^2} F(\mathrm{s}) = 2\,(p_1 + p_2)\,, \quad \forall i = 1, \ldots, n,$$

$$\frac{\partial^2}{\partial s_i s_j} F(\mathrm{s}) = 0.$$

And, its Hessian matrix,

$$H(F) = \begin{bmatrix} 2\,(p_1+p_2) & 0 & \cdot\ \cdot & 0 \\ 0 & 2\,(p_1+p_2)\ 0 & \cdot & \cdot \\ \cdot & 0 & \cdot\ 0 & \cdot \\ \cdot & \cdot & 0\ \cdot & 0 \\ 0 & \cdot & \cdot\ 0 & 2\,(p_1+p_2) \end{bmatrix}.$$

Since $H(F)$ is diagonal with $d_i = 2(p_1 + p_2) > 0, \forall i = 1, \ldots, n$, $F(s)$ is convex. Hence, we have the following result.

Theorem 3.6 *When $q \leq 1$, the optimal continuous and consistent sublot sizes, which minimize the sum of the weighted completion times of all sublots, are given by $s_i^* = U/n, i = 1, \ldots, n$, i.e., they are equal.*

Case 2. $q > 1$

Without loss of generality, let $p_1 = 1$, $p_2 = q$, and $U = 1$. For this case, it can be shown [31] that the optimal solution must necessarily satisfy the following condition:

$$qs_i \geq s_{i+1}, \quad \forall i = 1, \ldots, (n-1).$$

In lieu of this condition, the sublot completion times on machine 2, once again, can explicitly be expressed as,

$$C_{i2} = s_1 + q \sum_{j=1}^{i} s_j.$$

Consequently, we have the following model formulation for this case.

$$\text{Minimize:}\quad F(\mathrm{s}) \equiv \left(s_1 + q \sum_{i=1}^{n} s_i \sum_{j=1}^{i} s_j\right)$$

Subject to:

$$h(\mathrm{s}) \equiv \sum_{i=1}^{n} s_i = 1,$$

$$g(\mathrm{s}) \equiv s_{i+1} - qs_i \leq 0, \quad \forall i = 1, \ldots, (n-1), \text{ and}$$

$$s_i \geq 0, \quad \forall i = 1, \ldots, n.$$

Consider a solution with the sublot sizes given as follows:

$$s_1^* = \frac{\left(\dfrac{q^v - 1}{q - 1}\right) q - (n - v)}{\left(\dfrac{q^{2v} - 1}{q^v - 1}\right) q(n - v) + \left(\dfrac{q^v - 1}{q - 1}\right)^2 q},$$

$$s_i^* = q^{i-1} s_1^*, \quad \forall i = 1, \ldots, v,$$

and

$$s_i^* = \left(\frac{1 - s_1^* \sum_{j=1}^{v} q^{j-1}}{(n - v)}\right), \quad \forall i = (v + 1), \ldots, n,$$

where $v < n$ and $q s_v \geq s_{v+1} \geq s_v$.

This solution prescribes geometric sublot sizes until a sublot v and equal-size sublots for sublots $(v + 1), \ldots, n$. A typical schedule would appear as shown in Fig. 3.11.

As in Case 1, the objective function for this case can also be shown to be convex, and hence, in order to prove that the above solution is optimal, it will be sufficient to show that it satisfies the Karush–Kuhn–Tucker (KKT) conditions. The KKT conditions for the above model formulation are as follows:

$$\nabla F(s^*) + \sum_{i=1}^{v-1} \lambda_i \nabla g(s^*) + \delta \nabla h(s^*) = 0, \quad \text{and} \quad \lambda_i \geq 0, \forall i = 1, \ldots, (v - 1), \tag{3.12}$$

where $\lambda_i, i = 1, \cdots, v - 1$, are the Lagrange multipliers associated with the first $v - 1$ binding constraints (being geometric), and δ is the Lagrange multiplier associated with $h(s)$ above.

Using (3.12), we can show that the Lagrange multipliers $\lambda_i, i = 1, \ldots, (v - 1)$, are nonnegative and the resulting sublot sizes are identical to those stated above.

In case $v = n$, i.e., all sublot sizes are geometric, the resulting system of equations has a consistent solution, and hence, it is sufficient to show the nonnegativity

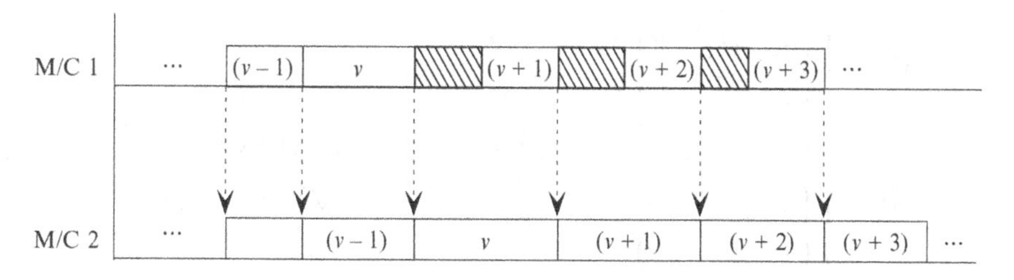

FIGURE 3.11. Optimal schedule for 2/1/C/II/CV/$\sum_{i=1}^{n} s_i C_{i2}$ when $q > 1$

of the Lagrange multipliers, λ_i, $i = 1, \ldots, (n-1)$. The Lagrange multipliers (see [31] for details) are given as,

$$\lambda_{n-1} = \frac{-q^{2n} + 2q^{n+1} + 2q^n - 2q - 1}{(q^n - 1)(q + 1)\sum_{j=0}^{n-1} q^j}$$

and

$$\lambda_i = \lambda_{n-1} \sum_{j=0}^{n-i-1} q^j + \sum_{j=i-1}^{n-1} (s_n - s_j)\, q^{j-i}, \quad i = 1, \ldots, n-2.$$

Since $\lambda_i > \lambda_{n-1}, i = 1, \ldots, (n-2)$, it is sufficient to check the nonnegativity of λ_{n-1}, which is ensured if the numerator (for a given q) is nonnegative as the denominator is always positive. The algorithm, thus, begins by checking this condition, and then, adjusts the sublot sizes accordingly, as follows:

Step 1. Initialize by setting $v = 0$.
Step 2. Is $f(q) > 0$, where $f(q) = -q^{2n} + 2q^{n+1} + 2q^n - 2q - 1$? If yes, then the geometric sublots are optimal and Stop; otherwise, go to Step 3.
Step 3. Increase v, i.e., $v = v + 1$. Calculate new sublot sizes as

$$\begin{aligned} s_1^* &= \frac{\left(\frac{q^v-1}{q-1}\right) q - (n-v)}{\left(\frac{q^{2v}-1}{q^v-1}\right) q(n-v) + \left(\frac{q^v-1}{q-1}\right)^2 q}, \\ s_i^* &= q^{i-1} s_1^*, \quad \forall i = 1, \ldots, v, \\ s_i^* &= \left(\frac{1 - s_1^* \sum_{j=1}^{v} q^{j-1}}{(n-v)} \right), \quad \forall i = (v+1), \ldots, n. \end{aligned}$$

Step 4. Is $qs_v \geq s_{v+1} \geq s_v$?. If yes, then the current sublot sizes are optimal and Stop; otherwise, go to Step 3.

It is interesting to note that the closed-form expressions for optimal sublot sizes can be determined when $n = 2$ and 3. These are given in [31] and are as follows:

$$(s_1^*, s_2^*) = \begin{cases} \left(\frac{1}{q+1}, \frac{q}{q+1}\right), & \text{if } 1 \leq q \leq 1 + \sqrt{2}, \\ \left(\frac{q-1}{2q}, \frac{q+1}{2q}\right), & \text{if } 1 + \sqrt{2} \leq q, \end{cases}$$

$$(s_1^*, s_2^*, s_3^*) = \begin{cases} \left(\frac{1}{q^2+q+1}, \frac{q}{q^2+q+1}, \frac{q^2}{q^2+q+1}\right), & \text{if } 1 \leq q \leq \left(1 + \sqrt{5}\right)/2, \\ \left(\frac{1}{2}\frac{q^2+q-1}{q^3+q^2+q}, \frac{1}{2}\frac{q^3+q^2-q}{q^3+q^2+q}, \frac{1}{2}\frac{q^3+2q+1}{q^3+q^2+q}\right), & \text{if } \left(1 + \sqrt{5}\right)/2 \leq q \leq \left(3 + \sqrt{13}\right)/2, \\ \left(\frac{q-2}{3q}, \frac{q+1}{3q}, \frac{q+1}{3q}\right), & \text{if } \left(3 + \sqrt{13}\right)/2 \leq q. \end{cases}$$

The above analysis holds for consistent sublot sizes. Unlike the makespan criterion, where no improvement in the objective function is possible under variable sublot sizes, one might expect decrement in the completion time-based objective function under variable sublot sizes (if each job, as it is completed on machine 1 can be reconfigured to form a new sublot (see Sect. 2.1.6)) since it is influenced by the completion time of each sublot. The following conjectures can be made for this problem [31].

Conjecture 3.1 *When $q \leq 1$, the total weighted sublot completion time value is minimized by equal sublots on both machines.*

Conjecture 3.2 *When $q > 1$, the total weighted sublot completion time value is minimized by geometric sublots on the first machine, and equal sublots on the second machine.*

These conjectures are based on the notion that the processing of sublots should be continuous on the dominant machine and should start as soon as possible. When $q \leq 1$, machine 1 is dominant and determines the sublot sizes. When $q > 1$, machine 2 is dominant and geometric sublot sizes allow optimal sublot processing to start as early as possible on machine 2. We illustrate this by using the following example.

Example 3.4 Let $U = 80$, $p_1 = 1$, $p_2 = 8$, and $n = 2$. Then, the optimal consistent sublot sizes can be obtained by using the algorithm (or the closed-form expression for $n = 2$) presented above. We obtain $s_1^* = 30$ and $s_2^* = 50$. The total weighted sublot completion time value for these sublot sizes is $(30 \times 150 + 50 \times 350) = 22,000$ (see Fig. 3.12a).

The geometric sublot sizes for the above data are $s_1 = 16$ and $s_2 = 64$ (see (3.2a) and (3.3)). In case geometric sublot sizes are used on machine 1 and equal sublot sizes ($= 40$) on machine 2 (see Fig. 3.12b), the resulting weighted sublot

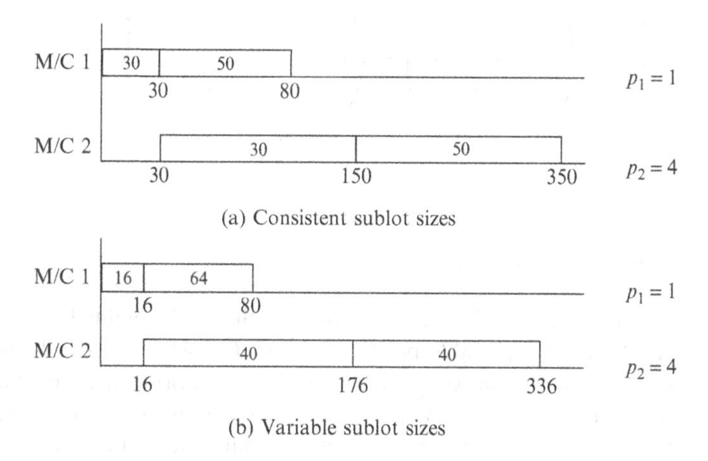

FIGURE 3.12. Nonoptimality of consistent sublot sizes for the total weighted sublot completion time performance measure

completion time = $(40 \times 176 + 40 \times 336) = 20,480$, which is smaller than that for the consistent sublots.

Hence, for the 2/1/{C,V}/II/CV/$\sum_{j=1}^{n} s_i C_{i2}$ problem, when $q \leq 1$, the optimal sublot sizes are equal for both consistent and variable sublot size scenarios. However, when $q > 1$, optimal consistent sublot sizes can be obtained by using the algorithm presented above. The optimal variable sublot sizes are geometric on machine 1 and equal on machine 2.

3.7 2/1/C/II/CV/$\sum_{i=1}^{n} s_i C_{i2}$, Sublot-Attached Setup Times

Next, we consider the problem discussed in Sect. 3.6 but add the following two features in it (1) the number of sublots is not known a priori but needs to be determined and (2) each sublot requires a sublot-attached setup before its processing. Let $\bar{n}$ be an upper bound on the number of sublots, and τ_1 and τ_2 be the sublot-attached setup times on machines 1 and 2, irrespective of the size of a sublot.

A mathematical formulation of this problem is as follows:

$$\text{P-1: Minimize: } F(s, C) \equiv \sum_{i=1}^{\bar{n}} s_i C_{i2} \tag{3.13}$$

Subject to:

$$C_{i2} = i\tau_1 + p_2 \sum_{j=1}^{i} s_j + I_i, \quad \forall i = 1, \ldots, \bar{n}, \tag{3.14}$$

$$I_1 = \tau_1 + p_1 s_1, \tag{3.15}$$

$$I_i = \max\left\{ I_{i-1}, \left(it_1 + p_1 \sum_{j=1}^{i} s_j \right) - \left((i-1)t_2 + p_2 \sum_{j=1}^{i-1} s_j \right) \right\}, \quad \forall i = 2, \ldots, \bar{n}, \tag{3.16}$$

$$\sum_{i=1}^{\bar{n}} s_i = U, \tag{3.17}$$

$$s_i \geq 0, \quad \forall i = 1, \ldots, \bar{n}. \tag{3.18}$$

The objective function $F(s, C)$ seeks to minimize the total weighted sublot completion time of all the $\bar{n}$ possible positive sublots. Constraint (3.14) defines the completion time of any sublot i on machine 2 as the sum of (a) setup times of all previous sublots including sublot i on machine 2, (b) processing times of all previous sublots including sublot i on machine 2, and (c) cumulative idle time appearing before sublot i on machine 2. Constraint (3.15) defines the idle time encountered ahead of sublot 1 on machine 2 and is the sum of its setup and processing time on machine 1. Constraint (3.16) defines the cumulative idle time on machine 2 for sublots $2, \ldots, \bar{n}$. Two cases are possible (1) the cumulative idle time remains the same, i.e., $I_i = I_{i-1}$, implying that sublot $(i-1)$ finishes processing on machine 2 later than the completion of sublot i on machine 1 or (2) the cumulative idle time increases, implying that sublot $(i - 1)$ finishes processing on machine 2 before the completion of sublot i on machine 1. Constraint (3.17) ensures that the sum of sublot sizes does not exceed the lot size. The last constraint defines the sublot sizes to be nonnegative.

When the idle time in constraint (3.16) is achieved by the first term of the max expression for all the sublots, the following inequality holds:

$$C_{i2} \geq C_{i+11}, \quad \forall i = 1, \ldots, \bar{n}. \tag{3.19}$$

This implies that machine 2 is the unique bottleneck. Alternatively, when constraint (3.16) is satisfied by the second term of the max expression for all the sublots, then the following inequality holds

$$C_{i2} < C_{i+11}, \quad \forall i = 1, \ldots, \bar{n}, \tag{3.20}$$

which implies that machine 1 is the unique bottleneck. We refer to this property as the Consistent Bottleneck Property (CBP) implying that the bottleneck remains the same in the optimal solution. However, one might expect that the inequalities (3.19) or (3.20) might not hold exclusively for all optimal solutions and there might exist an optimal solution in which some sublots satisfy (3.19) while the remaining satisfy (3.20). This is indeed true as is illustrated in the example below.

Example 3.5 (CBP is not satisfied in the optimal solution) Consider a two-machine flow shop and a single lot consisting of 25 jobs, each requiring 2.6 and 3 units for processing on machines 1 and 2, respectively. The setup times are 11 and 5 units on machines 1 and 2, respectively. The upper bound on the number of sublots is 5.

The optimal solution to this problem consists of five sublots of sizes 7.49, 6.34, 5, 3.72, and 2.45. The cumulative idle times are 30.4795, 30.4795, 30.4795, 31.1459, and 32.3312. The resulting objective function value is 2229.68. It can be seen that for the first three sublots, the first expression in constraint (3.16) achieves the maximum, and hence, machine 2 is the bottleneck. For sublots 4 and 5, the second expression in (3.16) achieves the maximum, and hence, machine 1 is the bottleneck.

When machine 1 is constrained to be the bottleneck, the optimal solution results in sublot sizes of 7.49, 6.34, 5.01, 3.72, and 2.44. The idle times are 30.4701, 30.4801, 30.4901, 31.1466, and 32.3318. The objective function value is 2229.69. When machine 2 is constrained to be the bottleneck, the optimal solution results in sublot sizes of 7.64, 6.51, 5.2, 3.7, and 1.95. The idle time is 30.8726 and remains the same for all the sublots. The objective function value is 2231.68.

Having made the above observation, a natural approach to develop a solution procedure for the problem under consideration would be to identify conditions under which:

A. Only machine 1 is the bottleneck in the optimal solution.
B. Only machine 2 is the bottleneck in the optimal solution.
C. Both machines 1 and 2 are the bottlenecks in the optimal solution.

We consider each of these cases below.

Case A. Machine 1 is the bottleneck

In the above example, if machine 1 is constrained to be the bottleneck, it can be seen that the following condition holds:

$$\tau_2 + p_2 s_i \leq \tau_1 + p_1 s_{i+1}, \quad \forall i = 1, \ldots, (\bar{n} - 1). \tag{3.21}$$

Under certain conditions, machine 1 is indeed the unique bottleneck in all optimal solutions. This result is formally stated below.

Theorem 3.7 (Bukchin et al. [5]) *If $p_1 \geq p_2$ and $\tau_1 \geq 2\tau_2$, then, in an optimal solution, machine 1 is the unique bottleneck.*

Since machine 1 is the unique bottleneck, the completion time of sublot i is now the sum of the setup and processing times of sublots 1 to i on machine 1 and its setup and processing time on machine 2, i.e.,

$$C_{i2} = \sum_{j=1}^{i} \left(\tau_1 + p_1 s_j\right) + \tau_2 + p_2 s_i.$$

The objective function $F(\boldsymbol{s}, \boldsymbol{C})$ can, now, be written as,

$$\begin{aligned} F(s, C) &= \sum_{i=1}^{\bar{n}} s_i \left(\sum_{j=1}^{i} \left(\tau_1 + p_1 s_j\right) + \tau_2 + p_2 s_i \right) \\ &= \tau_1 \sum_{i=1}^{\bar{n}} i s_i + p_1 \sum_{i=1}^{\bar{n}} s_i \sum_{i=1}^{i} s_j + \tau_2 \sum_{i=1}^{\bar{n}} s_i + p_2 \sum_{i=1}^{\bar{n}} s_i^2. \end{aligned}$$

Consider the second term in the above expression:

$$\begin{aligned} p_1 \sum_{i=1}^{\bar{n}} s_i \sum_{j=1}^{i} s_j &= p_1 \left\{ s_1^2 + s_2 \left(s_1 + s_2\right) + s_3 \left(s_1 + s_2 + s_3\right) + \cdots \right\} \\ &= p_1 \left\{ s_1^2 + s_2^2 + s_3^2 + \cdots + s_M^2 + s_1 s_2 + s_1 s_3 + s_1 s_4 + \cdots \right\} \\ &= \frac{p_1}{2} \left\{ \sum_{i=1}^{\bar{n}} s_i^2 + \left(s_1 + s_2 + s_3 + \cdots + s_M\right)^2 \right\} \\ &= \frac{p_1}{2} \left\{ \sum_{i=1}^{\bar{n}} s_i^2 + U^2 \right\}, \end{aligned}$$

$$F(\mathrm{s}, \mathrm{C}) = \tau_1 \sum_{i=1}^{\bar{n}} i s_i + \tau_2 \sum_{i=1}^{\bar{n}} s_i + \left(\frac{p_1}{2} + p_2\right) \sum_{i=1}^{\bar{n}} s_i^2 + \frac{p_1}{2} U^2.$$

Note that $\sum_{i=1}^{\bar{n}} s_i = U$. In addition, we need to enforce constraint (3.21). Since n, the number of sublots, is not known, (3.21) can be written as,

$$g(s) \equiv \left[\tau_2 + p_2 s_1 - (\tau_1 + p_1 s_{i+1})\right] s_i \leq 0, \quad \forall i = 1, \ldots, (\bar{n} - 1). \quad (3.22)$$

Hence, an equivalent formulation (CBP-1) to P-1, for this case, is as follows:

$$\text{CBP-1: Minimize:} \quad F^{\text{CBP-1}}(\text{s, C}) \equiv \tau_1 \sum_{i=1}^{\bar{n}} i s_i + \tau_2 U + \left(\frac{p_1}{2} + p_2\right) \sum_{i=1}^{\bar{n}} s_i^2 + \frac{p_1}{2} U^2$$

Subject to:

$$\left[\tau_2 + p_2 s_i - (\tau_1 + p_1 s_{i+1})\right] s_i \leq 0, \quad \forall i = 1, \ldots, (\bar{n} - 1),$$

$$\sum_{i=1}^{\bar{n}} s_i = U,$$

$$s_i \geq 0, \quad \forall i = 1, \ldots, \bar{n}.$$

In order to determine the convexity of $F^{\text{CBP-1}}$, we determine its partial derivatives as follows:

$$\frac{\partial}{\partial s_i} F^{\text{CBP-1}} = i\tau_1 + \tau_2 + 2s_i \left(\frac{p_1}{2} + p_2\right) + p_1 U, \quad \forall i = 1, \ldots, \bar{n},$$

$$\frac{\partial^2}{\partial s_i^2} F^{\text{CBP-1}} = 2(p_1 + p_2) > 0, \quad \forall i = 1, \ldots, \bar{n},$$

$$\frac{\partial^2}{\partial s_i s_j} F^{\text{CBP-1}} = p_1 > 0, \quad \forall i \neq j, i, j = 1, \ldots, \bar{n}.$$

Hence, the objective function is convex since its Hessian is positive definite. Similarly the partial derivatives of the left side of constraint (3.22) are as follows:

$$\frac{\partial}{\partial s_i} g = t_2 + 2p_2 s_i - t_1 - p_1 s_{i+1}, \quad \forall i = 1, \ldots, \overline{n},$$

$$\frac{\partial^2}{\partial s_i^2} g = 2p_2 > 0, \quad \forall i = 1, \ldots, \overline{n},$$

$$\frac{\partial^2}{\partial s_i s_j} g = -p_1 < 0, \quad \forall i \neq j, i, j = 1, \ldots, \overline{n}.$$

Hence, the left side of the above constraint is not necessary convex since its Hessian can be indefinite.

Thus, the CBP-1 formulation is not a convex program. However, if we relax constraint (3.22), then the resulting formulation is indeed a convex program. We refer to this formulation as the Relaxed CBP-1 (RCBP-1) formulation. Using RCBP-1, the closed-form expressions for the sublot sizes and the number of sublots [5] are as follows:

$$n^* = \left\lceil \sqrt{\frac{1}{4} + \frac{2U(p_1 + 2p_2)}{\tau_1}} - \frac{1}{2} \right\rceil,$$

$$s_i^* = \frac{U}{n^*} + \frac{\tau_1(n^* + 1 - 2i)}{2(p_1 + 2p_2)}, \quad \forall i = 1, \ldots, n^*,$$

$$= 0, \quad \forall i = (n^* + 1), \ldots, \overline{n}.$$

If this solution satisfies constraint (3.22), then it is an optimal solution of CBP-1. For $i > n^*$, this is satisfied trivially. For $i < n^*$, conditions under which the above solution satisfies constraint (3.22) can be identified as follows. Since $s_{i+1}^* - s_i^* = -\tau_1/(p_1 + 2p_2)$ (obtained by using the expression for s_i^* given above), constraint (3.21), for $i = 1, \ldots, \bar{n} - 1$, now becomes,

$$\tau_2 + p_2\left(s_{i+1}^* + \frac{t_1}{(p_1 + 2p_2)}\right) - \tau_1 - p_1 s_{i+1}^* \leq 0.$$

Therefore,

$$(\tau_2 - \tau_1) + (p_2 - p_1)\, s_{i+1}^* + \frac{p_2\tau_1}{(p_1 + 2p_2)} \leq 0$$

or

$$(\tau_1 - \tau_2) + (p_1 - p_2)\, s_{i+1}^* - \frac{p_2\tau_1}{(p_1 + 2p_2)} \geq 0.$$

Since $p_1 \geq p_2$, the above condition is satisfied when

$$\tau_1 - \tau_2 \geq \frac{p_2\tau_1}{(p_1 + 2p_2)} \quad \text{or} \quad \tau_1 \geq \left(\frac{p_1 + 2p_2}{p_1 + p_2}\right)\tau_2.$$

This result is formally stated below.

Theorem 3.8 (Bukchin et al. [5]) *If* $p_1 \geq p_2$ *and* $\tau_1 \geq \left(\frac{p_1+2p_2}{p_1+p_2}\right)\tau_2$, *then the solution of RCBP-1 is optimal for CBP-1.*

When $p_1 = p_2$, the above condition becomes $\tau_1 \geq 1.5\tau_2$ and when $p_1 \gg p_2$, it approaches $\tau_1 \geq \tau_2$. However, as stated in Theorem 3.7, both of these situations are already satisfied since $\tau_1 \geq 2\tau_2$ for machine 1 to be a bottleneck. This result is stated below.

Corollary 2 If $p_1 \geq p_2$ and $\tau_1 \geq 2\tau_2$, then the solution given by RCBP-1 is optimal for the original formulation P-1.

Hence, when $p_1 \geq p_2$ and $\tau_1 \geq 2\tau_2$, then all optimal solutions must have machine 1 as the unique bottleneck and CBP-1 is equivalent to P-1. However, CBP-1 is not a convex program, and hence, closed-form expressions for the sublot sizes and the number of sublots cannot be obtained. CBP-1 is a convex program if constraint (3.22) is relaxed, and the resulting formulation RCBP-1 can be solved easily by using the KKT conditions from which closed-form formulas can be obtained. This solution will satisfy the (sufficient) conditions of Theorem 3.8, which are weaker than those specified in Theorem 3.7. Consequently, the solution will be optimal to P-1. Hence, the solution of RCBP-1 is optimal to P-1.

Case B. Machine 2 is the bottleneck

The constraint that is required for machine 2 to be a bottleneck is as follows:

$$\left((i-1)t_2 + p_2\sum_{j=1}^{i-1} s_j\right) - \left((i-1)t_1 + p_1\sum_{j=2}^{i} s_j\right) \geq 0, \quad \forall i = 2, \ldots, n. \tag{3.23}$$

In fact, we can identify conditions under which machine 2 is the unique bottleneck in all optimal solutions. These are stated below.

Theorem 3.9 (Bukchin et al. [5]) *When $p_2 \geq p_1$ and $\tau_2 \geq \left(\frac{2p_2}{p_1+2p_2}\right)\tau_1$, then, in an optimal solution, machine 2 is the unique bottleneck.*

Since machine 2 is the unique bottleneck, the completion time of any sublot on machine 2 is the sum of the setup and processing times of the first sublot on machine 1 and the setup and processing times of sublots 1 to i on machine 2, i.e.,

$$C_{i2} = t_1 + p_1 s_1 + i t_2 + p_2 \sum_{j=1}^{i} s_j, \quad \forall i = 1, \ldots, n.$$

The objective function $F(\boldsymbol{s}, \boldsymbol{C})$ of P-1 can, now, be written as,

$$\begin{aligned} F^{\text{CBP-2}}(\text{s}, \text{C}) &= \sum_{i=1}^{\overline{n}} s_i C_{i2} \\ &= \sum_{i=1}^{\overline{n}} \left(\tau_1 + p_1 s_1 + i\tau_2 + p_2 \sum_{j=1}^{i} s_j\right) s_i, \end{aligned}$$

which, after simplification, becomes

$$F^{\text{CBP-2}}(\text{s}, \text{C}) = \tau_1 U + p_1 s_1 U + \tau_2 \sum_{i=1}^{\overline{n}} i s_i + \frac{p_2}{2} \sum_{i=1}^{\overline{n}} s_i^2 + \frac{p_2}{2} U^2.$$

In addition, we also need to enforce constraint (3.23). However, since n is a decision variable itself, we replace (3.23) by the following:

$$\left[\left((i-1)\tau_2 + p_2 \sum_{j=1}^{i-1} s_j\right) - \left((i-1)\tau_1 + p_1 \sum_{j=2}^{i} s_j\right)\right] s_i \geq 0, \quad \forall i = 2, \ldots, \overline{n}. \tag{3.24}$$

Hence, a formulation equivalent to P-1, when conditions in Theorem 3.9 are satisfied, is as follows:

CBP-2: Minimize: $F^{\text{CBP-2}}(\text{s}, \text{C}) \equiv \tau_1 U + p_1 s_1 U + \tau_2 \sum_{i=1}^{\overline{n}} i s_i + \frac{p_2}{2} \sum_{i=1}^{\overline{n}} s_i^2 + \frac{p_2}{2} U^2$

Subject to:

$$\left[\left((i-1)t_2 + p_2 \sum_{j=1}^{i-1} s_j\right) - \left((i-1)t_1 + p_1 \sum_{j=2}^{i} s_j\right)\right] s_i \geq 0,$$

$$\forall i = 2, \ldots, \overline{n},$$

$$\sum_{i=1}^{\overline{n}} s_i = U,$$

$$s_i \geq 0, \quad \forall i = 1, \ldots, \overline{n}.$$

By following an analysis similar to the one for the formulation CBP-1, it can be shown that $F^{\text{CBP-2}}(s, C)$ is convex. However, the left side of (3.24) is not concave, and consequently, CBP-2 is not a convex program. As before, by relaxing constraint (3.24), the resulting formulation (RCBP-2) is convex and the closed-form formulas for n^* and s_i^* can be obtained using the KKT conditions [5]. These are as follows:

$$\text{RCBP-2: Minimize: } F^{\text{RCBP}-2}(\text{s, C}) \equiv \tau_1 U + p_1 s_1 U + \tau_2 \sum_{i=1}^{\overline{n}} i s_i + \frac{p_2}{2} \sum_{i=1}^{\overline{n}} s_i^2 + \frac{p_2}{2} U^2$$

Subject to:

$$\sum_{i=1}^{\overline{n}} s_i = U,$$

$$s_i \geq 0, \quad \forall i = 1, \ldots, \overline{n}.$$

And, the closed-form expressions for n^* and s_i^*,

$$n^* = \left\lceil \sqrt{\frac{1}{4} + \frac{2U(p_1 + 2p_2)}{\tau_2}} - \frac{1}{2} \right\rceil,$$

$$\begin{aligned} s_i^* &= -\frac{p_1 U}{p_2} + \frac{U}{n^*}\left(\frac{p_1 + p_2}{p_2}\right) + \frac{\tau_2}{p_2}\left(\frac{n^*+1}{2} - i\right), & i = 1, \\ &= \frac{U}{n^*}\left(\frac{p_1 + p_2}{p_2}\right) + \frac{\tau_2}{p_2}\left(\frac{n^*+1}{2} - i\right), & \forall i = 2, \ldots, n^*. \\ &= 0, & \forall i = (n^*+1), \ldots, \overline{n}. \end{aligned}$$

Unlike the analysis done earlier for the case when machine 1 is the bottleneck, the conditions which guarantee that the above solution satisfies constraint (3.24) cannot be obtained in this case due to the complexity of the relevant expressions involved. Hence, we need to check if the solution to RCBP-2 satisfies constraint (3.24). If it does, then it is optimal to CBP-2 as well. However, we need to develop a different way to solve CBP-2 for the case in which it does not.

As mentioned earlier, the formulation CBP-2 is not a convex program. However, if the number of sublots is fixed, then it becomes a convex program. The resulting formulation CBP-2(n), for a fixed n, is as follows:

$$\text{CBP-2}(n)\text{: Minimize: } F^{\text{CBP-2}(n)}(\text{s, C}) \equiv \tau_1 U + p_1 s_1 U + \tau_2 \sum_{i=1}^{n} i s_i + \frac{p_2}{2} \sum_{i=1}^{n} s_i^2 + \frac{p_2}{2} U^2$$

Subject to:

$$\left((i-1)\tau_2 + p_2\sum_{j=1}^{i-1} s_j\right) - \left((i-1)\tau_1 + p_1\sum_{j=2}^{i} s_j\right) \geq 0,$$

$$\forall i = 2, \ldots, n,$$

$$\sum_{i=1}^{n} s_i = U,$$

$$s_i \geq 0, \quad \forall i = 1, \ldots, n.$$

Denote a solution consisting of n sublots by sublot–sizes $s_1(n), s_2(n), s_3(n), \ldots, s_n(n)$. Consider a corresponding solution for $\overline{n}$ ($\geq n$) sublots having sublot sizes $s_i(\overline{n}) = s_i(n), \forall i = 1, \ldots, n$ and $s_i(\overline{n}) = 0, \forall i = (n+1), \ldots, \overline{n}$. Then, this will be a feasible solution for the $\overline{n}$-sublot problem as long as $\tau_2 \geq \tau_1$ because, then, the constraint (3.23) will be satisfied for $i = n+1, \ldots, \bar{n}$. Thus, under this condition, any feasible solution of the n-sublot problem is a feasible solution of the $\overline{n}$-sublot problem and this must be true for the optimal solution of CBP-2, as well. Hence, when $\tau_2 \geq \tau_1$, an optimal solution to CBP-2 can be obtained by solving the formulation CBP-2($\overline{n}$). When $\tau_2 < \tau_1$, then the optimal solution to CBP-2 can be obtained by solving CBP-2(n) for all possible values of n, or by performing a binary search if $F^{\text{CBP-2}(n)}(s, C)$ is a unimodal function of n and picking the best solution. The resulting solution is optimal to CBP-2, and hence, to P-1 as well. The overall algorithm is shown in Fig. 3.13.

Case C. Both machines 1 and 2 are the bottlenecks

When the conditions specified by Theorems 3.7 and 3.9 do not hold, both machines 1 and 2 are bottlenecks. In that case, a heuristic procedure can be used, which is based on the analysis of the previous cases. The heuristic begins by solving RCBP-1 and RCBP-2:

1. If a solution to RCBP-1 satisfies the relaxed constraint (3.22), then it is optimal to CBP-1 but only an approximation to P-1. If it is not feasible, then we convert CBP-1 to a convex program CBP-1(n) by fixing the number of sublots. This formulation is given below:

$$\text{CBP-1}(n)\text{: Minimize: } F^{\text{CBP-1}(n)}(\text{s, C}) \equiv \tau_1\sum_{i=1}^{n} is_i + \tau_2\sum_{i=1}^{n} s_i + \left(\frac{p_1}{2} + p_2\right)\sum_{i=1}^{n} s_i^2 + \frac{p_1}{2}U^2$$

Subject to:

$$(\tau_2 + p_2 s_i) - (\tau_1 + p_1 s_{i+1}) \leq 0, \quad \forall i = 1, \ldots, (n-1),$$

$$\sum_{i=1}^{n} s_i = U,$$

$$s_i \geq 0, \quad \forall i = 1, \ldots, n.$$

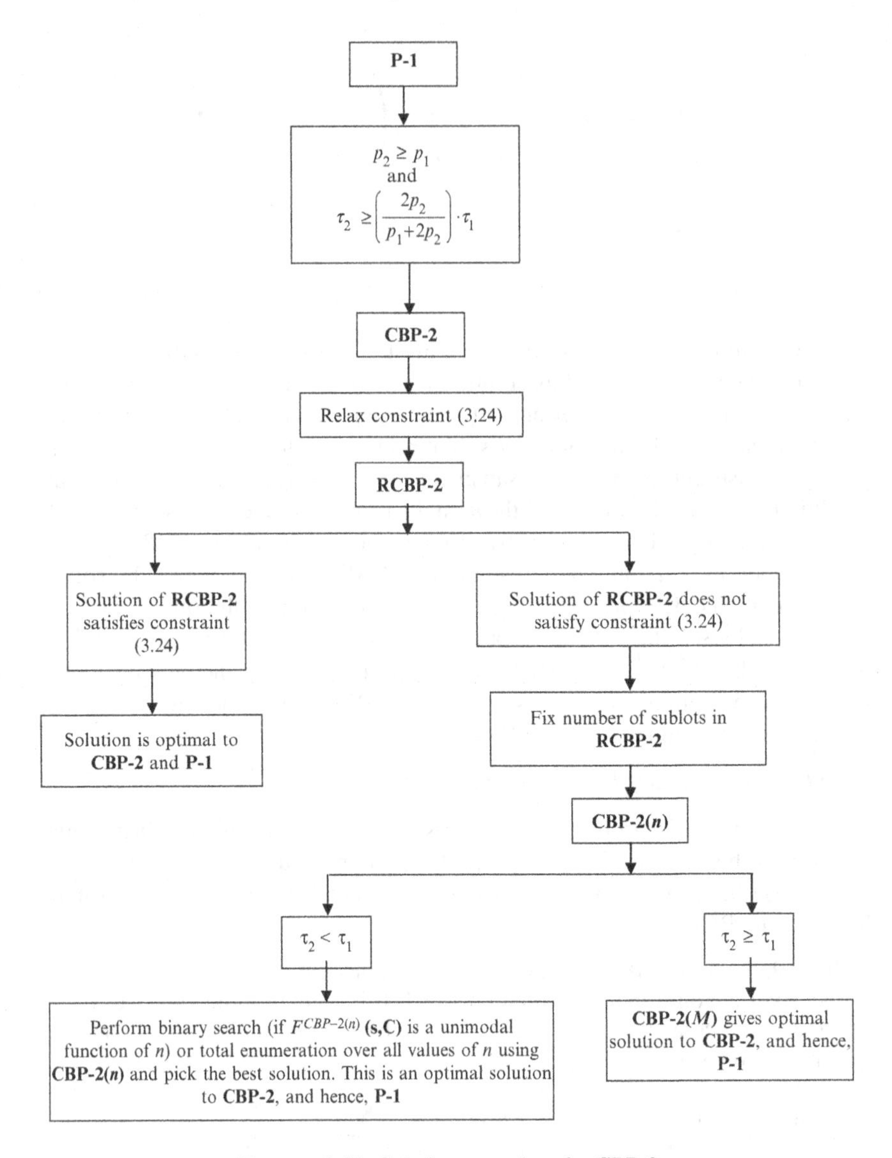

FIGURE 3.13. Solution procedure for CBP-2

The optimal value of n can be obtained by solving the formulation CBP-1(n) for all possible values of n or by performing a binary search if $F^{\text{CBP-1}(n)}(s, C)$ is a unimodal function of n. This solution is optimal to CBP-1 but not for P-1.

2. If a solution of RCBP-2 satisfies constraint (3.24), then this solution is optimal to CBP-2 but only an approximation to P-1. If it does not satisfy constraint (3.24), then we perform a binary search (if $F^{\text{CBP-2}(n)}(s, C)$ is a unimodal function of n) or total enumeration over all possible values of n using CBP-2(n)

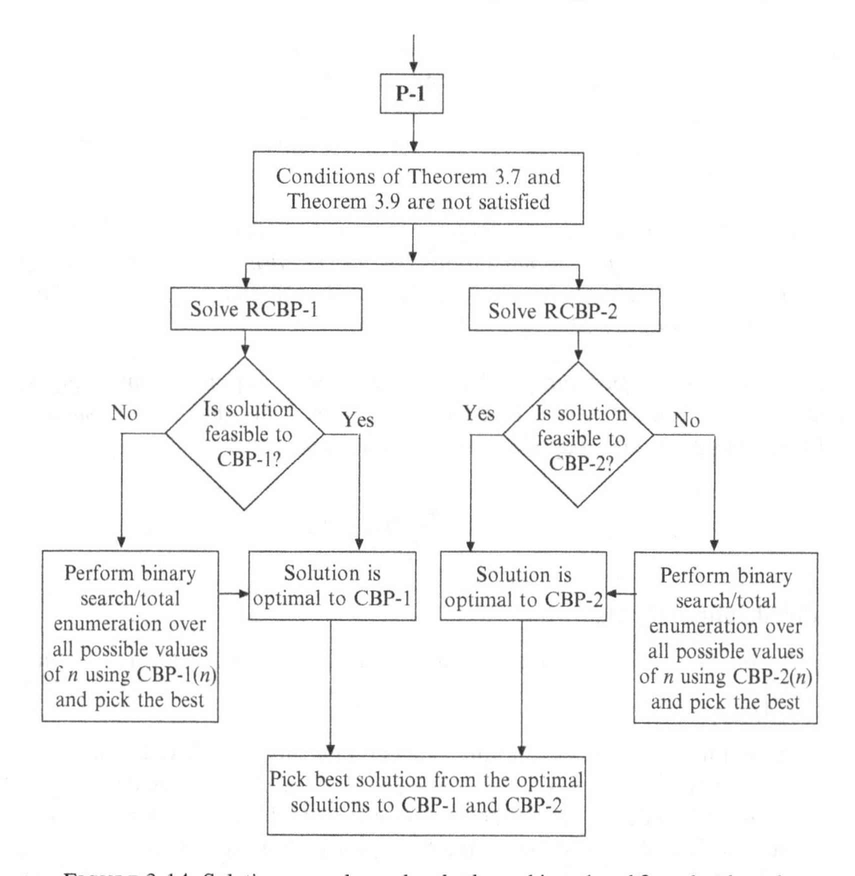

FIGURE 3.14. Solution procedure when both machines 1 and 2 are bottlenecks

and pick the best solution. This solution is, once again, an approximation to P-1. Having obtained optimal solutions to CBP-1 and CBP-2, the algorithm picks the best as an approximation to P-1. The corresponding algorithm is shown in Fig. 3.14.

Thus, the solution procedure to the 2/1/C/II/CV/{$\sum_{i=1}^{N} s_i C_{i2}$, Sublot-Attached Setup} problem first checks whether the optimal solution has a unique bottleneck. If so, it then uses the algorithm described in Case A or Case B above to find an optimal solution to CBP-1 or CBP-2, which are optimal to P-1. If not, then it uses the algorithm in Fig. 3.14 to obtain an approximate solution to P-1.

3.8 2/N/C/II/DV/{No Setup, Unit-Size Sublots}

Next, we consider the streaming of N lots. The simplest situation for this scenario is the one that involves no setup time and assumes unit-size sublots, i.e., $s_{ij} = 1$, $\forall i = 1, \ldots, n_j$ and $j = 1, \ldots, N$. The underlying issue is the determination of the sequence in which to process the lots. We address this problem first for the criterion of minimizing makespan. Clearly, unit-size sublots minimize the completion times of the lots in the absence of any setup time or restrictions that may arise during the transferring of the sublots, and hence, optimize a regular measure. We state this formally as follows.

Theorem 3.10 *Unit-size sublots are optimal for the N-lot, two-machine flow shop, lot streaming problem for the makespan criterion in case no setup and transfer times are incurred and no restrictions on the transferring of the sublots or a limit on queue size on any machine exist.*

The underlying sequencing problem can be solved to optimality by applying Johnson's algorithm to every sublot (or job) of a lot. Thus, there are N' sublots, and Johnson's algorithm is applied to these sublots, where

$$N' = \sum_{j=1}^{N} U_j.$$

Modified Johnson's Algorithm

Step 1. Find a sublot ℓ with the least per unit processing time, i.e., $\min_{1 \le \ell \le N'} \{p_{\ell 1}, p_{\ell 2}\}$:

(a) If the minimum processing time occurs on machine 1, schedule that sublot in the first available position in the sequence. Since ties may be broken arbitrarily, schedule any sublot of a lot requiring the least processing times on machine 1 in this position. Furthermore, since all n_j sublots of lot j would satisfy this criterion, the remaining $(n_j - 1)$ sublots of that lot may be scheduled after the first sublot. This implies that, in the solution to the $2/N'/C_{\max}$ problem, every sublot of lot j will precede all the remaining $N' - \{n_j\}$ sublots. Go to Step 2.

(b) If the minimum processing time occurs on machine 2, schedule the corresponding sublot in the last available position. Arbitrarily select a sublot of this lot and schedule all the remaining $n_j - 1$ sublots of that lot before it. Thus, the solution will have all the sublots of lot j processed consecutively and appearing after all the remaining $N' - \{n_j\}$ sublots. Go to Step 2.

Step 2. Delete all the sublots of the lots that have been sequenced from N'. If no more sublots are remaining, Stop; otherwise, go to Step 1.

Example 3.6 Consider a two-machine two-lot system with data as shown in Table 3.7.

The optimal sequence obtained from the modified Johnson's algorithm is 1–2, with a makespan of 22. Note that if we apply Johnson's algorithm to the aggregated problem, it may result in a different sequence. For example, in the above example, the optimal sequence for the aggregated problem is 2–1 with a makespan of 26.

TABLE 3.7. Data for the illustration of modified Johnson's algorithm

Lot	Unit processing time on machine 1 ($p_{\ell 1}$)	Unit processing time on machine 2 ($p_{\ell 2}$)	Number of jobs
1	2	3	4
2	3	4	2

Johnson's rule, which results in the optimal sequence of the lots in a two-machine flow shop lot streaming problem, can be stated as follows.

Theorem 3.11 *In a two-machine flow shop, lot i precedes lot j, in an optimal sequence, if*

$$\min\left\{p_{i1}, p_{j2}\right\} \leq \min\left\{p_{j1}, p_{i2}\right\}.$$

An easier way to implement the above rule is to partition the lots into two sets. The first set consists of lots for which $p_{i1} < p_{i2}$ while the second set consists of those lots for which $p_{i1} \geq p_{i2}$. The lots in the first set are arranged in the nondecreasing order of their processing times on machine 1 and those in the second set in the nonincreasing order of their processing times on machine 2. The optimal sequence is given by the lots in first set followed by those in the second set.

Since the Johnson's algorithm sequences the lots on machine 1 such that the shorter lots appear earlier in the sequence and the longer lots appear later, while the converse is true on machine 2, its effect is twofold:

1. It causes maximum overlap in the schedule (since longer lots appear in the middle).
2. It minimizes the idle time on machine 2 by sequencing the shortest lot first on machine 1 and then minimizes the tail (or the idle time on machine 1) by sequencing the shorter lots later on machine 2.

Sequencing of Lots Using Run-In and Run-Out Times

Recall that the run-in and run-out times of a job (see Sect. 1.2.6) represent the idle times on machines 2 and 1, respectively. Consequently, the Johnson's rule can also be stated as follows.

Theorem 3.12 *In a two-machine flow shop, lot i precedes lot j, in an optimal sequence, if the following is true*

$$\min\left\{\mathrm{RI}_{i1}, \mathrm{RO}_{j2}\right\} \leq \min\left\{\mathrm{RI}_{j1}, \mathrm{RO}_{i2}\right\}.$$

The corresponding implementation of this rule forms two groups of lots with $\mathrm{RI}_{i1} < \mathrm{RO}_{i2}$ and $\mathrm{RI}_{i1} \geq \mathrm{RO}_{i2}$ and proceeds in a similar fashion to that described above. The makespan expression in terms of the run-in and run-out times can be given as follows:

$$C_{\max} = \max_{1 \le u \le N} \left\{ \sum_{k=1}^{u} \mathrm{RI}_{k1} - \sum_{k=1}^{u-1} \mathrm{RO}_{k2} \right\} + \sum_{i=1}^{N} p_{i2} U_i.$$

In the presence of time lags, a job can begin its processing on machine 2 even when the processing on machine 1 is not completed. The start lag u_i and stop lag v_i represent the minimum delay required between the starts and completions, respectively, of that job on the machines. Then, a generalization of Johnson's rule can be stated as follows.

Theorem 3.13 *In a two-machine flow shop with time lags, job i precedes job j in the optimal schedule if*

$$\min\left\{p_{i1} + d_i, p_{j2} + d_j\right\} \le \min\left\{p_{j1} + d_j, p_{i2} + d_i\right\},$$

where $d_i = \max\{u_i - p_{i1}, v_i - p_{i2}\}$.

The transfer lag may be positive if there exists an idle time on machine 2, after the completion of job 1 on machine 1, or negative if the processing of the job on machines 1 and 2 overlap. In light of this, it is easy to realize that by adding the transfer lags to each processing time in the original Johnson's rule [19], we actually obtain the "idle time" on machine 1 and machine 2 in the presence of time lags. Thus, the concept of minimizing the idle times still holds even in this case. Therefore, we could either calculate the run-in and run-out times and use Theorem 3.12, or calculate the transfer lag and use Theorem 3.13. Both of these approaches are addressed next for the streaming of lots, where each lot contains identical jobs.

To develop expressions for the run-in and run-out times of a lot i in the presence of lot streaming, we consider the two cases of $p_{i1} < p_{i2}$ and $p_{i1} \ge p_{i2}$.

Case 1. $p_{i1} \le p_{i2}$

Figure 3.15 depicts the streaming of a lot containing five jobs with $p_{i1} \le p_{i2}$ for every job.

The run-in and run-out times for this lot are given below:

$$\mathrm{RI}_{i1} = p_{i1}, \tag{3.25}$$

$$\begin{aligned} \mathrm{RO}_{i2} &= U_i p_{i2} - (U_i - 1)\, p_{i1} \\ &= (U_i - 1)\,(p_{i2} - p_{i1}) + p_{i2}. \end{aligned} \tag{3.26}$$

Case 2. $p_{i1} > p_{i2}$

Figure 3.16 depicts the streaming of a lot containing five items with $p_{i1} > p_{i2}$ for every item.

The run-in and run-out times are given below:

$$\begin{aligned} \mathrm{RI}_{i1} &= U_i p_{i1} - (U_i - 1)\, p_{i2} \\ &= (U_i - 1)(p_{i1} - p_{i2}) + p_{i1}. \end{aligned} \tag{3.27}$$

$$\mathrm{RO}_{i2} = p_{i2}. \tag{3.28}$$

In view of both of the above cases, we have

$$\begin{aligned} \mathrm{RI}_{i1} &= \max\{p_{i1}, (U_i - 1)(p_{i1} - p_{i2}) + p_{i1}\}, \\ \mathrm{RO}_{i2} &= \max\{(U_i - 1)(p_{i2} - p_{i1}) + p_{i2}, p_{i2}\}. \end{aligned}$$

Note that when $p_{i1} < p_{i2}$, $\mathrm{RI}_{i1} < \mathrm{RO}_{i2}$, and when $p_{i1} \geq p_{i2}$, $\mathrm{RI}_{i1} \geq \mathrm{RO}_{i2}$. The optimal sequence among the lots can, then, be determined by using Theorem 3.12.

One issue that we need to deal with has to do with the sequencing of individual jobs on the two machines in the presence of arbitrary time lags. In this regard, Johnson [19] has shown the following result.

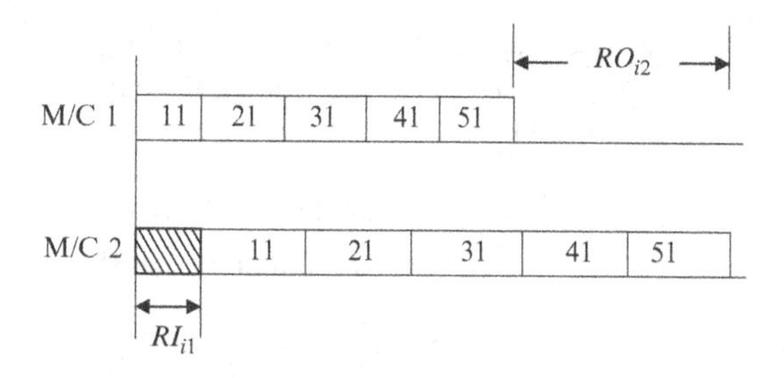

FIGURE 3.15. Run-in and run-out times of a lot when $p_{i1} \leq p_{i2}$

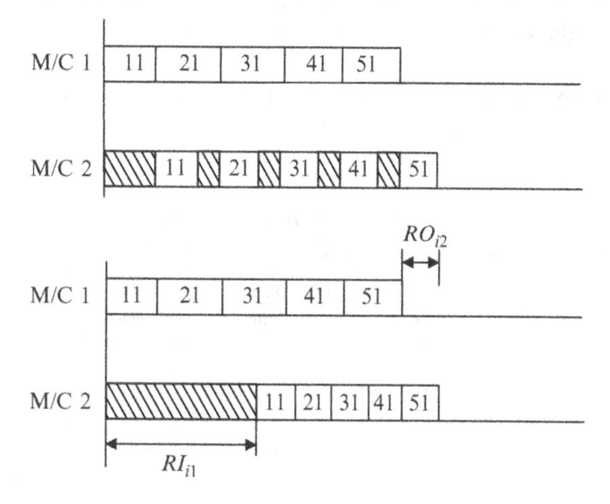

FIGURE 3.16. Run-in and run-out times of a lot when $p_{i1} > p_{i2}$

Theorem 3.14 *In a two-machine flow shop with time lags and makespan minimization objective, a permutation schedule is optimal if the following is satisfied for all consecutive pairs of jobs i and j*

$$d_i \leq d_j + \max\left\{p_{j1}, p_{j2}\right\},$$

where d_i is the transfer lag for job i (see Sect. 1.2.9).

Under lot streaming, the above condition always holds true for any two consecutive jobs. For instance, consider any two consecutive jobs belonging to the same lot i. In case $p_{i1} \geq p_{i2}$, the transfer lag is zero for all pairs of consecutive jobs, while if $p_{i1} < p_{i2}$, the transfer lag of the first job is zero, while those of the other jobs are nonnegative and increase with the position of the job in the lot, thereby, satisfying the above condition in both cases. However, for two consecutive lots, the above condition can, equivalently, be expressed as

$$D_i \leq D_j + \max\left(P_{j1}, P_{j2}\right),$$

where $P_{j1} = U_j p_{j1}$, $P_{j2} = U_j p_{j2}$, $D_i = \max\{p_{i1} - U_i p_{i1}, p_{i2} - U_i p_{i2}\}$, and $D_j = \max\{p_{j1} - U_j p_{j1}, p_{j2} - U_j p_{j2}\}$. Substituting the expressions for D_i and D_j in the above expression, we have

$$\max\left\{p_{i1} - U_i p_{i1}, p_{i2} - U_i p_{i2}\right\} \leq \max\left\{p_{j1} - U_j p_{j1}, p_{j2} - U_j p_{j2}\right\} + U_j \max\left\{p_{j1}, p_{j2}\right\}.$$

Note that when $p_{j1} > p_{j2}$, the right-hand side $p_{j2} - U_j p_{j2} + U_j p_{j1} > 0$. Also, when $p_{j1} < p_{j2}$, the right-hand side $p_{j1} - U_j p_{j1} + U_j p_{j2} > 0$. In other words, the right-hand side remains positive whether $p_{j1} > p_{j2}$ or $p_{j1} \leq p_{j2}$. Similarly, we can show that the left-hand side remains negative irrespective of whether $p_{j1} > p_{j2}$ or $p_{j1} \leq p_{j2}$. Hence, permutation schedules constitute a dominant set for the streaming of lots.

Example 3.7 Consider a five-lot, two-machine problem with data as given in Table 3.8.

TABLE 3.8. Data for the illustration of lot sequencing using run-in and run-out times

	Processing time on		Number
	Machine 1	Machine 2	of items
1	3	4	3
2	3	3	4
3	2	6	3
4	5	3	2
5	4	1	2

TABLE 3.9. Run-in and run-out times

	RI_{i1}	RO_{i2}
1	3	6
2	3	3
3	2	14
4	7	3
5	7	1

TABLE 3.10. Data for the illustration of lot sequencing using transfer lags

Lot	P_{i1}	P_{i2}	Start lag	Stop lag	Transfer lag d_i	$I_{i1} = P_{i1} + d_i$	$I_{i2} = P_{i2} + d_i$
1	9	12	3	4	−6	3	6
2	12	12	3	3	−9	3	3
3	6	18	2	6	−4	2	14
4	10	6	5	3	−3	7	3
5	8	2	4	1	−1	7	1

The run-in and run-out times are calculated and are shown below (Table 3.9).

Let Set $I = \{i : \mathrm{RI}_{i1} < \mathrm{RO}_{i2}\}$ and Set II $= \{i : \mathrm{RI}_{i1} \geq \mathrm{RO}_{i2}\}$. For the example on-hand, Set $I = \{1, 3\}$ and Set II $= \{2, 4, 5\}$. By sorting Set I in the nondecreasing order of RI_{i1} and Set II in the nonincreasing order of RO_{i2} and appending the resulting sequences, the optimal sequence is 3–1–5–4–2, with a makespan of 52.

Sequencing of Lots Using Transfer Lags

In the case of lot streaming, the start and stop lags for lot i are p_{i1} and p_{i2}, respectively. As $P_{i1}(= U_i p_{i1})$ and $P_{i2}(= U_i p_{i2})$ represent the total workload on machines 1 and 2, respectively, then, the transfer lag for lot i, D_i, is given by $\max\{p_{i1} - P_{i1}, p_{i2} - P_{i2}\}$. As mentioned earlier, the transfer lags when added to the processing times give us the idle time, which can be used to guide the sequencing of lots. The method is illustrated below in Table 3.10 using the example considered above.

Note that the numbers in column I_{i1} and I_{i2} of Table 3.10 are the same as in the columns RI_{i1} and RO_{i2} of Table 3.9. Hence, either approach yields the same result.

Next, we consider the situations involving setup times. We first develop general expressions for the run-in and run-out times by analyzing the various cases, and then, invoke Theorem 3.12 to sequence the lots. The cases of lot-detached and lot-attached setups are considered separately and are then illustrated with the help of examples.

3.9 2/N/C/II/DV/{Lot-Detached Setup, Unit-Size Sublots}

Case 1. Consider the lots for which $p_{i1} \le p_{i2}$

The setup on machine 2 may or may not delay the processing of the first sublot on machine 2. Hence, we consider the following subcases.

Case 1a. $t_{i1} + p_{i1} \ge t_{i2}$ (see Fig. 3.17)

The run-in and run-out times for this case are:

$$\begin{aligned} \mathrm{RI}_{i1} &= t_{i1} + p_{i1} - t_{i2}, \\ \mathrm{RO}_{i2} &= (\mathrm{RI}_{i1} + t_{i2} + U_i p_{i2}) - (t_{i1} + U_i p_{i1}) \\ &= p_{i1} + U_i (p_{i2} - p_{i1}). \end{aligned}$$

Case 1b. $t_{i1} + p_{i1} < t_{i2}$ (see Fig. 3.18)

The run-in and run-out times for this case are:

$$\begin{aligned} \mathrm{RI}_{i1} &= 0, \\ \mathrm{RO}_{i2} &= (t_{i2} + U_i p_{i2}) - (t_{i1} + U_i p_{i1}) \\ &= (t_{i2} - t_{i1}) + U_i (p_{i2} - p_{i1}). \end{aligned}$$

Therefore, for those lots with $p_{i1} \le p_{i2}$, we can summarize the run-in and run-out times as follows:

$$\mathrm{RI}_{i1} = \max\{t_{i1} + p_{i1} - t_{i2}, 0\}, \tag{3.29}$$

$$\mathrm{RO}_{i2} = \max\{p_{i1} + U_i (p_{i2} - p_{i1}), (t_{i2} - t_{i1}) + U_i (p_{i2} - p_{i1})\}. \tag{3.30}$$

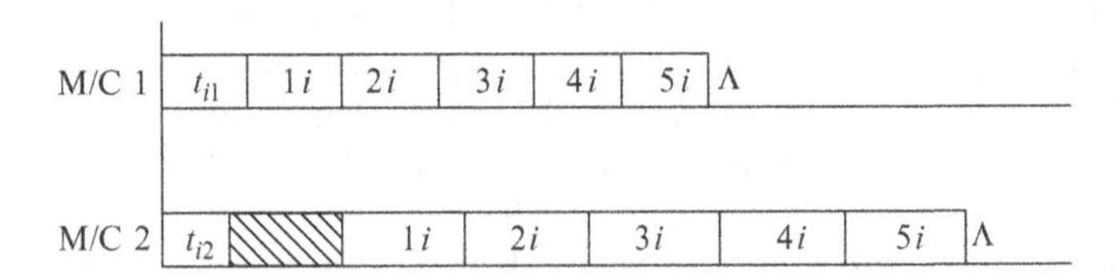

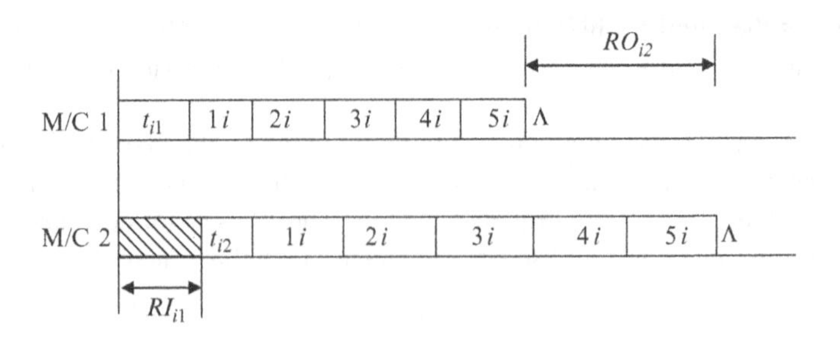

FIGURE 3.17. Lot-detached setup with $t_{i1} + p_{i1} \ge t_{i2}$

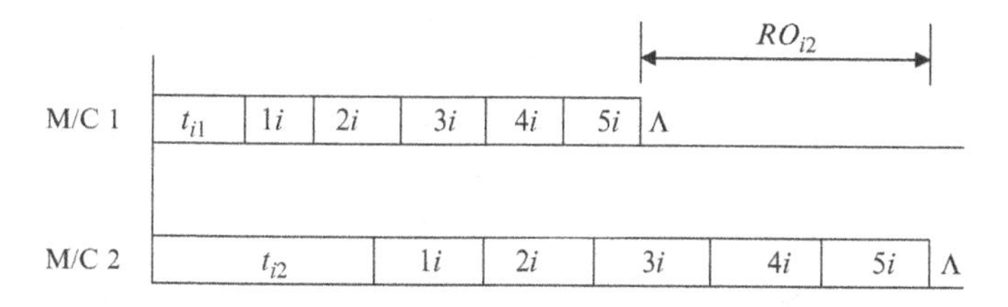

FIGURE 3.18. Lot-detached setups when $t_{i1} + p_{i1} < t_{i2}$

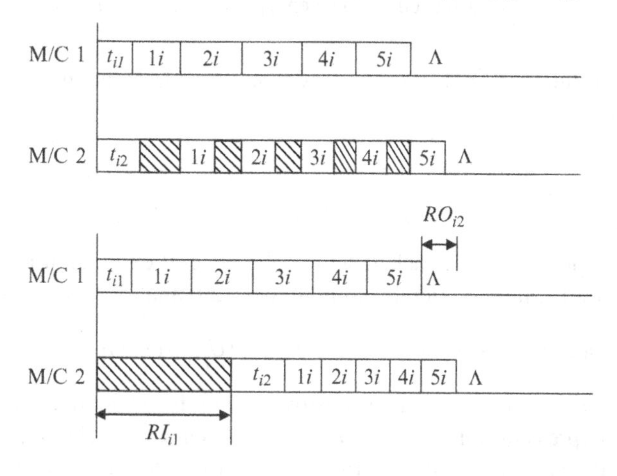

FIGURE 3.19. Lot-detached setup when $t_{i1} + U_i p_{i1} \geq t_{i2} + (U_i - 1) p_{i2}$

Case 2. Consider the lots for which $p_{i1} > p_{i2}$

As in the previous case, we consider the following two subcases:

Case 2a. This case refers to the situation in which the last sublot of lot i is scheduled continuously, i.e., $t_{i1} + U_i p_{i1} \geq t_{i2} + (U_i - 1) p_{i2}$ (like sublot 5 in Fig. 3.19)

We have

$$\begin{aligned} \text{RI}_{i1} &= (t_{i1} + U_i p_{i1}) - (t_{i2} + (U_i - 1)\, p_{i2}) \\ &= (t_{i1} - t_{i2}) + (U_i - 1)\,(p_{i1} - p_{i2}) + p_{i1}, \\ \text{RO}_{i2} &= p_{i2}. \end{aligned}$$

Case 2b. This case refers to the situation in which the last sublot of lot i is not scheduled continuously, i.e., $t_{i1} + U_i p_{i1} < t_{i2} + (U_i - 1) p_{i2}$ (see Fig. 3.20). This gives the following values of run-in and run-out times:

$$\begin{aligned} \text{RI}_{i1} &= 0, \\ \text{RO}_{i2} &= (t_{i2} + U_i p_{i2}) - (t_{i1} + U_i p_{i1}) \\ &= (t_{i2} - t_{i1}) + U_i\,(p_{i2} - p_{i1}). \end{aligned}$$

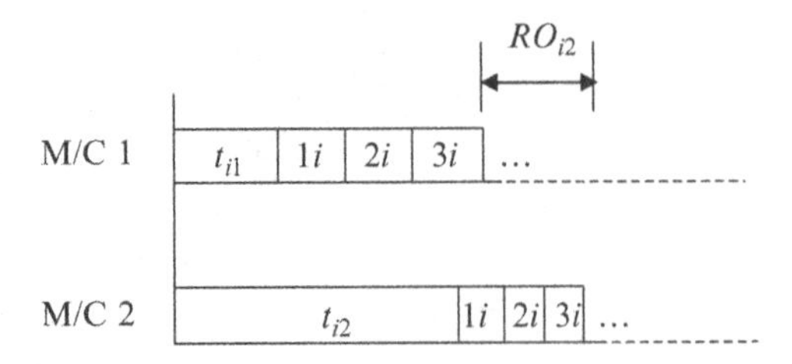

FIGURE 3.20. Lot-detached setup when $t_{i1} + U_i p_{i1} < t_{i2} + (U_i - 1) p_{i2}$

Therefore, for those lots with $p_{i1} > p_{i2}$, we can summarize the run-in and run-out times as follows:

$$\mathrm{RI}_{i1} = \max \left\{ (t_{i1} - t_{i2}) + (U_i - 1)(p_{i1} - p_{i2}) + p_{i1}, 0 \right\}, \tag{3.31}$$

$$\mathrm{RO}_{i2} = \max \left\{ p_{i2}, (t_{i2} - t_{i1}) + U_i (p_{i2} - p_{i1}) \right\}. \tag{3.32}$$

Hence, in the presence of detached setups, the general expression for the run-in time for any lot i can be written by combining (3.29) and (3.31) as follows:

$$\mathrm{RI}_{i1} = \max \left\{ 0, t_{i1} + p_{i1} - t_{i2}, (t_{i1} - t_{i2}) + (U_i - 1)(p_{i1} - p_{i2}) + p_{i1} \right\}.$$

When $p_{i1} \leq p_{i2}$, if $t_{i1} + p_{i1} - t_{i2}$ is positive, then it is greater than the third term in the above expression, since $(U_i - 1)(p_{i1} - p_{i2})$ is negative; and if $t_{i1} + p_{i1} - t_{i2}$ is negative, the second and third terms are negative and RI_{i1} is zero. Thus, the above expression is equivalent to (3.29). On the other hand, when $p_{i1} > p_{i2}$, if $t_{i1} + p_{i1} - t_{i2}$ is negative, then the above expression will reduce to (3.31); and if $t_{i1} + p_{i1} - t_{i2}$ is positive, then the second term in the above expression is redundant since it is dominated by the third term due to the fact that $(U_i - 1)(p_{i1} - p_{i2})$ is positive. Thus, in either case, the above expression is equivalent to those written separately for each case, and hence, is valid. It can also be rearranged and written as follows:

$$\mathrm{RI}_{i1} = (t_{i1} - t_{i2}) + \max \left\{ t_{i2} + t_{i1}, p_{i1}, (U_i - 1)(p_{i1} - p_{i2}) + p_{i1} \right\}. \tag{3.33}$$

Similarly, the run-out time for any lot i can be written by combining (3.30) and (3.32)

$$\mathrm{RO}_{i2} = \max \left\{ p_{i1} + U_i (p_{i2} - p_{i1}), p_{i2}, (t_{i2} - t_{i1}) + U_i (p_{i2} - p_{i1}) \right\}.$$

When $p_{i1} \leq p_{i2}$, the last sublot is never processed continuously, irrespective of the setup time on machine 2, implying that the first and third terms in the above expression dominate the second term, thus reducing it to (3.30). Similarly, when $p_{i1} > p_{i2}$, the first term in the above expression, which represents the difference of processing U_i items on machine 2 and $(U_i - 1)$ items on machine 1, is always negative (see Figs. 3.19 and 3.20), and hence, is dominated by the second and third terms, thereby, giving (3.32). Therefore, in either case, we get the individual expressions (3.30) and (3.32). Hence, the above expression is valid. It can be rearranged and rewritten as follows:

$$\mathrm{RO}_{i2} = U_i\,(p_{i2} - p_{i1}) + \max\{p_{i1}, t_{i2} - t_{i1}, p_{i2} - U_i\,(p_{i2} - p_{i1})\}. \quad (3.34)$$

Having determined run-in and run-out times for every lot, given by expressions (3.33) and (3.34) above, we can sequence the lots using Theorem 3.12.

Next, we consider the lot-attached setup case.

3.10 2/N/C/II/DV/{Lot-Attached Setup Times, Unit-Size Sublots}

We consider two cases as before depending on lots with $p_{i1} \leq p_{i2}$ or $p_{i1} > p_{i2}$.

Case 1. Consider the lots for which $p_{i1} \leq p_{i2}$

In this case, the value of t_{i2} is inconsequential and all sublots will be processed consecutively since $p_{i1} \leq p_{i2}$ for all sublots of lot i. We have (Fig. 3.21):

$$\mathrm{RI}_{i1} = t_{i1} + p_{i1}, \quad (3.35)$$

$$\begin{aligned} \mathrm{RO}_{i2} &= (\mathrm{RI}_{i1} + t_{i2} + U_i p_{i2}) - (t_{i1} + U_i p_{i1}) \\ &= t_{i2} + p_{i1} + U_i\,(p_{i2} - p_{i1})\,. \end{aligned} \quad (3.36)$$

Case 2. Consider the lots for which $p_{i1} > p_{i2}$

We have the following two subcases:

Case 2a. This case corresponds to the situation for which the last sublot is processed continuously implying that the following condition is satisfied (see Fig. 3.22)

$$(U_i - 1)\,p_{i1} \geq t_{i2} + (U_i - 1)\,p_{i2}.$$

The equality corresponds to the case where the setup time on machine 2 is equal to the difference between the processing times of $(U_i - 1)$ sublots on machine 1 and machine 2.

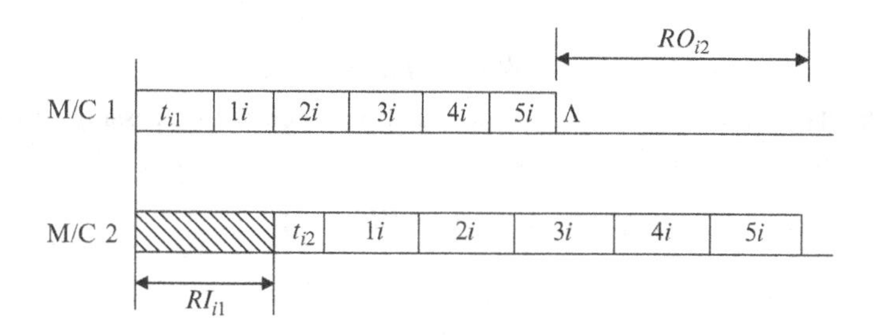

FIGURE 3.21. Lot-attached setup when $p_{i1} \leq p_{i2}$

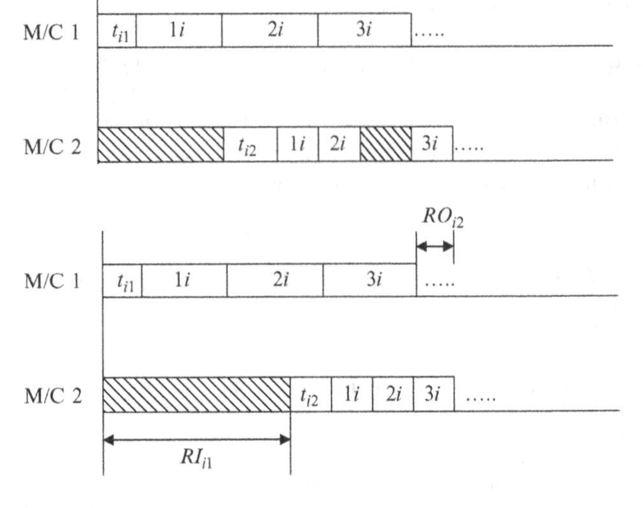

FIGURE 3.22. Lot-attached setup times when $(U_i - 1)p_{i1} \geq t_{i2} + (U_i - 1)p_{i2}$

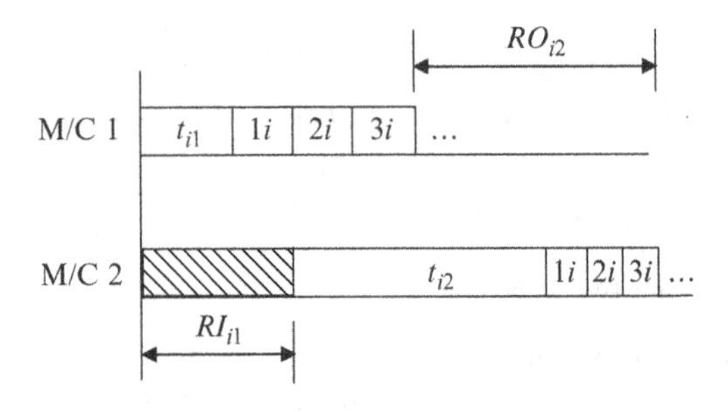

FIGURE 3.23. Lot-attached setup when $(U_i - 1)p_{i1} < t_{i2} + (U_i - 1)p_{i2}$

We have,

$$\begin{aligned} \mathrm{RI}_{i1} &= (t_{i1} + U_i p_{i1}) - (t_{i2} + (U_i - 1)\, p_{i2}) \\ &= t_{i1} + p_{i1} - t_{i2} + (U_i - 1)\,(p_{i1} - p_{i2}) \end{aligned} \tag{3.37}$$

and

$$\mathrm{RO}_{i2} = p_{i2}. \tag{3.38}$$

Case 2b. This case corresponds to the situation when the processing of the last sublot is delayed (see Fig. 3.23), i.e.,

$$(U_i - 1)\, p_{i1} < t_{i2} + (U_i - 1)\, p_{i2}.$$

In this case, we have

$$\mathrm{RI}_{i1} = t_{i1} + p_{i1}, \tag{3.39}$$

$$\begin{aligned} \mathrm{RO}_{i2} &= (t_{i2} + U_i p_{i2}) - (U_i - 1)\, p_{i1} \\ &= t_{i2} + p_{i1} + U_i\,(p_{i2} - p_{i1}). \end{aligned} \tag{3.40}$$

Thus, for a lot i, when $p_{i1} > p_{i2}$, the run-in time can be written by combining (3.37) and (3.39). That is,

$$\mathrm{RI}_{i1} = \max\left\{t_{i1} + p_{i1} - t_{i2} + (U_i - 1)(p_{i1} - p_{i2}),\, t_{i1} + p_{i1}\right\}. \tag{3.41}$$

Similarly, the run-out time can be written by combining (3.38) and (3.40) to obtain,

$$\mathrm{RO}_{i2} = \max\left\{p_{i2},\, t_{i2} + p_{i1} + U_i(p_{i2} - p_{i1})\right\}. \tag{3.42}$$

Therefore, for a lot i, in the presence of lot-attached setups, the run-in time can be written by combining (3.39) and (3.41) as follows,

$$\mathrm{RI}_{i1} = \max\left\{t_{i1} + p_{i1},\, t_{i1} + p_{i1} - t_{i2} + (U_i - 1)(p_{i1} - p_{i2})\right\}.$$

When $p_{i1} \leq p_{i2}$, the second term in the above expression will always be less than the first since $(U_i - 1)(p_{i1} - p_{i2})$ is negative thus giving (3.39). When $p_{i1} > p_{i2}$, the above expression is identical to (3.41). The above expression can be rearranged and rewritten as follows:

$$\mathrm{RI}_{i1} = t_{i1} + p_{i1} + \max\left\{0, (U_i - 1)(p_{i1} - p_{i2}) - t_{i2}\right\}. \tag{3.43}$$

Similarly, the run-out time for a lot with attached setups can be written by combining (3.40) and (3.42) to obtain,

$$\mathrm{RO}_{i2} = \max\left\{t_{i2} + p_{i1} + U_i(p_{i2} - p_{i1}),\, p_{i2}\right\}.$$

When $p_{i1} \leq p_{i2}$, the first term, which can be written as $t_{i2} + U_i p_{i2} - (U_i - 1)p_{i1}$, is always greater than p_{i2} (see Fig. 3.23). The higher value of p_{i2} does not permit the continuous processing of the last sublot. Hence, the second term is redundant and the expression is equivalent to (3.40). When $p_{i1} > p_{i2}$, the above expression is identical to (3.42). This expression can be rearranged and rewritten as follows:

$$\mathrm{RO}_{i2} = p_{i2} + \max\left\{t_{i2} + (U_i - 1)(p_{i2} - p_{i1}),\, 0\right\}. \tag{3.44}$$

Once again, after having determined the expressions of run-in and run-out times (given by expressions (3.43) and (3.44)), we can sequence the lots using Theorem 3.12.

Example 3.8 Consider the sequencing of five lots with the data as given below (Table 3.11).

TABLE 3.11. Data for the illustration of lot sequencing using run-in and run-out times in the presence of lot-attached/detached setups

	p_{i1}	p_{i2}	t_{i1}	t_{i2}	U_i
Lot 1	5	6	3	2	2
Lot 2	3	3	8	4	3
Lot 3	2	3	2	5	5
Lot 4	4	1	3	6	2
Lot 5	6	4	3	2	2

Lot-detached setup. The run-in and run-out times are calculated and are shown below:

	RI_{i1}	RO_{i2}
1	6	7
2	7	3
3	0	8
4	4	1
5	9	4

Set I = $\{1, 3\}$ and Set II = $\{2, 4, 5\}$. By sorting Set I in the nondecreasing order of RI_{i1} and Set II in the nonincreasing order of RO_{i2} and appending the sequences, the optimal sequence is 3–1–5–2–4 with a makespan of 74.

Lot-attached setup. The run-in and run-out times for this case are as follows:

	RI_{i1}	RO_{i2}
1	8	9
2	11	7
3	4	12
4	7	4
5	9	4

Set I = $\{1, 3\}$ and Set II = $\{2, 4, 5\}$. By sorting Set I in the nondecreasing order of RI_{i1} and Set II in the nonincreasing order of RO_{i2} and appending the sequences, the optimal sequence is 3–1–2–4–5 with a makespan of 78.

3.11 2/N/C/II/{CV,DV}/{Lot-Detached/Attached Setup and Lot-Attached Removal Times}

Next, we relax the assumption of unit lot sizes. The problem now is to determine sublot sizes in the presence of lot-detached/attached setup times. The number of sublots, n_j, for each lot j is known a priori. This may be the result of a restriction on the capacity of the material handling equipment or that of the container associated with the moving of the sublots. It turns out that the lot-attached removal time can also be handled conveniently in our analysis. The removal time arises due to the need to clear up the tools, jobs, fixtures, etc., after a lot has finished processing. The removal time is assumed to only delay the processing of the next lot on the current machine while the last sublot of the current lot can begin processing on the next machine (if available) as soon as it finishes processing. This is illustrated in Fig. 3.24, which depicts a schedule for a system consisting of two lots and for the data shown in Table 3.12.

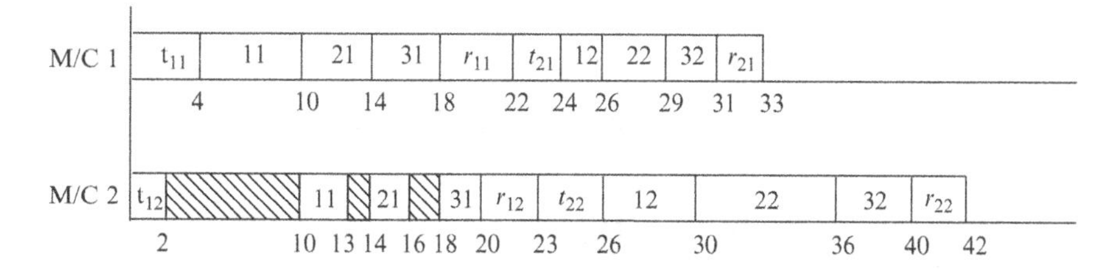

FIGURE 3.24. Schedule illustrating lot-detached setup and lot-attached removal times

TABLE 3.12. Data for illustration of lot-detached setup and lot-attached removal times

	t_{j1}	t_{j2}	r_{j1}	r_{j2}	p_{j1}	p_{j2}	s_{ij}
Lot 1	4	2	4	3	2	1	3, 2, 2
Lot 2	2	3	2	2	1	2	2, 3, 2

Consider the lot-detached case first. For a given sequence, the makespan expression for this case can be written as follows

$$C_{\max} = \max_{0 \leq u \leq N} \left\{ \sum_{k=1}^{u} \mathrm{RI}_{k1} - \sum_{k=1}^{u-1} \mathrm{RO}_{k2} \right\} + \sum_{j=1}^{N} \left(t_{j2} + p_{j2} U_j + r_{j2} \right),$$

where

$$\begin{aligned} \mathrm{RI}_{j1} &= t_{j1} - t_{j2} + I_j, \\ \mathrm{RO}_{j2} &= p_{j2} U_j + r_{j2} + I_j - \left(p_{j1} U_j + r_{j1} \right), \\ I_j &= \max_{1 \leq i \leq n_j} \left\{ p_{j1} \sum_{k=1}^{i} s_{kj} - p_{j2} \sum_{k=1}^{i-1} s_{kj} \right\}. \end{aligned}$$

It is clear from the above expressions that in order to minimize the makespan by choosing appropriate sublot sizes, for a given sequence of lots, it is sufficient to minimize the idle time, I_j, in between the processing of sublots on machine 2 since all the other terms are constants. In other words, the sublot sizes, which minimize the idle time for each lot separately, also minimize the makespan, and hence, the sublot sizing and lot sequencing procedures can be addressed independently.

The sublot sizes for each lot, j, can be determined by using the following linear program:

Minimize: I_j

Subject to:

$$I_j \geq p_{j1} s_{1j},$$
$$I_j \geq p_{j1} \left(s_{1j} + s_{2j} \right) - p_{j2} s_{1j},$$

$$\vdots$$

$$I_j \geq p_{j1} \sum_{k=1}^{n_j} s_{kj} - p_{j2} \sum_{k=1}^{n_j - 1} s_{kj},$$

$$\sum_{k=1}^{n_j} s_{kj} = U_j,$$

$$I_j, s_{kj} \geq 0.$$

The solution to this LP is trivial and it can be seen that the minimum value of I_j is $p_{j1}s_{1j}$. Hence, in the optimal solution, the inequalities must be satisfied as equalities implying that adjacent sublots must be in unique proportions.

Theorem 3.15 *For the makespan minimization criterion, the optimal sublot sizes for a lot in position j, for any arbitrary sequence, are given by*

$$s_{ij} = q^{i-1} s_{1j}, \quad \forall i = 2, 3, \ldots, n_j,$$

where

$$s_{1j} = \frac{U_j}{\sum_{i=1}^{n_j} q^{i-1}}.$$

The sublot sizes obtained by the application of Theorem 3.15 are continuous. To determine integer-size sublots, we proceed as follows

$$I_j \geq p_{j1} \sum_{k=1}^{i} s_{kj} - p_{j2} \sum_{k=1}^{i-1} s_{kj}, \quad \forall i = 1, 2, \ldots, n_j,$$

or,

$$I_j \geq p_{j1} s_{ij} + (p_{j1} - p_{j2}) \sum_{k=1}^{i-1} s_{kj}, \quad \forall i = 1, 2, \ldots, n_j.$$

Since we need integer sublot sizes, they must be the largest integer values satisfying this inequality. These can be found by rounding down the s_{kj} values as follows:

$$\hat{s}_{ij} = \left\lfloor \frac{I_j - (p_{j1} - p_{j2}) \sum_{k=1}^{i-1} s_{kj}}{p_{j1}} \right\rfloor.$$

Since the total items in all the sublots is restricted, their sizes can be calculated as follows:

$$\hat{s}_{ij} = \min \left\{ \left\lfloor \frac{I_j - (p_{j1} - p_{j2}) \sum_{k=1}^{i-1} s_{kj}}{p_{j1}} \right\rfloor, U_j - \sum_{k=1}^{i-1} \hat{s}_{kj} \right\}.$$

The corresponding algorithm for finding integer sublot sizes starts by solving the above linear program to obtain I_j. The sublot sizes are, then, calculated using the above equation. If $\sum_{k=1}^{n_j} \hat{s}_{kj} = U_j$, then the solution is optimal. If $\sum_{k=1}^{n_j} \hat{s}_{kj} < U_j$, then at least one of the sublots must contain an additional item implying that the Z_j value needs to be incremented as below

$$I_j = I_j + p_{j1} \min_{1 \leq i \leq n_j} (1 - d_{ij}),$$

where

$$d_{ij} = \left(\frac{I_j - (p_{j1} - p_{j2}) \sum_{k=1}^{i-1} s_{kj}}{p_{j1}} \right) - s_{ij}.$$

The new sublot sizes are re-calculated, and the procedure is repeated until the optimality condition is reached. The algorithm is stated below.

Algorithm for the Determination of Integer Sublot Sizes

Step 1. Calculate the sublot sizes for job j as

$$\hat{s}_{ij} = \min \left\{ \left\lfloor \frac{I_j - (p_{j1} - p_{j2}) \sum_{k=1}^{i-1} s_{kj}}{p_{j1}} \right\rfloor, U_j - \sum_{k=1}^{i-1} \hat{s}_{kj} \right\},$$

where

$$I_j = \max_{1 \leq i \leq n_j} \left\{ p_{j1} \sum_{k=1}^{i} s_{kj} - p_{j2} \sum_{k=1}^{i-1} s_{kj} \right\}.$$

Step 2. (a) If $\sum_{k=1}^{n_j} \hat{s}_{kj} = U_j$, the solution is optimal.

(b) Else, increment I_j as below

$$I_j = I_j + p_{j1} \min_{1 \leq i \leq n_j} \left(1 - d_{ij}\right),$$

where

$$d_{ij} = \left(\frac{I_j - \left(p_{j1} - p_{j2}\right) \sum_{k=1}^{i-1} s_{kj}}{p_{j1}} \right) - s_{ij}.$$

(c) Go to Step 1.

Having determined the optimal sublot sizes, Johnson's rule, in terms of run-in and run-out times, can be used to determine the optimal sequencing of lots. This is formally stated below.

Theorem 3.16 *In a two-machine flow shop with lot-detached setup and lot-attached removal times, lot i precedes lot j in an optimal sequence if the following is true*

$$\min \left\{ \mathrm{RI}_{i1}, \mathrm{RO}_{j2} \right\} \leq \min \left\{ \mathrm{RI}_{j1}, \mathrm{RO}_{i2} \right\},$$

where, for a given lot k,

$$\mathrm{RI}_{k1} = t_{k1} - t_{k2} + I_k,$$
$$\mathrm{RO}_{k2} = p_{k2}U_k + r_{k2} + I_k - (p_{k1}U_k + r_{k1}),$$
$$I_k = \max_{1\le i\le n_k}\left\{p_{k1}\sum_{\ell=1}^{i} s_{\ell k} - p_{k2}\sum_{\ell=1}^{i-1} s_{\ell k}\right\}.$$

We present the integrated sublot sizing and lot sequencing algorithm next.

Integrated Sublot Sizing and Lot Sequencing Algorithm

Sublot Sizing Procedure for Each Lot j

Step 1. Calculate the optimal continuous sublot sizes for all the jobs

(a) Unequal sublot sizes:

$$s_{ij} = q^{i-1}s_{1j}, \quad \forall i = 2, 3, \ldots, n_j,$$

where

$$s_{1j} = \frac{U_j}{\sum_{i=1}^{n_j} q^{i-1}}.$$

(b) Equal sublot sizes:

$$s_{ij} = U_j/n_j.$$

If s_{ij} is integer, go to *Lot Sequencing Procedure*; else, go to Step 2.

Step 2.

(a) Unequal sublot sizes: use the algorithm described above to obtain integer-size sublots. Let the resulting sublot sizes be represented as s_{ij}.

(b) Equal sublot sizes:

$$s_{ij} = U_j/\ell_j,$$

where ℓ_j is the largest integer less than or equal to n_j such that U_j/n_j is an integer.

Set $I_j = \max\{p_{j1}s_{1j} + p_{j2}U_j, p_{j1}U_j + p_{j2}s_{nj}\} - p_{j2}U_j$.

Lot Sequencing Procedure

Step 1. Calculate the following terms for each lot j:

$$\mathrm{RI}_{j1} = t_{j1} - t_{j2} + I_j,$$
$$\mathrm{RO}_{j2} = p_{j2}U_j + r_{j2} + I_j - (p_{j1}U_j + r_{j1}),$$
$$I_j = \max_{1\le k\le n_j}\left\{p_{j1}\sum_{k=1}^{i} s_{kj} - p_{j2}\sum_{k=1}^{i-1} s_{kj}\right\}.$$

Step 2. Separate the RI_{j1} and RO_{j2} into two sets:

$$\text{Set I}: \mathrm{RI}_{j1} \leq \mathrm{RO}_{j2} \quad \text{and} \quad \text{Set II}: \mathrm{RI}_{j1} > \mathrm{RO}_{j2}.$$

Step 3. Sort the jobs in Set I in nondecreasing order and those in Set II in nonincreasing order. The optimal sequence is given by $S^* = \{\text{Set I, Set II}\}$.

For the case of the lot-attached setup and lot-attached removal times, we can still use Theorem 3.16 to sequence the lots except that the expressions for RI_{j1}, RI_{j2}, and I_j for lot j are as follows:

$$\begin{aligned}
&\mathrm{RI}_{j1} = t_{j1} + I_j,\\
&\mathrm{RO}_{j2} = t_{j2} + p_{j2}U_j + r_{j2} + I_j - \left(p_{j1}U_j + r_{j1}\right),\\
&I_j = \max\left\{p_{j1}s_{1j}, \max_{2\leq i\leq n_j}\left\{p_{j1}\sum_{k=1}^{i} s_{kj} - p_{j2}\sum_{k=1}^{i-1} s_{kj}\right\} - t_{j2}\right\}.
\end{aligned}$$

Note that the geometric sublot sizes need not be optimal for this case. We can show this through a simple example.

Consider a lot with $U = 14$, $n = 3$, $p_1 = 1$, $p_2 = 2$, $t_1 = 4$, and $t_2 = 3$. The geometric sublot sizes are $s_1 = 2$, $s_2 = 4$, and $s_3 = 8$ with the value of 37. However, if we reduce the size of the first sublot by one unit and increase the size of the second sublot by the same amount, the idle time on the second machine reduces and we get the makespan value of 36.

The sublot sizing and lot sequencing problem are still independent for the lot-attached setup and lot-attached removal case since, in order to minimize the makespan, it is sufficient to minimize the idle time incurred in between the processing of the sublots of lot j on machine 2, given by I_j, as all other terms are constant. Therefore, to obtain optimal sublot sizes, we may solve a linear program for each lot j, $j = 1, \ldots, N$, independently. To that end, we can use the following linear program formulation:

$$\begin{aligned}
&\text{Minimize:} \quad I_j\\
&\text{Subject to:}\\
&\qquad I_j \geq p_{j1}s_{1j},\\
&\qquad I_j \geq p_{j1}\left(s_{1j} + s_{2j}\right) - p_{j2}s_{1j} - t_{j2},\\
&\qquad \vdots\\
&\qquad I_j \geq p_{j1}\sum_{k=1}^{n_j} s_{kj} - p_{j2}\sum_{k=1}^{n_j-1} s_{kj} - t_{j2},\\
&\qquad \sum_{k=1}^{n_j} s_{kj} = U_j,\\
&\qquad s_{1j} \geq 1,\\
&\qquad I_j, s_{kj} \geq 0.
\end{aligned}$$

TABLE 3.13. Data for the illustration of sublot sizing and lot sequencing algorithm in the presence of removal times and lot-detached setup times

Lot	p_{j1}	p_{j2}	t_{j1}	t_{j2}	r_{j1}	r_{j2}	n_j	U_j	$p_{j1}U_j$	$p_{j2}U_j$
1	3	2	3	4	1	2	4	10	30	20
2	2	4	4	2	2	2	3	15	30	60
3	3	4	1	3	2	1	5	20	60	80

We illustrate the above algorithm first for the case of the lot-detached setup, and then, for the lot-attached case with the help of an example problem below.

Example 3.9 Consider a three-job, two-machine flow shop with the details given in Table 3.13.

Sublot Sizing Procedure

Lot 1

Iteration 1:

The optimal continuous sublot sizes are {4.15, 2.77, 1.85, 1.23} with $I_1^* = 12.45$.

Integer sublot sizes	d_i	$1 - d_i$	$\min\{1 - d_i\}$	Incremented I_j value
$\hat{s}_1 = \lfloor 4.15 \rfloor = 4$	0.15	0.75		
$\hat{s}_2 = \lfloor 2.82 \rfloor = 2$	0.82	0.18	0.18	$12.45 + 3 \times (0.18) = 12.99$
$\hat{s}_3 = \lfloor 2.15 \rfloor = 2$	0.15	0.75		
$\hat{s}_4 = \lfloor 1.48 \rfloor = 1$	0.48	0.52		

Since $\sum_{i=1}^{n_j} \hat{s}_i < U_1$, the I_1 value is incremented.

Iteration 2:

Integer sublot sizes	d_i	$1 - d_i$	$\min\{1 - d_i\}$	Incremented I_j value
$\hat{s}_1 = \lfloor 3.99 \rfloor = 3$	0.99	0.01	0.01	$12.99 + 3 \times (0.01) = 13.02$
$\hat{s}_2 = \lfloor 3.33 \rfloor = 3$	0.33	0.67		
$\hat{s}_3 = \lfloor 2.33 \rfloor = 2$	0.33	0.67		
$\hat{s}_4 = \lfloor 1.66 \rfloor = 1$	0.66	0.34		

Since $\sum_{i=1}^{n_j} \hat{s}_i < U_1$, the I_1 value is incremented.

Iteration 3:

Integer sublot sizes
$\hat{s}_1 = \lfloor 4.01 \rfloor = 4$
$\hat{s}_2 = \lfloor 3.01 \rfloor = 3$
$\hat{s}_3 = \lfloor 2.01 \rfloor = 2$
$\hat{s}_4 = \lfloor 1.34 \rfloor = 1$

Since $\sum_{i=1}^{n_j} \hat{s}_i = U_1$, the solution is optimal. The sublot sizes are {4, 3, 2, 1}

Lot 2

Iteration 1:

The optimal continuous sublot sizes are {2.14, 4.29, 8.57} with $I_2^* = 4.28$.

Integer sublot sizes	d_i	$1 - d_i$	$\min\{1 - d_i\}$	Incremented I_j value
$\hat{s}_1 = \lfloor 2.14 \rfloor = 2$	0.14	0.86		
$\hat{s}_2 = \lfloor 4.14 \rfloor = 4$	0.14	0.86	0.18	$4.28 + 2 \times (0.86) = 6$
$\hat{s}_3 = \lfloor 5.14 \rfloor = 5$	0.14	0.86		

Since $\sum_{i=1}^{n_j} \hat{s}_i < U_1$, the I_2 value is incremented.

Iteration 2:

Integer sublot sizes
$\hat{s}_1 = \lfloor 4 \rfloor = 4$
$\hat{s}_2 = \lfloor 7 \rfloor = 7$
$\hat{s}_3 = \min\{\lfloor 14 \rfloor, 15 - 11\} = 4$

Since $\sum_{i=1}^{n_j} \hat{s}_i = U_2$, the solution is optimal. The sublot sizes are {4, 7, 4}.

Lot 3

Iteration 1:

The optimal continuous sublot sizes are {2.07, 2.77, 3.69, 4.92, 6.56} with $Z_3^* = 6.21$.

Integer sublot sizes	d_i	$1-d_i$	$\min\{1-d_i\}$	Incremented I_j value
$\hat{s}_1 = \lfloor 2.07 \rfloor = 2$	0.07	0.93		
$\hat{s}_2 = \lfloor 2.74 \rfloor = 2$	0.74	0.26	0.26	$6.21 + 3 \times (0.26) = 6.99$
$\hat{s}_3 = \lfloor 3.4 \rfloor = 3$	0.4	0.6		
$\hat{s}_4 = \lfloor 4.4 \rfloor = 4$	0.4	0.6		
$\hat{s}_5 = \lfloor 5.74 \rfloor = 5$	0.74	0.26		

Since $\sum_{i=1}^{n_j} \hat{s}_i < U_3$, the I_3 value is incremented.

Iteration 2:

Integer sublot sizes
$\hat{s}_1 = \lfloor 2.66 \rfloor = 2$
$\hat{s}_2 = \lfloor 3 \rfloor = 3$
$\hat{s}_3 = \lfloor 4 \rfloor = 4$
$\hat{s}_4 = \lfloor 5 \rfloor = 5$
$\hat{s}_5 = \min\{\lfloor 7 \rfloor, 6\} = 6$

Since $\sum_{i=1}^{n_j} \hat{s}_i = U_3$, the solution is optimal. The sublot sizes are {2, 3, 4, 5, 6}.

Lot Sequencing Procedure

The run-in time, run-out time, and idle time for all the lots are shown below:

Lot	RI_{j1}	RO_{j2}	I_j
1	12	4	13
2	10	38	8
3	5	26	7

The optimal sequence is 3–2–1 with a makespan of 179. The final results for both the procedures are summarized below (Table 3.14).

As an illustration for the lot-attached case, consider the following data:

Lot	t_{j1}	t_{j2}	r_{j1}	r_{j2}	U_j	p_{j1}	p_{j2}	$p_{j1}U_j$	$p_{j2}U_j$
1	3	4	1	2	10	3	2	30	20
2	4	2	2	2	15	2	4	30	60
3	1	3	2	1	20	3	4	60	80

TABLE 3.14. Solution for sublot sizing and lot sequencing in the presence of removal times and lot-detached setup times

Lot	Sublot sizes	I_j
1	4, 3, 2, 1	13
2	4, 7, 4	8
3	2, 3, 4, 5, 6	7
Optimal sequence		3–2–1
Makespan		179

By using the linear program presented above for the lot-attached case (while including the integrality constraints to obtain integer sublot sizes), we obtain the following sublot sizes:

Lot	Sublot 1	Sublot 2	Sublot 3	Sublot 4	Sublot 5
1	3	3	2	2	–
2	2	5	8	–	–
3	1	3	4	5	7

The run-in time, run-out time, idle time for all the lots, and the optimal sequence as well as the makespan value are summarized below:

Lot	RI_{j1}	RO_{j2}	I_j
1	13	5	10
2	8	36	4
3	6	27	5
Optimal sequence		3–2–1	
Makespan		166	

3.12 2/N/C/II/{CV,DV}/{Lot-Detached/Attached Setup and Sublot-Attached Transfer Times}

This problem involves the sublot-attached transfer time, which represents the time by which a sublot is delayed in its delivery from machine 1 to machine 2. This transfer time is made up of a fixed part f_j and a variable part v_j, which depends on the size of each sublot of lot j, $j = 1, \ldots, N$. Thus, the sublot transfer time of a sublot i of lot j can be represented as $f_j + v_j s_{ij}$. Note that the sublot-attached transfer time is different from the sublot-attached setup time in that the transfer time can be incurred while the next machine is occupied. The number of sublots for each lot is known a priori, and serves as an upper bound for the actual number of sublots used. We analyze both lot-detached and lot-attached setup cases.

Lot-Detached Setup

If I_{kj} is the idle time before the kth sublot of lot j on machine 2, then the total idle time of lot j on machine 2 is

$$I_j = \sum_{k=1}^{n_j} I_{kj}, \quad \forall j = 1, 2, \ldots, N.$$

Without loss of generality, assume that the lots are arranged so that the sequence is $1, 2, 3, \ldots, N$. Consider the first lot. The delivery of any sublot from machine 1 to machine 2 is delayed by $f_1 + v_1 s_{k1}$. The idle time preceding the first sublot of this lot on machine 2 is given by:

$$I_{11} = \max\{0, t_{11} + p_{11}s_{11} + f_1 + v_1 s_{11} - t_{12}\}.$$

If we let $p'_{j1} = p_{j1} + v_j$, $t'_{j1} = t_{j1} + f_j$, $p'_{j2} = p_{j2} + v_j$, and $t'_{j2} = t_{j2}$, we have:

$$I_{11} = \max\left\{0, t'_{11} - t'_{12} + p'_{11}s_{11}\right\}.$$

For $2 \leq k \leq n_j$, the idle time can be expressed as

$$I_{k1} = \max\left\{0, t_{11} + p_{11}\sum_{u=1}^{k} s_{u1} + f_1 + v_1 s_{k1} - t_{12} - p_{12}\sum_{u=1}^{k-1} s_{u1} - \sum_{u=1}^{k-1} I_{u1}\right\}.$$

But, $v_1 s_{k1} = v_1 \sum_{u=1}^{k} s_{u1} - v_1 \sum_{u=1}^{k-1} s_{u1}$. Therefore,

$$I_{k1} = \max\left\{0, t_{11} + p_{11}\sum_{u=1}^{k} s_{u1} + f_1 + v_1\sum_{u=1}^{k} s_{u1} - v_1\sum_{u=1}^{k-1} s_{u1} - t_{12} \right.$$
$$\left. - p_{12}\sum_{u=1}^{k-1} s_{u1} - \sum_{u=1}^{k-1} I_{u1}\right\},$$

or,

$$I_{k1} = \max\left\{0, t'_{11} + p'_{11}\sum_{u=1}^{k} s_{u1} - t'_{12} - p'_{12}\sum_{u=1}^{k-1} s_{u1} - \sum_{u=1}^{k-1} I_{u1}\right\}.$$

Also, it is easy to see that the total idle time on machine 2, up until the start of the kth sublot of lot 1, is given by:

$$\sum_{i=1}^{k} I_{i1} = \max\left\{0, t'_{11} - t'_{12} + \max_{1\leq i\leq k}\left(p'_{11}\sum_{u=1}^{i} s_{u1} - p'_{12}\sum_{u=1}^{i-1} s_{u1}\right)\right\}.$$

Note that, when $i = 1$, the above expression is identical to I_{11} mentioned above. When $k = n_1$, we obtain the total idle time on machine 2 inserted among the sublots of lot 1, designated by $z_1 (\equiv I_1)$ as follows:

$$z_1 = \sum_{i=1}^{n_1} I_{i1} = \max\left\{0, t'_{11} - t'_{12} + \max_{1 \le k \le n_1}\left(p'_{11}\sum_{u=1}^{k} s_{u1} - p'_{12}\sum_{u=1}^{k-1} s_{u1}\right)\right\}.$$

Consider the second lot. The idle time on machine 2 preceding the first sublot of this lot can be obtained as follows:

$$I_{12} = \max\{0, t_{11} + p_{11}U_1 + t_{21} + p_{21}s_{12} + f_2 + v_2 s_{12} - t_{22} - t_{12} - p_{12}U_1 - I_1\}.$$

If we let $y_1 = t_{11} - t_{12} + p_{11}U_1 - p_{12}U_1$, we have

$$I_{12} = \max\left\{0, y_1 + t'_{21} + p'_{21}s_{12} - t'_{22} - I_1\right\}.$$

Similarly,

$$I_{k2} = \max\left\{0, y_1 + t'_{21} + p'_{21}\sum_{u=1}^{k} s_{u2} - p'_{22}\sum_{u=1}^{k-1} s_{u2} - t'_{22} - I_1 - \sum_{u=1}^{k-1} I_{u2}\right\}, \quad k \ge 2.$$

As before, the total idle time on machine 2, up until the start of the kth sublot of lot 2, is given by:

$$\sum_{i=1}^{k} I_{i2} = \max\left\{0, y_1 + t'_{21} - t'_{22} - I_1 + \max_{1 \le i \le k}\left(p'_{21}\sum_{u=1}^{i} s_{u2} - p'_{22}\sum_{u=1}^{i-1} s_{u2}\right)\right\}.$$

For $k = n_2$, we now have

$$I_2 = \sum_{i=1}^{n_2} I_{i2} = \max\left\{0, y_1 + t'_{21} - t'_{22} - I_1 + \max_{1 \le i \le n_2}\left(p'_{21}\sum_{u=1}^{i} s_{u2} - p'_{22}\sum_{u=1}^{i-1} s_{u2}\right)\right\}.$$

Hence,

$$I_1 + I_2 = \max\left\{z_1, y_1 + t'_{21} - t'_{22} + \max_{1 \le i \le n_2}\left(p'_{21}\sum_{u=1}^{i} s_{u2} - p'_{22}\sum_{u=1}^{i-1} s_{u2}\right)\right\}.$$

We know that

$$\begin{aligned} y_1 &= t_{11} - t_{12} + p_{11}\sum_{u=1}^{n_1} s_{u1} - p_{12}\sum_{u=1}^{n_1} s_{u1} \quad \text{(from above)} \\ &\le t'_{11} - t'_{12} + p'_{11}\sum_{u=1}^{n_1} s_{u1} - p'_{12}\sum_{u=1}^{n_1-1} s_{u1} \\ &\le z_1. \end{aligned}$$

Hence, an additional y_1 can be inserted in the above expression of $I_1 + I_2$ to obtain

$$I_1 + I_2 = \max\left\{z_1, y_1, y_1 + t'_{12} - t'_{22} + \max_{1 \le i \le n_2}\left(p'_{21}\sum_{u=1}^{i} s_{u2} - p'_{22}\sum_{u=1}^{i-1} s_{u2}\right)\right\}$$

$$= \max\left(z_1, y_1 + \max\left(0, t'_{21} - t'_{22} + \max_{1 \le i \le n_2}\left(p'_{21}\sum_{u=1}^{i} s_{u2} - p'_{22}\sum_{u=1}^{i-1} s_{u2}\right)\right)\right).$$

Let

$$z_2 = \max\left(0, t'_{21} - t'_{22} + \max_{1 \le i \le n_2}\left(p'_{21}\sum_{u=1}^{i} s_{u2} - p'_{22}\sum_{u=1}^{i-1} s_{u2}\right)\right).$$

Then,

$$I_1 + I_2 = \max\{z_1, y_1 + z_2\}.$$

By repeating the same argument, it can be shown that

$$\sum_{k=1}^{j} I_k = \max\{z_1, y_1 + z_2, \ldots, y_1 + \cdots + y_{j-1} + z_j\}, \quad \forall j = 1, 2, \ldots, N,$$

where

$$z_j = \max\left\{0, t'_{j1} - t'_{j2} + \max_{1 \le k \le n_j}\left(p'_{j1}\sum_{u=1}^{k} s_{uj} - p'_{j2}\sum_{u=1}^{k-1} s_{uj}\right)\right\}$$

and

$$y_j = t_{j1} - t_{j2} + p_{j1}U_j - p_{j2}U_j.$$

Remark 3.1 *The above expression holds true even if the number of sublots, m_j, is less than n_j, in which case we can set $s_{kj} = 0$, $k = m_j + 1, \ldots, n_j$.*

Note that, y_j, $\forall j = 1, \ldots, N$, do not contain n_j, and z_j, $\forall j = 1, \ldots, N$, are independent of one another. Hence, the following result holds.

Theorem 3.17 *The optimal sublot sizes for a lot j can be obtained by minimizing z_j irrespective of the lot sequence and sublot sizes of the other lots.*

In lieu of Theorem 3.17, the optimal number of sublots for each lot j, which minimizes z_j, also minimizes the overall makespan value. This problem for each lot j can be formulated as a problem of determining the optimal number of sublots of a lot, which minimizes the idle time appearing in between its sublots as follows:

$$
\begin{aligned}
&\text{Minimize:} \quad z_j \\
&\text{Subject to:} \\
&\qquad z_j \ge t'_{j1} + p'_{j1}s_{1j} - t'_{j2}, \\
&\qquad z_j \ge \left\{t'_{j1} + p'_{j1}\left(s_{1j} + s_{2j}\right)\right\} - \left\{t'_{j2} + p'_{j2}s_{1j}\right\}, \\
&\qquad \vdots \\
&\qquad z_j \ge \left\{t'_{j1} + p'_{j1}\sum_{u=1}^{n_j} s_{uj}\right\} - \left\{t'_{j2} + p'_{j2}\sum_{u=1}^{n_j-1} s_{uj}\right\}, \\
&\qquad \sum_{u=1}^{n_j} s_{uj} = U_j, \\
&\qquad z_j \ge 0, \\
&\qquad s_{ij} \ge 0, \quad \forall i = 1, 2, \ldots, n_j.
\end{aligned}
$$

The constraints essentially follow the definition of z_j, and the fact that the sum of all sublots of lot j is U_j.

Remark 3.1 ensures the validity of this formulation even if $s_i j = 0$, $i = m, \ldots, n_j, m < n_j$.

For any lot j, let $q = p'_{j2}/p'_{j1}$, as before, represent the ratio of the processing time on machine 2 to that on machine 1, and $\xi_j = z_j + t'_{j2} - t'_{j1}$ represent the maximum idle time appearing in between the sublots of lot j in the absence of setup times (but in the presence of sublot transfer times). The three possible cases along with their optimal solutions are presented below.

Case 1. $\frac{p'_{j1}U}{\left(1-q_j^n\right)/\left(1-q_j\right)} + t'_{j1} \ge t'_{j2}$

Note that $\frac{p'_{j1}U}{\left(1-q_j^n\right)/\left(1-q_j\right)}$ represents processing time of the first of the geometric sublot sizes (see Sect. 3.2). If the above relationship holds, then t'_{j2} does not interfere with the geometric sublot size and they are optimal for lot j. Hence,

$$
s_{ij} = \frac{q_j^{i-1}U}{\left(1-q_j^n\right)/\left(1-q_j\right)}, \quad \forall i = 1, \ldots, n_j.
$$

In the absence of setup times, the idle time, that appears only in front of the first sublot on machine 2,

$$
\xi_j^* = \frac{p'_{j1}U}{\left(1-q_j^n\right)/\left(1-q_j\right)} = p'_{j1}s_{1j}.
$$

Hence,

$$
z_j = p'_{j1}s_{1j} + t'_{j1} - t'_{j2}.
$$

Case 2. $\frac{p'_{j1}U}{\left(1-q_j^n\right)/\left(1-q_j\right)} + t'_{j1} < t'_{j2}$, but $p'_{j1}U_j + t'_{j1} > t'_{j2}$

In this case, the processing time of the first of the geometric sublots plus the setup time on machine 1 $\left(t'_{j1} = t_{j1} + f_j\right)$ is less than t'_{j2} $(= t_{j2})$, but the processing of the entire lot requires more time than t'_{j2}.

In order to minimize the idle time on machine 2, the first sublot size, s_{1j}, is such that,

$$s_{1j}p'_{j1} = t'_{j2} - t'_{j1} \quad \text{or} \quad s_{1j} = \frac{t'_{j2} - t'_{j1}}{p'_{j1}}.$$

The remaining sublot sizes are geometric. Since $\sum_{i=1}^{n_j} s_{ij} = U_j$, we need to determine the number of sublots, the largest integer m, such that

$$\sum_{i=1}^{m}\left(\frac{t'_{j2} - t'_{j1}}{p'_{j1}}\right)q_j^{i-1} \le U_j,$$

or,

$$\left(\frac{t'_{j2} - t'_{j1}}{p'_{j1}}\right)\left(\frac{1 - q_j^m}{1 - q_j}\right) \le U_j.$$

After simplification, this results in,

$$m = \left\lfloor \frac{\log\left\{1 - \frac{p'_{j1}U_j(1 - q_j)}{t'_{j2} - t'_{j1}}\right\}}{\log q_j} \right\rfloor$$

with the last sublot of size,

$$s_{m+1j} = U_j - \left(\frac{t'_{j2} - t'_{j1}}{p'_{j1}}\right)\left(\frac{1 - q_j^m}{1 - q_j}\right),$$

while

$$s_{ij} = 0, \quad \forall i \ge m + 2.$$

Also,

$$\xi_j^* = t'_{j2} - t'_{j1} \quad \text{and} \quad z_j^* = 0.$$

Case 3. $p'_{j1}U_j + t'_{j1} \le t'_{j2}$

This case corresponds to the situation in which the setup and total processing time of lot j on machine 1 is less than its setup time on machine 2. Under this case, the entire lot can be completed on machine 1 before the setup on machine 2 is incurred. Hence, lot streaming gains no advantage and the entire lot can be treated as a single transfer lot.

Once the optimal sublot sizes s_{ij}^* and idle time z_j^* have been obtained for each lot j, the next question is to determine an optimal sequence among the lots that satisfies the optimal z_j^* for every lot j. We address this problem next.

The minimization of makespan is equivalent to minimizing the total idle time

$$I = \sum_{k=1}^{N} I_k = \max_{1 \le k \le N} \left\{ \sum_{i=1}^{k-1} y_i + z_k \right\}.$$

Let $z_j'^* = z_j^* - y_j$. Then

$$I = \max \left\{ z_1^*, z_1^* + z_2^* - z_j'^*, \ldots, \sum_{i=1}^{N} z_i^* - \sum_{i=1}^{N-1} z_i'^* \right\}.$$

This is the expression for the total idle time on machine 2 for a normal flow shop with processing times $p_{j1} = z_j^*$ and $p_{j2} = z_j'^*$, and hence, Johnson's algorithm can be used to sequence these lots. This follows since Johnson's algorithm is a procedure which seeks to find a permutation that minimizes an expression of the type

$$\max_{1 \le k \le N} \left\{ \sum_{i=1}^{k} p_{i1} - \sum_{i=1}^{k-1} p_{i2} \right\}.$$

Thus, when lot-detached setup times and sublot transfer times are present, the optimal continuous sublot sizes are obtained for each lot independently and the lots can be sequenced by Johnson's algorithm with processing times of z_j^* and $z_j'^*$ on machines 1 and 2, respectively.

For the case of integer sublot sizes, we can use the following algorithm.

Algorithm for Integer Sublots Sizes

Let $S_{1j} = s_{1j}$, $\xi_j^* = p'_{j1} S_{1j}$, and $S_{kj} = S_{1j} + \sum_{i=2}^{k} s_{ij}$.

The sublot sizes must satisfy the relation $S_{1j} \le S_{2j} \le S_{3j} \le \cdots \le S_{nj}$ and the constraints

$$S_{1j} \le \frac{\xi_j}{p'_{j1}},$$
$$S_{k+1j} \le \frac{\xi_j + p'_{j2} S_{kj}}{p'_{j1}}, \quad \forall k = 1, \ldots, n-1,$$
$$S_{nj} = U_j.$$

Step 1. Initialize lower bound on $\xi_{j\text{INT}}^*$ to $\underline{\xi}_j = \lceil \xi_j^* \rceil$:

$$\text{Set } v = \underline{\xi}_j.$$

Step 2. Compute $S_{1j}, S_{2j}, S_{3j}, \ldots, S_{nj}$ as follows:

$$S_{1j} = \left\lfloor \frac{v}{p'_{j1}} \right\rfloor,$$
$$S_{kj} = \left\lfloor \frac{v + p'_{j2} S_{k-1j}}{p'_{j1}} \right\rfloor, \quad \forall k = 2, \ldots, n.$$

Step 3.

(a) If $S_{nj} < U_j$, the value of the integer sublot sizes are too small and must be incremented. Update the new lower bound to $\underline{\xi}_j = \upsilon + 1$. For the first iteration, calculate an upper bound using the Upper Bound Procedure mentioned below.

(b) If $S_{mj} \geq U_j$, for the first $m(\leq n)$ that violates this constraint, set $S_{mj} = U_j$. Also, set $S_{kj} = 0, k = m+1, \ldots, n$, if $m < n$. Since, this yields a feasible solution, the υ value used can serve as a valid upper bound on integer $\xi^*_{j\mathrm{INT}}$, i.e., $\bar{\xi}_j = \upsilon$.

Step 4. If $\bar{\xi}_j = \underline{\xi}_j$, the upper bound solution is optimal. Else, let $\upsilon = \left\lfloor (\bar{\xi}_j + \underline{\xi}_j)/2 \right\rfloor$ and go to Step 2.

Upper Bound Procedure

A valid upper bound can be obtained by rounding up the continuous optimal solution. The cumulative sublot sizes are

$$\hat{S}_{kj} = \min\left\{U_j, \left\lceil S^*_{1j} \right\rceil + \sum_{j=2}^{k} \left\lceil S^*_{k'_j} \right\rceil \right\}, \quad \forall k = 2, \ldots, n.$$

The desired upper bound on ξ_j is given by

$$\bar{\xi}_j = \max\left\{p'_{j1}\hat{S}_{1j}, \max_{2\leq k\leq n}\left(p'_{j1}\hat{S}_{kj} - p'_{j2}\cdot \hat{S}_{k-1j}\right)\right\}.$$

Lot-Attached Setup

Similar to the case of lot-detached setups, we let I_j define the total idle time appearing on machine 2 in front of the sublots of lot j, in the presence of attached setups, i.e.,

$$I_j = \max\left\{t'_{j1} + p'_{j1}s_{1j}, t'_{j1} - t'_{j2} + \mu_j\right\},$$

where

$$\mu_j = \max_{2\leq k\leq n_j}\left\{p'_{j1}\sum_{u=1}^{k} s_{uj} - p'_{j2}\sum_{u=1}^{k-1} s_{uj}\right\}.$$

Theorem 3.17 can once again be invoked to determine the sublot sizes of each lot.

The linear programming formulation below minimizes the maximum total idle time appearing on machine 2 in the presence of lot-attached setups and is solved for each lot separately.

$$
\begin{aligned}
&\text{Minimize:} \quad I_j \\
&\text{Subject to:} \\
&\qquad I_j \geq t'_{j1} + p'_{j1} s_{1j}, \\
&\qquad \vdots \\
&\qquad I_j \geq t'_{j1} - t'_{j2} + p'_{j1} \sum_{u=1}^{n_j} s_{uj} - p'_{j2} \sum_{u=1}^{n_j - 1} s_{uj}, \\
&\qquad \sum_{u=1}^{n_j} s_{uj} = U_j, \\
&\qquad I_j \geq 0, \\
&\qquad s_{1j} \geq 1, \\
&\qquad s_{ij} \geq 0, \quad \forall i = 2, \ldots, n_j.
\end{aligned}
$$

In the absence of a constraint on the size of the first sublot, we might get a solution with $s_{1j} = 0$ (irrespective of the continuous or discrete version of the problem), implying that we must transfer an infinitesimal amount of work from machine 1 to enable the setup on machine 2. Since the smallest sublot size is 1, we add the constraint $s_{1j} \geq 1$. To ensure that all constraints are in the same format, we use the following notation.

Let $s_{1j} = 1 + s''_{1j}$, $t''_{j1} = t'_{j1} + p'_{j1}$, $t''_{j2} = t'_{j2} + p'_{j2}$, and $U''_j = U_j - 1$. Then, the above formulation becomes

$$
\begin{aligned}
&\text{Minimize:} \quad I_j \\
&\text{Subject to:} \\
&\qquad I_j \geq t''_{j1} + p'_{j1} s''_{1j}, \\
&\qquad I_j \geq t''_{j1} + p'_{j1} \left(s''_{1j} + s_{2j} \right) - t''_{j2} + p'_{j2} s''_{1j}, \\
&\qquad \vdots \\
&\qquad I_j \geq t''_{j1} - t''_{j2} + \left(p'_{j1} - p'_{j2} \right) s''_{1j} + p'_{j1} \sum_{u=2}^{n_j} s_{uj} + p'_{j2} \sum_{u=2}^{n_j - 1} s_{uj}, \\
&\qquad s''_{1j} + \sum_{u=2}^{n_j} s_{uj} = U''_j, \\
&\qquad I_j \geq 0, \\
&\qquad s''_{1j} \geq 0, \\
&\qquad s_{kj} \geq 0, \quad \forall k = 2, \ldots, n_j.
\end{aligned}
$$

Once the optimal I^*_j values have been obtained, Johnson's algorithm can be used to sequence the lots by using $p_{j1} = I^*_j$ and $p_{j2} = I'^*_j$, as before, where $I'^*_j \equiv I^*_j - y_j$.

TABLE 3.15. Illustration of sublot sizing in the presence of sublot transfer times and lot-attached setup times

Lot j	t_{j1}	t_{j2}	p_{j1}	p_{j2}	n_j	U_j	$p_{j1}U_j$	$p_{j2}U_j$	y_j	q_j	$p_{j1}U_j/\left(1-q_j^n/1-q_j\right)$
1	2	3	3	2	4	10	30	20	9	0.67	12.46
2	4	2	5	4	5	15	75	60	17	0.8	22.31
3	3	8	3	6	6	20	60	120	−65	2	0.95

The algorithm for finding the integer sublot sizes is also similar to the one presented earlier, except for the following changes.
Step 2. Define $\tau_j = t''_{j2}$, then

$$S_{1j} = \left\lfloor \frac{\upsilon - \tau_j}{p'_{j1}} \right\rfloor.$$

Replace all U_j with $U''_j - 1$.

Example 3.10 To illustrate the algorithm presented above for the case of lot-attached setup, consider an example with relevant data given in Table 3.15. For the sake of simplicity, we assume that the transfer times are zero.

Sublot Sizing Procedure

Continuous Optimal Sublot Sizes

Lot 1. Case 1 applies here. The optimal sublot sizes are {4.2, 2.8, 1.8, 1.2} with $\xi_1^* = 12.46$ and $z_1^* = 11.46$.

Lot 2. Case 1 applies here also. The optimal sublot sizes are {4.5, 3.6, 2.9, 2.3, 1.8} with $\xi_2^* = 22.31$ and $z_2^* = 24.31$.

Lot 3. Case 2 applies here with $m = \lfloor 3.7 \rfloor = 3$. The sublot sizes are {1.67, 3.33, 6.67, 8.33, 0, 0} with $\xi_3^* = 5$ and $z_3^* = 0$.

Integer Sublot Sizes

Lot 1.

Step 1. Initial lower bound, $\underline{\xi}_1 = \lceil \xi_1^* \rceil = 13$.
Step 2. The cumulative sublot sizes are {4, 7, 9, 10}.
Step 3. Since $S_{41} = U_1$, the solution is feasible and serves as a valid upper bound. Hence, we set $\bar{\xi}_1 = 13$.
Step 4. Since UB = LB, the solution is optimal.

The integer sublot sizes, $\hat{s}^*$, are {4, 3, 2, 1} with $\hat{\xi}_1^* = 13$ and $\hat{z}_1^* = 12$ (where $\hat{s}$, $\hat{\xi}_1$, and $\hat{z}_1$ represent integer values).

Lot 2.

Iteration 1:

Step 1. Initial lower bound, $\underline{\xi}_2 = \lceil \xi_2^* \rceil = 23 = \upsilon$.
Step 2. The cumulative sublot sizes are $\{4, 7, 10, 12, 14\}$.
Step 3. Since $S_{52} < U_2$, increment $\underline{\xi}_2 = 23 + 1 = 24 = \upsilon$.

Upper Bound Procedure

The cumulative sublot sizes obtained from rounding up the continuous optimal solution are $\{5, 9, 12, 15\}$. The upper bound $\bar{\xi}_2$ is $\max\{25, 25, 24, 27\} = 27$.

Step 4. Since UB $\neq$ LB, the new value for υ is calculated as $\lfloor (24 + 27)/2 \rfloor = 25$.

Iteration 2:

Step 2. The new cumulative sublot sizes are $\{5, 9, 12, 14, 16\}$.
Step 3. Since $S_{52} > U_2$, redefine $S_{52} = 15$. Update the upper bound to 25.
Step 4. Since UB $\neq$ LB, the new value for υ is calculated as $\lfloor (24 + 25)/2 \rfloor = 24$.

Iteration 3:

Step 2. The new cumulative sublot sizes are $\{4, 8, 11, 13, 15\}$.
Step 3. Since $S_{52} = U_2$, this is a feasible solution and yields a valid upper bound. Update the upper bound to 24.
Step 4. Since LB = UB, the solution is optimal.

The sublot sizes, $\hat{s}^*$, are $\{4, 4, 3, 2, 2\}$ with $\hat{\xi}_2^* = 24$ and $\hat{z}_2^* = 26$.

Lot 3.

Step 1. Initial lower bound, $\bar{\xi}_3 = \lceil \xi_3^* \rceil = 5 = \upsilon$.
Step 2. The cumulative sublot sizes are $\{1, 3, 7, 15, 31, 31\}$.
Step 3. Since $S_{53} = U_3$, redefine $S_{53} = 20$, $S_{63} = 20$. This yields a feasible solution and acts as a valid upper bound. Hence, we set $\underline{\xi}_3 = 5$.
Step 4. Since LB = UB, the solution is optimal.

The integer sublot sizes, $\hat{s}^*$, are $\{1, 2, 4, 5, 0\}$ with $\hat{\xi}_3^* = 5$ and $\hat{z}_3^* = 0$.
The results are summarized below in Table 3.16.

TABLE 3.16. Solution for sublot sizing in presence of sublot transfer times and lot-attached setup times

Lot j	Integer sublot sizes	$\hat{z}_j^*$	$\hat{\xi}_j^*$
1	4, 3, 2, 1	12	13
2	4, 4, 3, 2, 2	26	24
3	1, 2, 4, 8, 5, 0	0	5

TABLE 3.17. Solution for lot sequencing in presence of sublot transfer times and lot-attached setup times

Job j	$\hat{z}_j^*$	y_j	$z_j'^* = \hat{z}_j^* - y_j$
1	12	9	3
2	26	17	9
3	0	−65	65

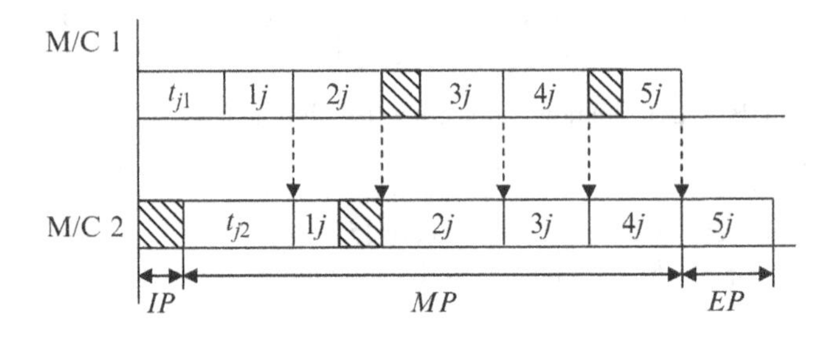

FIGURE 3.25. Notation for a schedule in a no-wait flow shop

Lot Sequencing Procedure

The lots can be sequenced via Johnson's algorithm by using $\hat{z}_j^*$ and $z_j'^*$ as the processing times. The calculation of $z_j'^*$ is shown in Table 3.17.

The optimal sequence is 3–2–1 with a total idle time of $\max\{0, -39, -36\} = 0$.

3.13 2/N/C/II/{CV,DV}/{Lot-Detached Setup Times, No-Wait}

This is an extension of the single lot, no-wait flow shop (Sects. 3.4 and 3.5) to N lots. Note that a schedule of a lot for processing in seclusion under lot streaming can be split into three parts as follows:

Initial part. This refers to the beginning part of the schedule where machine 2 is idle. It is equal to $\max\{0, t_1 + p_1 s_1 - t_2\}$ and is represented as IP.

End part. This refers to the tail part of the schedule where machine 1 is idle. We denote it as EP.

Middle part. This refers to the part of the schedule that is between the initial and end parts. These are illustrated in Fig. 3.25. Note that the initial part and end part have connotations of run-in and run-out times.

The optimal sublot sizes for the processing of a lot under lot streaming can be determined by using the results of Theorem 3.4. These optimal sublot sizes for each lot are such that the length of each section is less than or equal to that obtained by any other sublot sizes. This is formally stated in the theorem below.

Theorem 3.18 *For the set of optimal continuous sublot sizes $\hat{S}^*$ of a lot j, obtained using Theorem 3.4,*

$$\begin{aligned} \mathrm{MP}^* &\leq \mathrm{MP}, \\ \mathrm{IP}^* + \mathrm{MP}^* &\leq \mathrm{IP} + \mathrm{MP}, \\ \mathrm{EP}^* + \mathrm{MP}^* &\leq \mathrm{EP} + \mathrm{MP}, \end{aligned}$$

where IP^*, MP^*, *and* EP^* *are the values corresponding to optimal sublot sizes, and IP, MP, and EP are for any other sizes of sublots.*

The following question arises at this point for the N-lot problem in hand: do the optimal sublot sizes, determined for each lot individually using Theorem 3.4, remain optimal as the N-lots are sequenced to obtain an optimal solution? The answer to this question, as shown by Sriskandarajah and Wagneur [32], is yes for continuous sublot sizes but not necessarily for the discrete case.

Consider the continuous sublot size case first. Once the sublot sizes have been obtained for each lot, the parts IP^*, MP^*, and EP^* can be calculated. For any general sequence of lots, the makespan can be written as follows

$$C_{\max} = \sum_{j=1}^{N+1} \max\left\{\mathrm{IP}_j^*, \mathrm{EP}_{j-1}^*\right\} + \sum_{j=1}^{N+1} \mathrm{MP}_j^*,$$

where dummy lots 0 and $(N+1)$ are added in the beginning and at the end of the schedule with IP_j^*, EP_j^*, and $\mathrm{MP}_j^* = 0$, for $j = 0$ and $j = N + 1$. Since the last term is a constant, it is sufficient to minimize the first term only, i.e.,

$$C_{\max} = \sum_{j=1}^{N+1} \max\left\{\mathrm{IP}_j^*, \mathrm{EP}_{j-1}^*\right\}.$$

This expression can be viewed as minimizing the makespan in a two-machine flow shop where lot j has processing times IP_j^* and EP_j^* on machines 1 and 2, respectively.

Before proceeding further, we first show that the problem of sequencing the lots for the no-wait flow shop is different from the case when the no-wait assumption is not present, and hence, Johnson's algorithm need not give an optimal solution. Consider, for instance, the data shown in Table 3.18 for three lots.

If we apply the Johnson's algorithm to this data, the sequence obtained is 1–2–3. The makespan for this sequence is 22, as shown in Fig. 3.26a. However, if we process the lots in the sequence 1–3–2, which is also the optimal sequence, the makespan obtained is 21. Note that the Johnson's algorithm results in an idle time on machine 2 due to the no-wait processing of the lots, which the optimal sequence avoids. Similarly, it can be shown that the no-wait optimal sequence need not be optimal for the case when the no-wait assumption is not present. Consider the data given in Table 3.19.

TABLE 3.18. Data for the no-wait problem

Lot	IP	EP
1	1	9
2	3	8
3	9	3

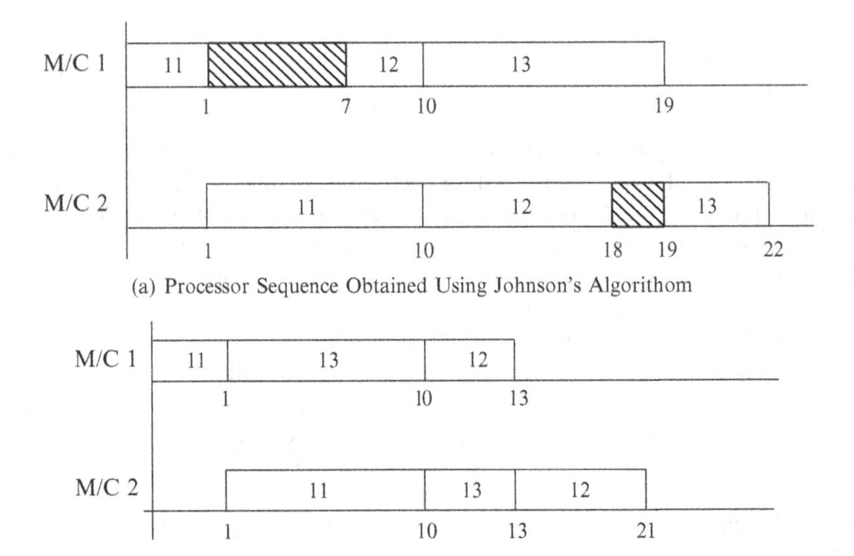

FIGURE 3.26. Sequence of lots for the no-wait problem obtained using Johnson's algorithm and optimally

TABLE 3.19. Data for the problem when the no-wait assumption is not present

Lot	IP	EP
1	1	9
2	3	7
3	9	2

The optimal no-wait sequence (1–3–2) is shown in Fig. 3.27a while the optimal sequence (1–2–3) for the situation when the no-wait assumption is not present (obtained using the Johnson algorithm) is shown in Fig. 3.27b.

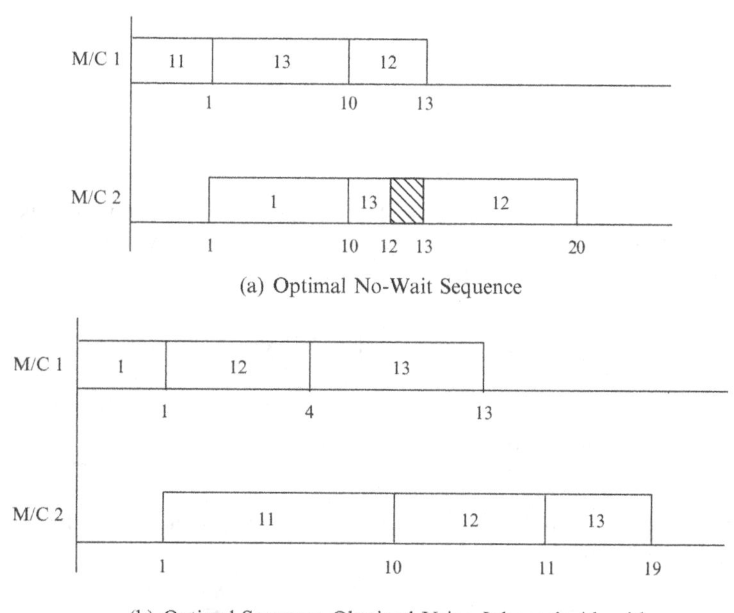

FIGURE 3.27. Optimal no-wait sequence and optimal sequence obtained using Johnson's rule for the situation when the no-wait assumption is not present

The problem of determining an optimal sequence for the no-wait flow shop is identical to the special case of the traveling salesman problem addressed by Gilmore and Gomory [15]. They present a polynomial time algorithm ($O(n \log N)$) to obtain an optimal solution for this problem. Next, we describe this algorithm in the context of our no-wait flow shop problem.

Let x_{ij} be the relative size of the ith sublot of lot j. As the sublot sizes of a lot are geometric in nature, we have

$$x_{1j} = \begin{cases} \dfrac{(1-q_j)}{1-q_j^{n_j}}, & q_j \neq 1, \forall j = 1, \ldots, N, \\ \dfrac{1}{n_j}, & q_j = 1, \ \forall j = 1, \ldots, N, \end{cases}$$

where $q_j = p_{2j}/p_{1j}$ and $x_{ij} = x_{i-1,j}q_j$ for $i = 2, \ldots, n_j$.

Having determined x_{1j} for every j, $j = 1, \ldots, N$, we have

$$\begin{aligned} \mathrm{IP}_j^* &= \max\left\{0, t_{j1} - t_{j2} + p_{j1}U_j x_{1j}\right\}, \\ \mathrm{EP}_j^* &= p_{j2}U_j x_{n_j j}, \end{aligned}$$

and

$$\mathrm{MP}_j^* = C_j^* - \mathrm{IP}_j^* - \mathrm{EP}_j^*,$$

where C_j^* is the optimal makespan of lot j in the single-lot case, and is given by,

$$C_j^* = t_{j2} + p_{j2}U_j + \mathrm{IP}_j^*.$$

This gives

$$\mathrm{MP}_j^* = t_{j2} + p_{j2}U_j \left(1 - x_{n_j j}\right).$$

Let a sequence of lots be denoted by σ. For dummy starting and ending lots 0 and $N+1$, we have $\mathrm{IP}_{\sigma(i)}^* = \mathrm{MP}_{\sigma(i)}^* = \mathrm{EP}_{\sigma(i)}^* = 0$ for $i = 0$ and $N+1$. Note that for a given σ, we can express $C_{\max}(\sigma)$ as follows:

$$\begin{aligned} C_{\max}(\sigma) &= \sum_{i=1}^{N+1} \max\left\{\mathrm{IP}_{\sigma(i)}^*, \mathrm{EP}_{\sigma(i-1)}^*\right\} + \sum_{i=1}^{N+1} \mathrm{MP}_{\sigma(i)}^* \\ &= \sum_{i=1}^{N+1} \max\left\{\mathrm{IP}_{\sigma(i)}^* - \mathrm{EP}_{\sigma(i-1)}^*, 0\right\} + \sum_{i=1}^{N+1} \mathrm{EP}_{\sigma(i-1)}^* + \sum_{i=1}^{N+1} \mathrm{MP}_{\sigma(i)}^*. \end{aligned}$$

Since the second and third terms are constants for given sublot sizes, the problem is to minimize the first term

$$F(\sigma) = \sum_{i=1}^{N+1} \max\left\{\mathrm{IP}_{\sigma(i)}^* - \mathrm{EP}_{\sigma(i-1)}^*, 0\right\} = \sum_{i=1}^{N+1} c_{\sigma(i-1)\sigma(i)},$$

where

$$c_{ij} = \max\left\{\mathrm{IP}_j^* - \mathrm{EP}_i^*, 0\right\}.$$

Therefore, in order to determine an optimal sequence in which to process the lots, it is enough to minimize $F(\sum) = \sum_{i=1}^{N+1} c_{\sigma(i-1)\sigma(i)}$.

Gilmore and Gomory [15] define the function c_{ij}, in general, to be as follows

$$\begin{cases} c_{ij} = \int_{B_i}^{A_j} f(x)\mathrm{d}x & \text{if } A_j \geq B_i, \\ c_{ij} = \int_{A_j}^{B_i} g(x)\mathrm{d}x & \text{if } A_j < B_i, \end{cases}$$

where $f(x)$ and $g(x)$ are any integrable functions satisfying $f(x) + g(x) \geq 0$. In view of our definition of c_{ij}, we can replace A_j with IP_j^* and B_i with EP_i^*. We can now present the steps of the Gilmore and Gomory's algorithm as follows.

Step 1. Index the lots in the nondecreasing order of EP_i^* values, i.e., $\mathrm{EP}_i^* \leq \mathrm{EP}_{i+1}^*$.

Step 2. Arrange IP_j^* in the nondecreasing order of its values as well.

Step 3. Find the permutation $\varphi(p)$ such that $\varphi(p) = q$, where q is a lot with the pth smallest IP^* value.

Step 4. Compute $c_\varphi(\alpha_{ii+1})$, for $i = 1, \ldots, N-1$, as follows:

$$c_\varphi(\alpha_{ii+1}) = 0 \quad \text{if} \quad \max\left(\mathrm{EP}_i, \mathrm{IP}_{\varphi(i)}\right) \geq \min\left(\mathrm{EP}_{i+1}, \mathrm{IP}_{\varphi(i+1)}\right);$$

otherwise

$$c_\varphi(\alpha_{ii+1}) = \min\left(\mathrm{EP}_{i+1}, \mathrm{IP}_{\varphi(i+1)}\right) - \max\left(\mathrm{EP}_i, \mathrm{IP}_{\varphi(i)}\right).$$

TABLE 3.20. Data for Example 3.11

Lot j	p_{j1}	p_{j2}	t_{j1}	t_{j2}	n_j	U_j
1	3	7	1	2	1	13
2	7	1	1	3	1	12
3	9	4	1	2	2	10
4	2	6	1	2	3	18
5	3	4	3	2	2	17
6	8	4	1	2	3	6
7	3	2	2	3	3	16

Step 5. Form an undirected graph with N nodes and undirected arcs connecting ith and $\varphi(i)$th nodes, $i = 1, \ldots, N$. If the resultant graph has only one connected component, go to Step 8; otherwise continue.

Step 6. Choose the smallest $c_\varphi(\alpha_{ii+1})$ such that i is in one connected component and $i+1$ in another. In case of a tie, choose any.

Step 7. Add the undirected arc A_{ii+1} to the graph using the i value selected in Step 6. Go to Step 6.

Step 8. Divide the arcs added in Step 7 into two groups. Those A_{ii+1} for which $I^*_{\varphi(i)} \geq E^*_i$ go in Group 1 and those for which $I^*_{\varphi(i)} < E_i$ go in Group 2.

Step 9. Arrange the arcs A_{ii+1} in Group 1 in the nonincreasing order of index i. In case there are ℓ arcs in Group 1, we have $i_1 \geq i_2 \geq \cdots \geq i_\ell$.

Step 10. Arrange the arcs A_{ii+1} in Group 2 in the nondecreasing order of index j. In case there are k arcs in Group 2, we have $j_1 \leq j_2 \leq \cdots \leq j_k$.

Step 11. The optimal sequence that minimizes $\sum_{i=1}^{N+1} \max\left\{\text{IP}^*_{\sigma(i)}, \text{EP}^*_{\sigma(i-1)}\right\}$ is obtained by following the ith lot with lot $\psi^*(i)$, where

$$\psi^* = \varphi\alpha^1_{i_1 i_1+1}\alpha^1_{i_2 i_2+1}\cdots\alpha^1_{i_\ell i_\ell+1}\cdots\alpha^2_{j_1 j_1+1}\alpha^2_{j_2 j_2+1}\cdots\alpha^2_{j_\ell j_\ell+1},$$

and α_{pq} is defined as follows:

$$\begin{cases} \alpha_{pq}(p) = q, \\ \alpha_{pq}(q) = p, \\ \alpha_{pq}(i) = i \quad (i \neq p, q). \end{cases}$$

Next, we illustrate the above algorithm through an example problem.

Example 3.11 Suppose there are seven lots with processing times (p_{jk}), setup times (t_{jk}), maximum allowed number of sublots (n_j), and lot size (U_j) as shown in Table 3.20.

The problem is to determine optimal sublot sizes and a sequence in which to process these lots so as to minimize the makespan for a no-wait flow shop.

Calculate the optimal sublot sizes for each lot. The first sublot x_{1j} and the last sublot size x_{nj} for each lot j are as shown below:

Lot j	1	2	3	4	5	6	7
x_{1j}	0.57	0.47	1.00	0.69	0.43	0.08	1.00
x_{nj}	0.14	0.21	1.00	0.31	0.57	0.69	1.00

Calculate IP_j^* and EP_j^*. Add a dummy lot 0, and index the lots in the nondecreasing order of EP_j^* values.

Lot j	EP_j^*	IP_j^*	$\varphi(j)$
0	0	0	0
1	3.43	1.77	6
2	6.74	21.74	2
3	12.00	22.86	5
4	12.31	26.43	1
5	38.86	38.00	7
6	74.77	61.31	4
7	91.00	82.00	3

Step 1. Rank the lots in the nondecreasing order of EP_j^* values. We have a permutation of $\varphi = (0, 6, 2, 5, 1, 7, 4, 3)$.
Step 2. Arrange IP_j^* in the nondecreasing order of its values as well.
Step 3. $\varphi(j)$ is given in the fourth column of the above table.
Step 4. Calculate the numbers $c_\varphi(\alpha_{i,i+1})$ for $i = 0, \ldots, 6$. We have $c_\varphi = (1.77, 3.31, 0, 0, 11.57, 22.45, 7.23)$, and EP_i and $\mathrm{IP}_{\varphi(i)}$, as follows:

Lot i	7	6	5	4	3	2	1	0
EP_i	91.00	74.77	38.86	12.31	12.00	6.74	3.43	0
$\mathrm{IP}_{\varphi(i)}$	82.00	61.31	38.00	26.43	22.86	21.74	1.77	0

Step 5. The undirected graph with arc $A_{i\varphi(i)}$ is shown in Fig. 3.28.
Step 6. There are four connected components in this graph: namely, {0}, {1, 4, 6}, {2}, and {3, 5, 7}. The $(i, i+1)$ pairs such that i is in one connected component and $i+1$ in another are (0, 1), (1, 2), (2, 3), (3, 4), (4, 5), (5, 6), and (6, 7). Since $c_\varphi(\alpha_{23}) = c_\varphi(\alpha_{34}) = 0$ are the smallest, arbitrarily choose (2, 3).
Step 7. Connect nodes 2 and 3 as shown in Fig. 3.29. Repeat Step 6 until we have one connected component. This is shown in Fig. 3.29.

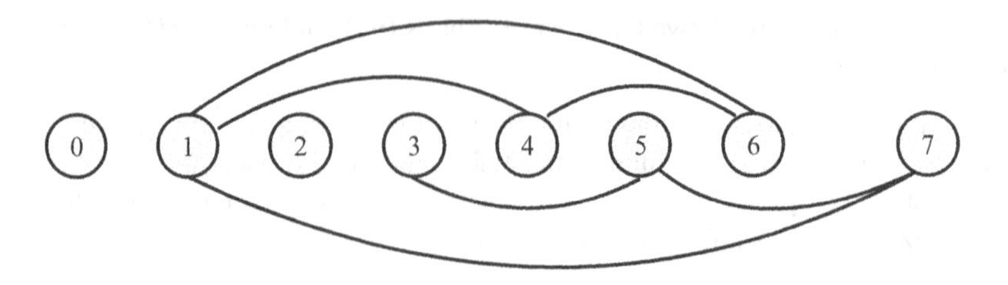

FIGURE 3.28. Undirectional graph at the start

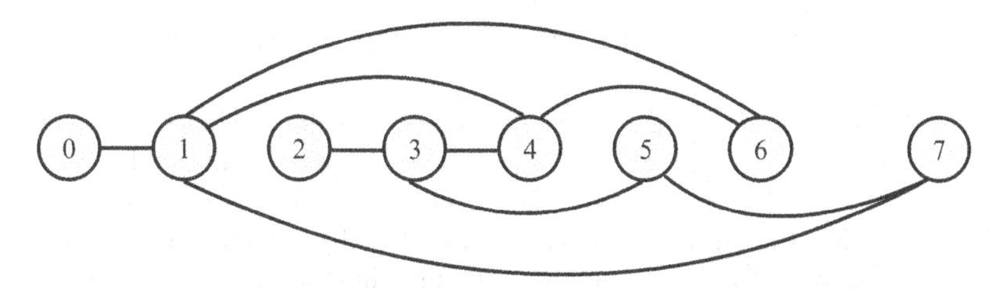

FIGURE 3.29. Undirectional graph with added arcs

Step 8. The arcs added in Step 7 are A_{23}, A_{34}, and A_{01}. Divide them into two groups. Group 1 has lots i with $\mathrm{IP}^*_{\varphi(i)} \geq \mathrm{EP}_i$, and the rest belong to Group 2. All arcs fall in Group 1.

Step 9. Arranging the arcs (A_{ii+1}) in Group 1 in nonincreasing order of index i, we have (A_{34}, A_{23}, A_{01}).

Step 10. This is not applicable.

Step 11. The optimal permutations is $\psi^* = \varphi\alpha_{34}\alpha_{23}\alpha_{01}$. It is obtained as follows:

$$\varphi\alpha_{34} = (0, 6, 2, 1, 5, 7, 4, 3),$$
$$\varphi\alpha_{34}\alpha_{23} = (0, 6, 1, 2, 5, 7, 4, 3),$$
$$\psi^* = \varphi\alpha_{34}\alpha_{23}\alpha_{01} = (6, 0, 1, 2, 5, 7, 4, 3).$$

The optimal sequence is obtained by following the ith lot with lot $\psi^*(i)$, which for this case is 0–6–4–5–7–3–2–1–0. The minimum cost of this sequence (without the dummy lot) is 280.85.

The integrated heuristic procedure for the integer sublot sizes and their optimal sequence is given below. Since the optimal integer sublot sizes for each individual lot need not remain optimal as the lots are sequenced to obtain an optimal solution, the following heuristic could be used, which essentially relies on the continuous sublot size solution.

Integrated Algorithm for Integer Sublot Sizing and Lot Sequencing in a No-Wait Flow Shop

Sublot Sizing

Step 1. For all lots, find the continuous optimal sublot sizes using Theorem 3.4.

Step 2. For all lots, find the integer sublot sizes using the algorithm for 2/1/C/II/DV/{Lot-Detached Setup, No-Wait} (see Sect. 3.5).

Step 3. For all lots, calculate IP^*_j and EP^*_j.

Sequencing Procedure

Use the algorithm of Gilmore and Gomory [15] to determine a sequence.

3.14 2/N/E/II/DV/Sublot-Attached Setup Times

Next, we address the situation in which the number of sublots of a lot is not known a priori and needs to be determined. However, we assume that the sublots of a lot are of the same size, take on integer values, and also incur sublot-attached setups. This problem is different from that discussed in Sect. 3.7 in that, here, we consider more than one lot and also the objective of minimizing the makespan as compared to the consideration of a single lot and the objective of minimizing the sum of the weighted sublot completion times for the problem discussed in Sect. 3.7. The following two special cases are in order:

1. Sublot sizes of all the lots are the same, i.e., $s_i = s, \forall i = 1, \ldots, N$. Hence, in this case, we need to find only one value of s, which serves as a sublot size for all the lots.
2. Sublot sizes of different lots are different, but sublot sizes of any particular lot are equal. Hence, in this case, we need to find $s_i, \forall i = 1, \ldots, N$.

Case 1. $s_i = s, \forall i = 1, \ldots, N$

Recall that the traditional n-job two-machine flow shop problem can be solved optimally using Johnson's rule. For the problem in hand, the lots cannot be sequenced until we have determined their sublot sizes. To that end, the proposed algorithm iterates over the sublot sizes. At each iteration of this procedure, with a fixed sublot size for all lots, an optimal sequence is obtained by using Johnson's rule [19]; the sublot-attached setup times are included in the sublot processing times. Clearly, as a result of this procedure, the sublot sizes and the corresponding sequence obtained are optimal. This follows from the fact that the procedure iterates over the entire set of feasible solutions. A significant amount of computations is, however, saved at each stage since the optimal sequence is obtained via Johnson's rule (as opposed to complete enumeration). The algorithm is given below. Here, τ_{ik} is the setup time of sublot i on machine k.

Algorithm for 2/N/E/DV/$C_{\max}$, Sublot-Attached Setup, $s_i = s, \forall i = 1, \ldots, N$

Step 1. Input $\tau_{i1}, \tau_{i2}, p_{i1}, p_{i2}, U_i, \ \forall i = 1, \ldots, N$ and an UB value (use $\text{UB} = \min_{1 \leq i \leq N} \{U_i\}$, in the absence of a tighter value). Initialize sublot size, $s = \text{LB}$ (Use $\text{LB} = 1$ in the absence of a tighter value).

Step 2. For each lot i, compute the number of "jobs" to be considered as

$$n_i = \left\lfloor \frac{U_i}{s} \right\rfloor.$$

Consider the last job to be of size, $R_i = U_i - n_i s$, the remainder sublot. With each of these jobs, associate the following "job" processing times on the two machines

$$A_i = \tau_{i1} + s p_{i1} \quad \text{and} \quad B_i = \tau_{i2} + s p_{i2},$$

except for the last sublots for which we replace s by R_i.

TABLE 3.21. Data for Example 3.12

Lot i	Processing times		Setup times		Lot size
	p_{j1}	p_{j2}	τ_{i1}	τ_{i2}	U_i
1	3	4	2	1	4
2	4	3	0	3	3
3	4	4	2	0	2

TABLE 3.22. Solution summary for Example 3.12

s	Lot	n_i	A_i	B_i	R_i	A_i	B_i	$S^*(s)$	$C^*_{\max}(S^*(s))$
	1	4	5	5	0				
1	2	3	4	6	0			{2, 2, 2, 1, 1, 1, 1, 3, 3}	50
	3	2	6	4	0				
	1	2	8	9	0				
2	2	1	8	9	1	4	6	$\{\underline{2}, 2, 1, 1, 3\}$	47
	3	1	10	8	0				
	1	1	11	13	1	5	5		
3	2	1	12	12	0			$\{\underline{1}, 1, 2, 3\}$	49
	3	0			1	10	8		
	1	1	14	17	0				
4	2	0			1	12	12	$\{\underline{1}, \underline{2}, \underline{3}\}$	51
	3	0			1	10	8		

Step 3. Apply Johnson's rule on the entire set of jobs formed in Step 2 to obtain a sequence, $S(s)$. Compute the resultant makespan of this sequence, $C_{\max}(S(s))$. Save the best solution so far as s^*, $S^*(s^*)$, $C^*_{\max}(S^*(s^*))$.

Step 4. If $s = \text{UB}$, then stop and output the best solution so far. This is the optimal solution. Else set $s = s + 1$, and go to Step 2.

Example 3.12 Consider a two-machine flow shop with three lots. The relevant data are given in Table 3.21.

Initially, the sublot size s is set at 1. For a sublot size, the number of full sublots of each lot-type, n_i, is computed and the size of the remainder sublot, R_i, is determined. Next, the job processing times A_i and B_i are computed for these sublots, and the optimal sequence $S^*(s)$ is established using Johnson's rule. Lastly, the corresponding makespan is calculated. These calculations, for every value of s, are summarized in Table 3.22.

Note that in specifying a sequence, an underscore is used to differentiate between full sublots and remainder sublots, which have been underscored. It can be seen that a sublot size of 2 is optimal. In this optimal solution, lot 1 is split into exactly two full sublots. Lot 2, on the other hand, is split into a full and a remainder sublot. Lot 3 is not split at all and is taken in its entirety. The optimal makespan for this sublot size is 47.

The overall complexity of this algorithm is $O(N \log(N) U_{\max})$. The algorithm, therefore, qualifies as pseudopolynomial, since it depends on the lot size(s).

Case 2. $s_i \neq s, \forall i = 1, \ldots, N$

Clearly, this case is more complicated as compared to the one considered above. However, a pseudo-polynomial procedure can still be constructed to obtain an optimal solution by modifying the above algorithm. This new algorithm iterates over the individual lot sizes. Hence, the modified algorithm extends the search from a single-dimensional variable (namely, the identical sublot size) to a search over a multidimensional variable (namely, the lot-based sublot sizes). The iterative process is repeated for every combination of values of sublot sizes. Although optimal, this algorithm is very cumbersome and practically inefficient when the number of sublots and the lot sizes are relatively large. The algorithm iterates $\prod_{i=1}^{N} U_i$ times using Johnson's rule, to identify the optimal solution consisting of the sublot sizes for all the lots and the sequence in which to process them. Hence, the complexity of this algorithm is of the order $O\left(N \log(N) \prod_{i=1}^{N} U_i\right)$. The expression $\prod_{i=1}^{N} U_i$ is problematic, from a computational viewpoint, since it is exponential in the number of lots. However, we can use a quick and efficient heuristic procedure as follows.

This heuristic procedure consists of two phases (1) a *construction phase*, which obtains near optimal sequence of the lots and an (2) *improvement phase*, which optimizes the sublot sizes of the lots, given the sequence obtained in the construction phase.

In the construction phase, the single lot, two-machine problem is solved optimally for all the lots individually. Johnson's rule is then applied to the sublots of the lots in order to obtain a sequence. In all likelihood, the first lot (denoted as f) in this sequence is indeed the best lot to sequence first. This is due to two reasons. First, the processing time of the first sublot of this lot on machine 1 must be the smallest from among all the lots in accordance with Johnson's rule. Hence, this minimizes the start time (or the lag time) on machine 2. Second, it is most likely that lot f will not cause any idle time in between the processing of its sublots on machine 2. This follows by the fact that the Johnson's rule schedules the jobs with $A_f \leq B_f$ first, i.e., the sublot processing time on machine 2 is not less than that on machine 1. The only special case when the first lot would cause some idle time in between the processing of its sublots on machine 2 is when $A_i > B_i$ for all lots. However, in that case, every lot would cause some unavoidable idle time on machine 2.

Once this initial sequence has been obtained, the improvement phase is implemented wherein the sublot sizes of the lots are reoptimized. The idea is to reduce the setup times by increasing their sublot sizes. If the net decrement in setup times as a result of sublot size increment is greater than an increment in idle time, then an improvement in makespan value will be achieved. To that end, the lots are considered, except for the first lot, in the nonincreasing order of the total setup times required on machine 2 in their solution obtained in the construction phase. Then, starting with the first lot on the list, the sublot size is iteratively incremented until a net reduction in the makespan value is not obtained. This process is repeated for all the lots on the list. While iterations are carried out to reduce the makespan with respect to a lot, the heuristic accepts a larger sublot size even if that sublot size does not result in a strict reduction of makespan. This is due to the fact that a larger sublot size implies a smaller run-in time due to a lower total setup time. Hence, between the two sublot sizes resulting in the same makespan, the larger one is chosen. While ordering the lots based on the total setup time on machine 2, the first lot is not considered since its sublot size, as determined in the construction phase, is already optimal. However, after all the lots in the list have been reoptimized, it may still be beneficial to reoptimize the size of the first sublot. Hence, it is placed last in LIST.

It should be apparent that the re-optimized sublot sizes need not be the same as those computed in the construction phase. In fact, they will always be greater than or equal to the sublot size obtained earlier, implying that some setup time is saved. Hence, the iterations made over the sublot sizes start from the sublot sizes obtained in the construction phase. It is followed by their increments in steps of one, to obtain the re-optimized sublot sizes. This algorithm is formally presented next.

Algorithm for 2/N/E/DV/Sublot-Attached Setup Times, $n_i \neq n, \forall i = 1, \ldots, N$

Construction Phase

Step 1. Input $N, m, \tau_{i1}, \tau_{i2}, p_{i1}, p_{i2}, U_i, \ \forall i = 1, \ldots, N$.

Step 2. For each lot i, solve the single lot, two-machine problem. The optimal solution is given by

$$s_i^* = \arg \min_{s_i} \left\{ C_{\max}(s_i) = \left(\frac{U_i}{s_i} - 1 \right) \max_{k=1,2} \{\tau_{ik} + s_i p_{ik}\} + \sum_{k=1}^{2} (\tau_{ik} + s_i p_{ik}) \, | s_i : \left[\sqrt{\frac{U_i \tau_{i1}}{p_{i2}}} \right]^{\pm}, \left[\sqrt{\frac{U_i \tau_{i2}}{p_{i1}}} \right]^{\pm}, \left[\frac{\tau_{i2} - \tau_{i1}}{p_{i1} - p_{i2}} \right], 1, U_i \right\}.$$

The above expression states that, in the case of two machines, the optimal sublot size for a single lot is either the rounded-up or rounded-down value of one of the following possibilities:

(a) If machine 1 dictates the makespan, then the optimal sublot size is $\sqrt{U_i \tau_{i1} / p_{i2}}$.

(b) If machine 2 dictates the makespan, then the optimal sublot size is $\sqrt{U_i \tau_{i2} / p_{i1}}$.

(c) If both the machines dictate the makespan, i.e., the critical path contains some sublots on machine 1 and some sublots on machine 2, then the intersection point corresponds to the optimal solution, i.e., the optimal continuous sublot size is $(\tau_{i2} - \tau_{i1})/(p_{i1} - p_{i2})$.

(d) All of the above result in a sublot size which is either greater than the lot size U or smaller than the discrete sublot size of 1. In that case, the solution will be either U or 1, respectively.

Consider each lot to be a job. With each of these jobs, associate the following "job" processing times:

$$A_i = A_i = \tau_{i1} + s_i p_{i1} \quad \text{and} \quad B_i = \tau_{i2} + s_i p_{i2}.$$

Step 3. Apply Johnson's rule on the entire set of jobs formed in Step 2 to obtain a sequence, S. This sequence remains unchanged in the remaining steps.

Step 4. Schedule the lots according to this sequence (with size computed in Step 2). Compute the makespan $C_{\max}(S, \underline{s})$; Set $C^*_{\max}(S, \underline{s})$.

Improvement Phase

Step 5. Compute the total setup time associated with each lot on machine 2. Form a list, LIST, in which all the lots (except the first one) are arranged in nonincreasing order of the total setup time. Place the first lot last in LIST.

Step 6. For any lot in LIST, say j, reoptimize its sublot size as follows:

(a) $s_j = s_j + 1$; Compute the resulting makespan $C^t_{\max}\left(S, \underline{s^{\text{new}}}\right)$.
(b) If $C^t_{\max}\left(S, \underline{s^{\text{new}}}\right) \le C^*_{\max}\left(S, \underline{s}\right)$, then set $C^*_{\max}\left(S, \underline{s}\right) = C^t_{\max}\left(S, \underline{s^{\text{new}}}\right)$; repeat Step 6(a). Otherwise, go to Step 7.

Step 7. Remove lot j from LIST. If LIST is empty, then go to Step 8. Otherwise, repeat Step 6.

Step 8. The final solution is obtained with the makespan values of $C^*_{\max}(S, \underline{s})$, Stop.

The proposed heuristic overcomes the shortcoming of the optimal solution algorithm by first fixing a sequence of lots and then iterating, in accordance with this sequence, to determine sublot sizes. Consequently, the order of complexity, $O(U^N)$, of the original optimal solution algorithm reduces to $O(UN)$, due to the substitution of the exponential search in the optimal solution algorithm by a sequential search in the proposed heuristic. The overall order of complexity of the proposed heuristic is $O(N^2 \log(N)U)$. Although still pseudopolynomial, it is not exponential in U and computational experiments [21] have shown that the solutions via this heuristic are obtained in a matter of few seconds. Example 3.13 illustrates the working of the heuristic.

Example 3.13 Consider a two-machine flow shop with five lots. The data are given in Table 3.23. For simplicity, all the lot sizes are set to 4 and only integer number of sublots of a lot are evaluated, i.e., $s = 3$ is not considered. The total number of combinations that would have to be evaluated under total enumeration is $3^5 = 243$.

The computations of the construction phase of the heuristic are shown in the Table 3.24. For each i, the values of A_i and B_i and the resultant single-lot makespan are shown. The optimal (unconstrained) sublot size for the single-lot problem is indicated by "*." The resultant sequence, determined in Step 3, is {2–5–4–3–1} and the overall makespan is 85, as shown in Fig. 3.30.

The improvement phase of the heuristic begins by computing the total setup time associated with each lot on machine 2, in order to determine the LIST, which

TABLE 3.23. Data for Example 3.13

Lot i	Processing times		Setup times		Lot size
	p_{i1}	p_{i2}	τ_{i1}	τ_{i2}	U_i
1	4	2	2	1	4
2	2	4	1	2	4
3	3	3	2	1	4
4	3	3	1	2	4
5	4	3	1	2	4

TABLE 3.24. Computations performed in Step 2

Lot	Sublot size	A_i	B_i	Makespan
	1	$2+4=6$	$1+2=3$	27
1	2	$2+8=10$	$1+4=5$	25(*)
	4	$2+16=18$	$1+8=9$	27
	1	3	6	27
2	2	5	10	25(*)
	4	9	18	27
	1	5	4	24
3	2	8	7	23(*)
	4	14	13	27
	1	4	5	24
4	2	7	8	23(*)
	4	13	14	27
	1	5	5	20(*)
5	2	9	8	26
	4	17	14	31

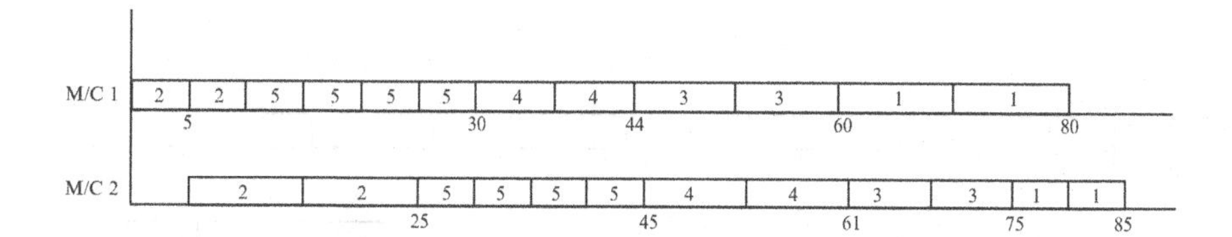

FIGURE 3.30. Schedule after construction phase

consists of lots arranged in nonincreasing order of their setup times. The total setup time associated with the sublots of lots 1, 3, 4, and 5 is 2, 2, 4, and 8, respectively. Consequently, the LIST $= \{5, 4, 3, 1, 2\}$. Note that lot 2, being first in the sequence, is placed last in LIST. The reoptimization of the sublot sizes starts with lot 5. Its current sublot size is 1. When the sublot size is incremented to 2, the makespan value reduces from 85 to 83, despite the creation of an idle time slot of 2 units on machine 2. The resultant schedule is shown in Fig. 3.31.

Incrementing the sublot size to 4 further decreases the makespan to 82. This is shown in Fig. 3.32.

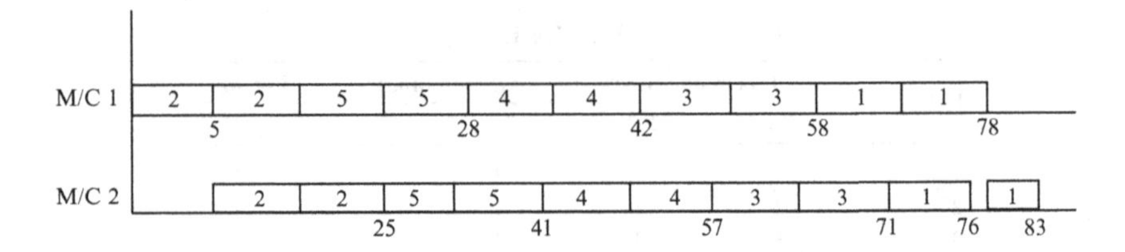

FIGURE 3.31. Improved schedule after iteration 1

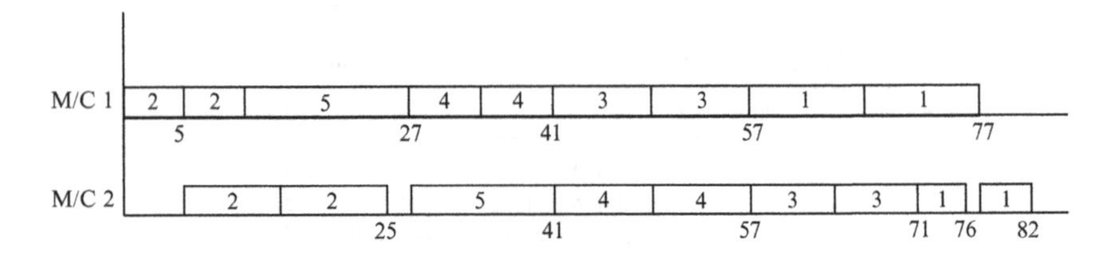

FIGURE 3.32. Improved schedule after iteration 2

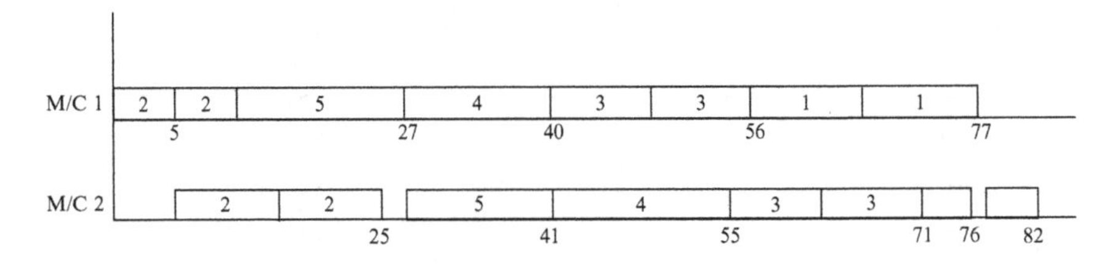

FIGURE 3.33. Improved schedule after iteration 3

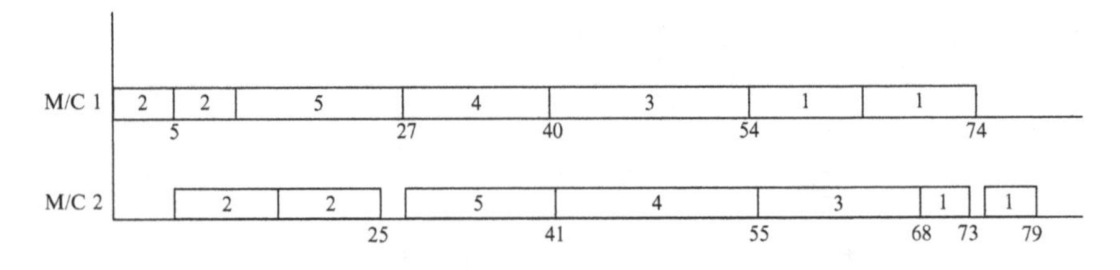

FIGURE 3.34. Improved schedule after iteration 1

Lot 4 is next on the LIST, with a sublot size of 2, which when incremented to 4, decreases the makespan value to 81, although the idle time on machine 2 increases to 4 units. This is depicted in Fig. 3.33.

The process is repeated for lot 3, which is next on the LIST. By incrementing its current sublot size from 2 to 4 also results in a reduction in the makespan value (to 79). This schedule is shown in Fig. 3.34.

The sublot size of lot 1 does not change, since an increment in its sublot size from 2 to 4 results in an increase in makespan by 2 units. Lastly, the current sublot

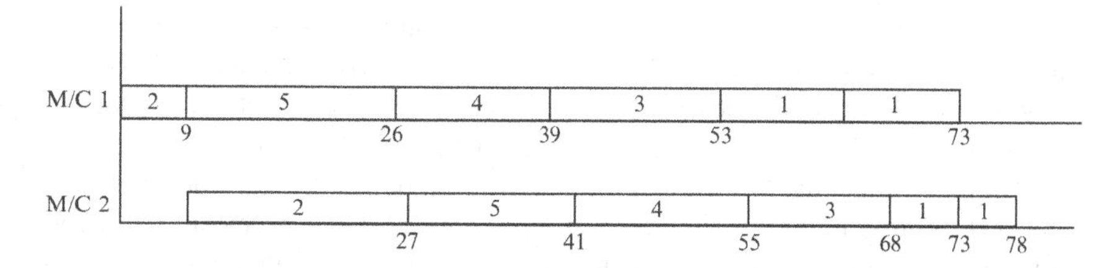

FIGURE 3.35. Final and optimal schedule for Example 3.13

size of lot 2 is incremented to 4, which results in a decrement of makespan. This is the final solution of the heuristic and is shown in Fig. 3.35.

The final solution can be summarized as below:

$$s_1 = 2, s_2 = 1, s_3 = 1, s_4 = 1, s_5 = 1, S = \{2-5-4-3-1\};\ C_{\max}\left(S, \underline{s}\right) = 78.$$

An evaluation of all possible solutions for the given problem (via total enumeration) shows that the solution found by the heuristic is indeed optimal for this case.

3.15 Chapter Summary

The generic mathematical model for the lot streaming problem presented in Chap. 2 reduces to a linear programming model if the integrality of the sublot sizes is dropped and the sequence in which to process the lots is assumed known a priori. The resulting model would, therefore, be relatively easier to solve. However, under some special conditions, the optimal sublot sizes can be shown to be related to each other, and thus, are easy to determine. The single lot, two-machine lot streaming problem, involving a given number (n) of continuous and consistent sublot sizes and involving no setups, is such a case. For this problem, the sizes of the n sublots can be shown to be geometric in q, where q is the ratio of the unit processing time on machine 2 to that on machine 1. The sublots are increasing in size if $q > 1$, and decreasing in size if $q < 1$. The resulting solution is compact in the sense that there is no idle time in between the sublots and that every sublot lies on a critical path that traverses from the first sublot on the first machine to the last sublot on the second machine. To determine integer sublot sizes, a polynomial time algorithm of $O(n^2)$ can be used, which begins from the optimal, continuous sublot sizes and iteratively converts them to optimal, integer values by causing the least amount of increment in the makespan. This is interesting because we are able to solve the underlying integer program (using this algorithm) in polynomial time.

The notion of a critical path (as well as that of a bottleneck or dominant machine) has been exploited quite effectively to address the lot streaming problems. As alluded to above, a critical path begins from the processing of the first sublot on the first machine and ends at the processing of the last sublot on the second machine; and it comprises of the sublots that are processed continuously, i.e., with no idle time among them. If the processing of the sublots on the machines is viewed as a directed network consisting of the sublots as nodes and a directed arc between every two adjacent nodes representing the precedence order in which to process the sublots, then the critical path is the largest path from the start node to the end node. Some interesting structural properties can be developed by studying such a network representation of the above problem, which help in devising a better algorithm (of complexity $O(n)$) to determine optimal integer sublot sizes.

The consideration of setup, removal, and transfer times generally makes the problem more difficult to solve; however, for some cases the optimal sublot sizes can be obtained easily. Recall that during removal time, the current machine remains occupied and the next lot or sublot, as the case may be, cannot begin processing until the removal process has been completed. The transfer time represents the time in transit from one machine to another.

In the presence of the lot-attached or lot-detached setups, geometric sublot sizes are still optimal for the case of continuous sublot sizes. The optimal discrete sublot sizes can be obtained in polynomial time by iteratively modifying the optimal continuous sublot sizes. In case the sublot-attached transfer time is also present, along with the lot-attached or lot-detached setup, the linear programming model of the underlying lot streaming problem can be used to obtain optimal continuous sublot sizes. For the case of lot-detached setup, it turns out that the optimal lot sizes are geometric in nature except for the first sublot. Algorithms are also available to obtain integer sublot sizes, starting from the optimal continuous sublot sizes.

The geometric sublot sizes remain optimal for the case when the lot-detached setup is involved and the sublots are processed in a no-wait fashion, i.e., each sublot is processed continuously on both the machines. In case the size of each sublot is required to take on integer value, the problem is difficult to solve and a heuristic solution procedure is available that does not guarantee attainment of an optimal solution.

For the makespan objective function, it is easy to see that the optimal sublot sizes are those that are consistent. However, this need not be the case for other objective functions like, for instance, the minimization of the sum of the weighted completion times of the sublots. When $q \leq 1$, the optimal sublot sizes are equal for both the consistent and variable sublot size scenarios. However, when $q > 1$, for the case of consistent sublots, the optimal sublot sizes are geometric until a sublot $\upsilon \leq n$, and equal after that. For the variable sublot case, when $q > 1$, the optimal sublot sizes are geometric on the first machine and equal on the second machine. This follows by the fact that the processing of the sublots should be continuous on the dominant machine, and hence, should start as soon as possible. Thus, when $q \leq 1$, machine 1 is dominant and determines the sublot sizes. However, when $q > 1$, machine 2 is dominant, and geometric sublot sizes on machine 1 allow processing to begin as early as possible on machine 2. In the presence of setups, the above solution does not change if the setup is lot-attached or lot-detached. However, for sublot-attached setups, the optimal sublot sizes are dictated by the nature of the bottleneck encountered. In case the bottleneck is unique, due to either machine 1 or machine 2, algorithms exist to obtain optimal sublot sizes. However, in case the bottleneck is not unique, then only a heuristic procedure is available for the determination of sublot sizes.

The presence of multiple lots adds another aspect to the lot streaming problem and that is of sequencing the lots. Consider the simplest case which involves unit-sized sublots and no setups. The optimal sequence in which to process the lots can be determined by applying Johnson's algorithm to individual sublots, or equivalently, to run-in and run-out times of the lots. In case lot-detached or lot-attached setups are present, but still involving unit-size sublots, run-in and run-out times can be determined accordingly, and the optimal sequence in which to process the lots can be determined using Johnson's algorithm. When the assumption of unit-size sublots is dropped, a question arises as to the relationship between the sublot sizes and the sequence in which to process the lots. It turns out that, in many cases, the sublot sizing and lot sequencing problems are independent, i.e., the sublot sizes for each lot can be determined optimally irrespective of the positions of that lot in the sequence. For the case of lot-detached or lot-attached setup and lot-attached removal times, the optimal continuous sublot sizes are geometric, and then, the lots can be sequenced optimally using Johnson's algorithm, applied to run-in and run-out times that are computed appropriately in the presence of pertinent setup times and removal times. Discrete sublot sizes can be obtained iteratively, starting from the optimal continuous sublot sizes. The sublot sizing and lot sequencing problems are also independent in the case of lot-detached or lot-attached setup and sublot-attached transfer times. The optimal continuous sublot sizes can be obtained by solving a linear program. The optimal sequence in which to process the lots is determined by applying Johnson's algorithm to properly defined lot processing times. Discrete sublot sizes can be obtained iteratively starting from the optimal continuous sublot sizes. The lot sizing and lot sequencing decisions are still independent for the no-wait flow shops in the presence of lot-detached setups. However, these decisions do not remain independent in the presence of sublot-attached setups. We have shown this for the case of same size sublots, where the problem is to determine an optimal number of sublots for each lot. The problem is easy to solve when the sublot size is the same for all the lots. The optimal solution can be obtained by enumerating over all possible sublot sizes, and for each sublot size, obtaining the optimal sequence by using Johnson's algorithm. For the case of different sublot sizes for different lots, the possible combinations of sublot sizes of different lots grow exponentially, even though for each combination, the optimal sequence of the lots can be determined by using Johnson's algorithm. A heuristic procedure is presented for this problem that relies on obtaining an optimal, equal sublot size for each lot, sequencing the lots by using Johnson's algorithm, and then, adjusting sublot sizes of individual lots to obtain a better makespan value while maintaining the same lot sequence. This as well as other problems involving additional features can be solved by using the formulations of the corresponding mathematical model from the generic model presented in Chap. 2.

4
Three-Machine Lot Streaming Models

4.1 A Brief Overview

As alluded to earlier, the three-machine lot streaming problems form a stepping stone for the study of general m-machine lot streaming problems. For the case of continuous and consistent sublot sizes and makespan objective, the single lot, three-machine lot streaming problem can be modeled as a linear program for minimizing the completion time of the last sublot on machine 3 or, equivalently, the total idle time appearing on machine 3. The requirements of no idling among the sublots and integer sublot sizes can be captured by making appropriate changes to the linear programming formulation. Alternatively, the continuous and consistent optimal sublot sizes in the presence of intermittent idleness among sublots can be obtained by analyzing the structural properties of a network representation of the problem.

In the absence of setup times, if the sublot sizes are variable and no idling is permitted among the sublots, then the problem can be decomposed into two subproblems consisting of machines 1 and 2, and machines 2 and 3. Each of these problems can be solved by using the methodology for the 2/1/NI/CV problem. However, when the sublot sizes are variable and intermittent idling is permitted, then the solution procedure depends on the processing time of machine 2. If $p_2^2 > p_1 p_3$, i.e., machine 2 dominates, then the problem can be solved to optimality by decomposing it into 2 two-machine problems and by solving each of these problems by using the two-machine procedures developed in Chap. 3. However, if $(p_2)^2 \leq p_1 p_2$, i.e., machine 2 is dominated, then consistent sublot sizes are optimal and they are geometric in the ratio $(p_2 + p_3) : (p_1 + p_2)$.

In the case of consistent sublot sizes and lot-detached/attached setups, the optimal sublot sizes can, once again, be obtained by using the network representation of the problem.

A schematic representation of the work reported on the three-machine lot streaming problems is given in Fig. 4.1. Table 4.1 shows the problem status for the various three-machine single and multiple-lot streaming problems. Table 4.2 gives a brief summary of the major results and/or solution methodologies used to solve these problems.

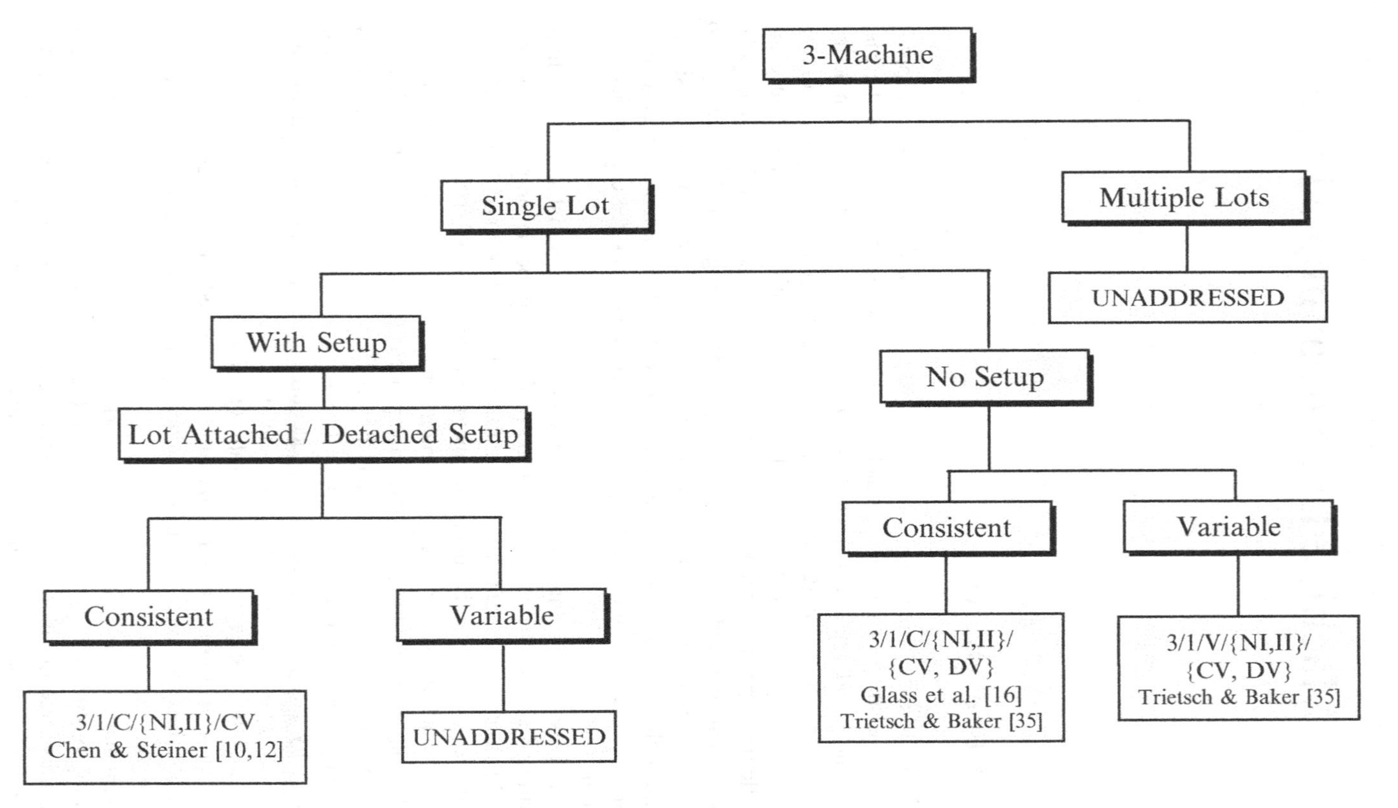

FIGURE 4.1. Research on three-machine lot streaming problems

TABLE 4.1. Problem status for three-machine lot streaming problems

No.	Problem	Problem type	Status
A. Single lot			
A1. No setup			
1.	3/1/C/{NI, II}/{CV, DV}	$\Im^L(s)$	Section 4.2
2.	3/1/V/{NI, II}/{CV, DV}	$\Im^L(s)$	Section 4.3
A2. With setup			
1.	3/1/C/II/CV/{Lot-Detached Setup}	$\Im^L(s)$	Section 4.4
2.	3/1/C/II/CV/{Lot-Attached Setup}	$\Im^L(s)$	Section 4.5
3.	3/1/C/NI/DV/{Lot-Detached/Attached Setup}	$\Im^L(s)$	Open
4.	3/1/V/{NI,II}/{CV,DV}/{Lot-Detached/Attached Setup}	$\Im^L(s)$	Open
B. Multiple lots			
1.	3/N/{C, V}/{II, NI}/{CV, DV}/{Lot/Sublot-Detached/Attached Setup Times, Transfer Times, Removal Times, Release Times}	$\Im^L(s, S)$	Open

4.2 3/1/C/{NI,II}/{CV,DV}

The subscript j is dropped here from the relevant notation since there is only one lot. For the case when only two sublots are present, the optimal solution can be obtained by analyzing the various cases depending on the relationships among the unit processing times of the jobs on the machines. Since machine 1 operates continuously, the idle time on this machine is zero. Similarly, right shifting the start times can eliminate the idle time on machine 3. Hence, the analysis focuses on machine 2 where we try to eliminate the inserted idle time by sizing the sublots such that the makespan is minimized. We have the following two cases:

1. $p_2^2 > p_1 p_3$
2. $p_2^2 \leq p_1 p_3$

Furthermore, for Case 1, the determination of sublot sizes depends on whether $p_1 \geq p_3$ or $p_1 < p_3$. We discuss these cases next.

Case 1a. $p_2^2 > p_1 p_3,\ p_1 \geq p_3$

In this case, machines 1 and 2 are critical (do not encounter idle time) and the optimal solution must be compact on these machines. Hence,

$$p_1 s_2 = p_2 s_1.$$

Rearranging gives

$$p_2 s_1 - p_1 s_2 = 0. \tag{4.1}$$

Also,

$$s_1 + s_2 = U. \tag{4.2}$$

TABLE 4.2. Summary of major results/solution methodologies for the three-machine lot streaming problems

No.	Problem	Major result/solution methodology	Complexity $[\Im^L(s)]$
1.	3/1/C/{NI, II}/{CV, DV} (Sect. 4.2) [16,35]	(a) Continuous optimal sizes can be obtained by LP formulations, which minimize the completion time of the last sublot on machine 3 or, equivalently, the total idle time on machine 3 (appearing in between the sublots).	Polynomial
		(b) Integer-sized sublots and the no-idling version of the problem can be obtained by making appropriate changes to the LP formulation.	Exponential
		(c) Alternatively, the continuous and consistent optimal sublot sizes in the presence of intermittent idleness among the sublots can also be obtained by analyzing the structural properties of a network representation of the problem.	Polynomial
2.	3/1/V/NI/{CV, DV} (Sect. 4.3) [35]	This problem can be solved by decomposing it into 2 two-machine problems and using the procedures presented in Sects. 3.2 and 3.3 to obtain the continuous and integer sublot sizes.	Polynomial
3.	3/1/V/II/{CV, DV} (Sect. 4.3) [35]	The optimal schedule is a no-wait schedule and the solution procedure depends on the processing time on machines 2. If $(p_2)^2 \geq p_1 p_3$, then we decompose the problem and obtain sublot sizes as in (2) above. If $(p_2)^2 < p_1 p_3$, then consistent sublot sizes are optimal and are geometric with ratio $(p_2+p_3) : (p_1+p_2)$. This result does not extend to the case of integer-sized sublots. These can be obtained by using an algorithm that iteratively rounds up the continuous optimal sublot sizes.	$O(\log n)$
4.	3/1/C/{NI, II}/CV/ { Lot-Detached Setup} (Sect. 4.4) [12]	Optimal sublot sizes can be obtained by analyzing a network representation of the problem.	$O(\log n)$
5.	3/1/C/II/CV/{Lot-Attached Setup} (Sect. 4.5) [10]	Optimal sublot sizes can be obtained by analyzing a network representation of the problem.	$O(n)$

By solving (4.1) and (4.2) for s_1 and s_2, we have

$$s_1 = \left(\frac{p_1}{p_1 + p_2}\right) U \quad \text{and} \quad s_2 = \left(\frac{p_2}{p_1 + p_2}\right) U.$$

Thus, the optimal sublot sizes are in the ratio $p_1 : p_2$, and there is no intermittent idling on machine 2.

Case 1b. $p_2^2 > p_1 p_3$, $p_1 < p_3$

In this case, machines 2 and 3 are the critical machines and the optimal solution must be compact on these machines. Hence,

$$p_3 s_1 = p_2 s_2.$$

Rearranging gives

$$p_3 s_1 - p_2 s_2 = 0. \tag{4.3}$$

Also,

$$s_1 + s_2 = U. \tag{4.4}$$

By solving (4.3) and (4.4) for s_1 and s_2, we have

$$s_1 = \left(\frac{p_2}{p_2 + p_3}\right) U \quad \text{and} \quad s_2 = \left(\frac{p_3}{p_2 + p_3}\right) U.$$

Hence, the optimal sublot sizes are in the ratio $p_2 : p_3$, and there is no intermittent idling on machine 2.

Case 2. $p_2^2 \leq p_1 p_3$

In this case, idle time may be present on machine 2. The optimal schedule will be obtained when there is maximum overlap, i.e., $C_{1,3} = C_{2,2}$. This can be written as

$$s_1 (p_1 + p_2 + p_3) = p_1 s_1 + s_2 (p_1 + p_2),$$

i.e.,

$$s_1 (p_2 + p_3) = s_2 (p_1 + p_2).$$

By rearranging, we have

$$s_1 (p_2 + p_3) - s_2 (p_1 + p_2) = 0. \tag{4.5}$$

Again,

$$s_1 + s_2 = U. \tag{4.6}$$

By solving (4.5) and (4.6) for s_1 and s_2, we have

$$s_1 = \left(\frac{p_1 + p_2}{p_1 + 2p_2 + p_3}\right) U \quad \text{and} \quad s_2 = \left(\frac{p_2 + p_3}{p_1 + 2p_2 + p_3}\right) U.$$

Hence, the optimal sublot sizes are in the ratio $(p_1 + p_2) : (p_2 + p_3)$, and there is intermittent idling on machine 2. These results are listed in Table 4.3.

For a more general case, when the number of sublots, $n > 2$, the problem of finding the optimal consistent sublot sizes can be formulated as a linear program as follows:

Minimize: C_{n3}

Subject to:

$$C_{11} \geq s_1 p_1, \tag{4.7}$$

$$C_{ik} \geq C_{i,(k-1)} + p_k s_i, \quad \forall i = 1, \ldots, n, k = 1, 2, 3, \tag{4.8}$$

TABLE 4.3. Sublot sizes for 3/1/II/CV/{No Setup} problem

	s_1	s_2	C_{max}
$p_2^2 > p_1p_3, p_1 \geq p_3$	$\left(\frac{p_1}{p_1+p_2}\right)U$	$\left(\frac{p_2}{p_1+p_2}\right)U$	$s_1(p_1+p_2)+s_2(p_2+p_3)$
$p_2^2 > p_1p_3, p_1 < p_3$	$\left(\frac{p_2}{p_2+p_3}\right)U$	$\left(\frac{p_3}{p_2+p_3}\right)U$	$(p_1+p_2+p_3)s_1+p_3s_2$
$p_2^2 \leq p_1p_3$	$\left(\frac{p_1+p_2}{p_1+2p_2+p_3}\right)U$	$\left(\frac{p_2+p_3}{p_1+2p_2+p_3}\right)U$	$p_1s_1+(p_1+p_2+p_3)s_2$

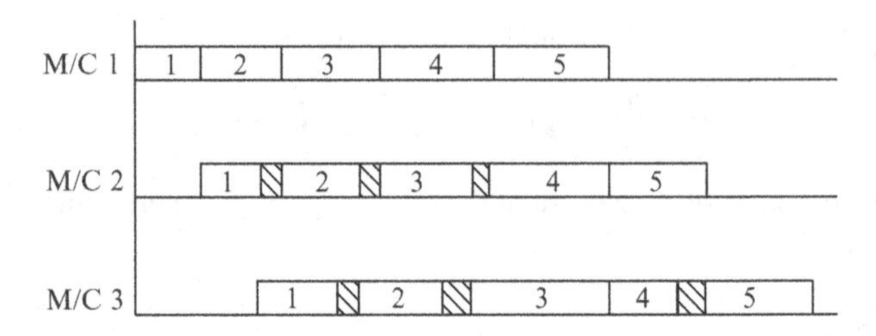

FIGURE 4.2. Occurence of idle time in a three-machine lot streaming problem

$$C_{ik} \geq C_{(i-1),k} + p_k s_i, \quad \forall i = 2, \ldots, n, k = 1, 2, 3, \tag{4.9}$$

$$\sum_{i=1}^{n} s_i = U, \quad s_i \geq 0, \forall i = 1, 2, \ldots, n. \tag{4.10}$$

Constraints (4.8) and (4.9) ensure that any sublot i begins processing on machine k after its completion on the previous machine or the processing of the $(i-1)$th sublot on machine k, whichever is maximum. Constraint (4.10) ensures that the total number of items in all the sublots equals U. The requirements of no idling and discrete sublot sizes can be incorporated by replacing the inequalities by equalities and by restricting the sublot sizes to take on integer values, respectively. This formulation has $2n + 3(n-1) + 2 = (5n-1)$ constraints and $3n + n = 4n$ variables.

An alternative formulation can be developed by taking into account the idle time, z_{ik}, which appears before the ith sublot on machine k (Fig. 4.2).

In general, this idle time can be expressed as

$$z_{ik} = \max(0, C_{i,k-1} - C_{i-1,k}). \tag{4.11}$$

Also, the expression for the completion time, C_{ik}, is as follows:

$$C_{ik} = p_k \sum_{u=1}^{i} s_u + \sum_{u=1}^{i} z_{uk}, \quad k = 1, 2, 3, i = 1, \ldots, n. \tag{4.12}$$

Note that $C_{10} = 0$ and $C_{01} = 0$

$$z_{i1} = 0, \quad \forall i = 1, \ldots, n \quad \text{and} \quad z_{lk} = s_1 \sum_{u=1}^{k-1} p_k, \quad \forall k = 1, 2, 3.$$

By substituting (4.12) in (4.11), we have

$$\begin{aligned} z_{ik} &= \max\left\{0, \, p_{k-1} \sum_{u=1}^{i} s_u + \sum_{u=1}^{i} z_{u,k-1} - p_k \sum_{u=1}^{i-1} s_u - \sum_{u=1}^{i-1} z_{uk}\right\} \\ &= \max\left\{0 \, (p_{k-1} - p_k) \sum_{u=1}^{i-1} s_u + p_{k-1} s_i + \sum_{u=1}^{i} z_{u,k-1} - \sum_{u=1}^{i-1} z_{uk}\right\}. \end{aligned}$$

This is equivalent to

$$z_{ik} \geq (p_{k-1} - p_k) \sum_{u=1}^{i-1} s_u + p_{k-1} s_i + \sum_{u=1}^{i} z_{u,k-1} - \sum_{u=1}^{i-1} z_{uk}$$

or,

$$(p_k - p_{k-1}) \sum_{u=1}^{i-1} s_u - p_{k-1} s_i - \sum_{u=1}^{i} z_{u,k-1} + \sum_{u=1}^{i} z_{uk} \geq 0,$$

and

$$z_{ik} \geq 0, \quad \forall k = 1, 2, 3, \forall i = 1, \ldots, n.$$

In the corresponding linear programming formulation, it is sufficient to minimize the total idle time on machine 3 since the total processing time, $p_3 U$, remains constant. We have

$$\text{Minimize:} \quad \sum_{u=1}^{n} z_{u3}$$

Subject to:

$$(p_k - p_{k-1}) \sum_{u=1}^{i-1} s_u - p_{k-1} s_i - \sum_{u=1}^{i} z_{u,k-1} + \sum_{u=1}^{i} z_{uk} \geq 0,$$
$$\forall i = 2, \ldots, n, \forall k = 1, 2, 3,$$
$$\sum_{i=1}^{n} s_i = U,$$
$$z_{ik} \geq 0, \quad \forall k = 1, 2, 3, \forall i = 1, \ldots, n,$$
$$s_i \geq 0, \quad \forall i = 1, 2, \ldots, n.$$

This formulation has $(2(n-1)+1)$ constraints and $2n + n = 3n$ variables. The reduction in size, when compared with the formulation presented earlier, is due to the elimination of constraints for the first machine. The above formulation can be appropriately altered, as before, to achieve the no idling and discrete sublot size requirements.

The continuous optimal sublot sizes without no-idling constraint can also be obtained by analyzing the structural properties of an optimal solution using a network representation. In this network, designated $\mathcal{G}(N, A)$, a node (i, k), represents the processing time of the ith sublot of size x_i on machine k with weight $p_k x_i$. A directed arc from node (i, k) to $(i, k + 1)$ captures the requirement that sublot i cannot begin processing on machine $(k + 1)$ unless it finishes processing on machine k. Similarly, a directed arc from (i, k) to $(i + 1, k)$ represents the constraint that machine k cannot start processing the $(i + 1)$th sublot until it has finished processing the ith sublot. The length of a path is the sum of the weights of the nodes that lie on that path. The longest path from (1, 1) to (n, m), where m is the number of machines, is known as the *critical path* and defines the makespan. The problem of finding the optimal sublot sizes corresponds to assigning weights to the nodes in the network representation such that the length of the longest path is minimized. A subpath of the critical path is referred to as a *critical segment.* A sublot i is said to be critical if $(i, k) - (i, k + 1)$ is a critical segment for some machine $k, k \in \{1, \ldots, m - 1\}$. A node (i, k) is said to be an *upper critical corner* (UCC) if there exists a critical segment $(i', k) - (i, k) - (i, k + 1)$, where $i' < i$. We say that the upper critical corner *extends from* node (i', k). Node (i, k) is called a *lower critical corner* (LCC) if there exists a critical segment $(i, k - 1) - (i, k) - (i', k), i' > i$. We say that the lower critical corner *extends to* node (i', k). Finally, we refer to nodes (i, ℓ) and (j, k) as *matching critical corners* (MCCs) (shown below) if $(i, k) - (j, k) - (j, \ell)$ and $(i, k) - (i, \ell) - (j, \ell)$ are both critical segments.

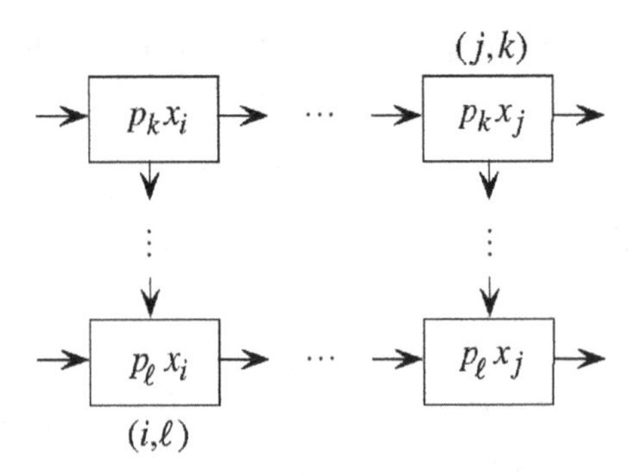

Lemma 4.1 defines the properties of the upper critical, lower critical, and matching critical corners.

Lemma 4.1 (Glass et al. [16]) *For a network, if (i, k) and $(j, k+1)$ are UCC and LCC extending from node (i', k) and to node $(j', k + 1)$, respectively, then*

$$p_k \sum_{u=i'+1}^{i} x_u \geq p_{k+1} \sum_{u=i'}^{i-1} x_u$$

and

$$p_{k+1} \sum_{u=j}^{j'-1} x_u \geq p_k \sum_{u=j+1}^{j'} x_u.$$

Furthermore, if (i, ℓ) *and* (j, k) *are MCC, then*

$$p_k \sum_{u=i+1}^{j-1} x_u + x_j \sum_{v=k}^{\ell-1} p_u = x_i \sum_{u=k+1}^{\ell} p_u + p_\ell \sum_{v=i+1}^{j-1} x_v.$$

These relations essentially follow from the definitions of UCC and LCC.

As before, the sublot sizes depend upon the relative values of p_1, p_2, and p_3. In case machine 2 is dominated by machines 1 and 3, then the critical path will not pass through machine 2, and we can transform the three-machine problem to an equivalent two-machine problem, with processing times of p_1+p_2 on one of these machines and $p_2 + p_3$ on the other. These two machines, then, dictate the sublot sizes. However, if machine 2 is dominant, then the critical path passes through machine 2, and the sublot sizes depend upon the pairs of machines 1 and 2, 2 and 3, or both.

We have the following three cases: $p_2^2 < p_1 p_3$, $p_2^2 = p_1 p_3$, and $p_2^2 > p_1 p_3$.

Case 1. $p_2^2 < p_1 p_3$

Theorem 4.1 (Glass et al. [16]) *When* $p_2^2 < p_1 p_3$*, the critical paths are of the form* $(1, 1) \cdots (i, 1) - (i, 2) - (i, 3) \cdots (n, 3)$ *for all sublots* $i, i = 1, \ldots, n$.

These are shown in Fig. 4.3. A bold line between the nodes of the network indicates that it is on the critical path. Note that in this case, a critical path does not pass through machine 2.

Case 2. $p_2^2 = p_1 p_3$

In this case, the ratio of the processing times on the first two and the last two machines are identical. Hence, the minimum makespan will be obtained when the schedule is compact. Consequently, the idle time on all the machines will be zero. Hence, all the paths in the network are critical and we obtain the critical network structure as shown in Fig. 4.4.

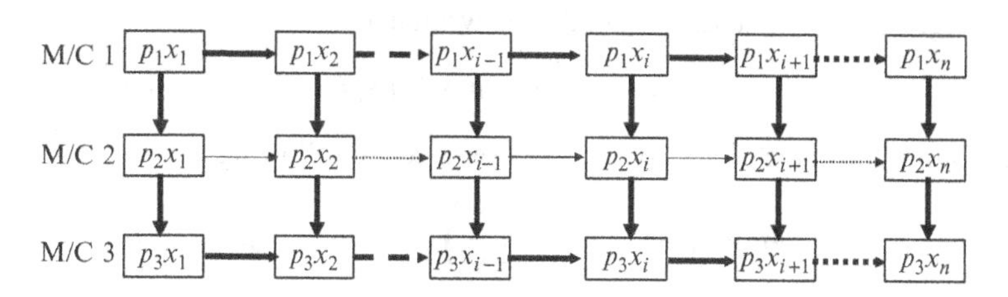

FIGURE 4.3. Optimal sublot structure when $p_2^2 < p_1 p_3$

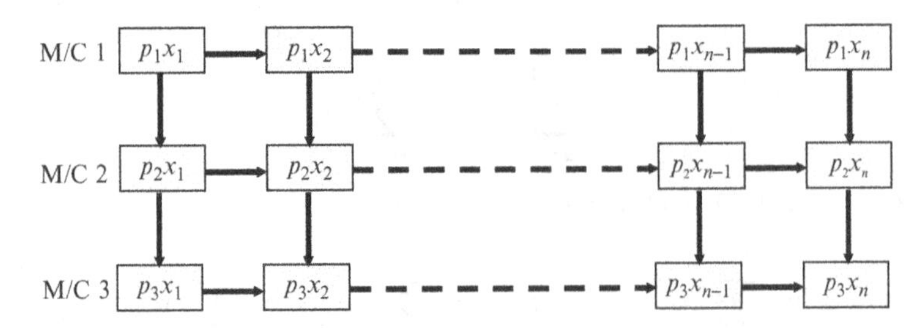

FIGURE 4.4. Optimal sublot structure when $p_2^2 = p_1p_3$

Theorem 4.2 *When $p_2 = p_1$ and $p_1 = p_3$, all the paths in the network are critical.*

Determination of Optimal Sublot Sizes When $p_2^2 \leq p_1p_3$

Since all the paths for the case $p_2^2 < p_1p_3$ are also common to those for $p_2^2 = p_1p_3$, we consider these cases together.

Note that node $(i, 1)$ is an UCC extending from node (1, 1), and node $(i - 1, 3)$ is a LCC extending to node $(n, 3)$, for all i, $i = 2, \ldots, n$. Hence, by Lemma 4.1, nodes $(i, 1)$ and $(i - 1, 3)$ for all i, $i = 2, \ldots, n$, are MCCs. Therefore,

$$p_1x_i + p_2x_i = p_2x_{i-1} + p_3x_{i-1} \quad \text{or} \quad x_i = \left(\frac{p_2 + p_3}{p_2 + p_1}\right)x_{i-1}.$$

Let

$$q = \left(\frac{p_2 + p_3}{p_2 + p_1}\right).$$

We have

$$x_i = qx_{i-1},$$

which, in terms of x_1, can be written as follows:

$$x_i = q^{i-1}x_1.$$

Since $\sum_{i=1}^{n} x_i = 1$, substituting in it x_i from above and simplifying, we obtain x_1 to be

$$\begin{aligned} x_1 &= \frac{q-1}{q^n - 1}, \quad p_1 \neq p_3, \\ &= 1/n, \qquad p_1 = p_3. \end{aligned}$$

From Fig. 4.3 or 4.4, the makespan can be written as

$$M = \{x_1(p_1 + p_2) + p_3\}U.$$

Substituting the value of x_1 from above, we have the following:

$$\begin{aligned} M &= \left(p_3 + \left(\frac{p_3 - p_1}{q^n - 1}\right)\right)U, \quad p_1 \neq p_3, \\ &= \left(p_3 + \left(\frac{p_1 + p_2}{n}\right)\right)U, \quad p_1 \neq p_3. \end{aligned}$$

Example 4.1 Consider an example where $p_1 = 1$, $p_2 = 2$, and $p_3 = 4$; $U = 56$ and $n = 3$.

We have

$$q = \frac{p_2 + p_3}{p_1 + p_2} = \frac{2+4}{1+2} = 2.$$

Also,

$$x_1 = \frac{2-1}{2^3-1} = \frac{1}{7},$$
$$x_2 = q x_1 = \frac{2}{7},$$
$$x_3 = q^2 x_1 = \frac{4}{7}.$$

$M = \{(1/7)(1+2)+4\}\,56 = 248$. Since $p_2^2 = p_1 p_3$ for this example, the schedule is compact and is shown in the following figure.

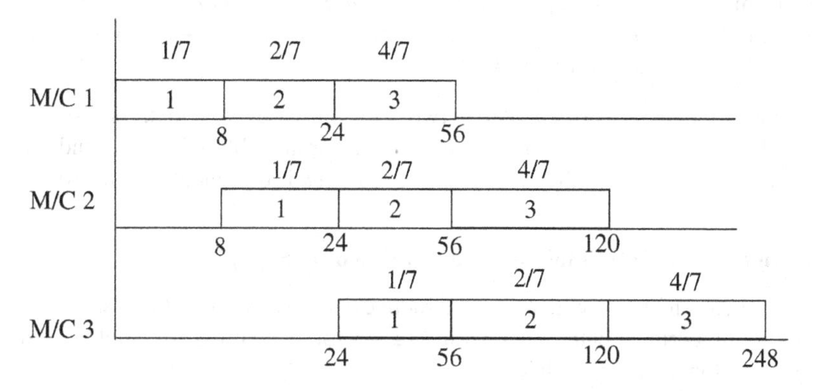

Let us change p_2 to 1.5 with p_1 and p_3 as before to obtain $p_2^2 < p_1 p_3$. We have

$$q = \frac{1.5+4}{1+1.5} = 2.2,$$
$$x_1 = \frac{2.2-1}{2.2^2-1} \simeq \frac{1}{8},$$
$$x_2 = 2.2 \times \frac{1}{8} = \frac{2.2}{8},$$
$$x_3 = (2.2)^2 \times \frac{1}{8} = \frac{4.84}{8} \simeq \frac{4.8}{8} \quad \text{(to ensure } (x_1 + x_2 + x_3 = 1)\text{)},$$
$$M = \left\{\frac{1}{8}(1+1.5)+4\right\} 56 = 241.5.$$

The resulting schedule is shown on the next page. Note that, since $p_2^2 < p_1 p_3$, the sublot sizes are dictated by machines 1 and 3, and an intermediate idle time is encountered on machine 2. Also, note that the completion time of the third sublot

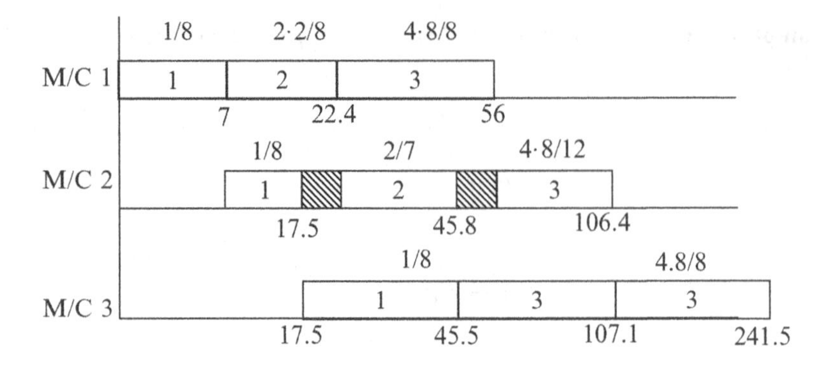

on machine 2 and that of the second sublot on machine 2 are slightly different due to truncation error.

Case 3. $p_2^2 > p_1 p_3$

Theorem 4.3 (Glass et al. [16]) *When* $p_2^2 > p_1 p_3$, *for a sublot* $k, k \in \{1, \ldots, n\}$, *the critical paths are of the form* $(1, 1) \cdots (i, 1) - (i, 2) - \cdots - (j, 2) - (j, 3) \cdots (n, 3), \forall i, i = 1, \ldots, k$ *and* $j = k, \ldots, n$.

Figure 4.5 depicts a critical path network for this case. Note that, for a sublot k, critical paths are dictated by machines 1 and 2 for all sublots $i, i \leq k$, and, then, by machines 2 and 3 for all sublot $j, j \geq k$. Sublot k is designated as the crossover sublot.

Determination of Optimal Sublot Sizes When $p_2^2 > p_1 p_3$

The implication of Theorem 4.3 is that there exists a vector of optimal sublot sizes resulting in a critical path as shown in Fig. 4.5 for some crossover sublot k. Note that the following are the MCCs:

$$
\begin{aligned}
&(i+1, 1) \text{ and } (i, 2), \quad \forall i, i = 1, \ldots, k-1, \\
&(i, 2) \text{ and } (i-1, 3), \quad \forall i, i = k+1, \ldots, n.
\end{aligned}
$$

Using Lemma 4.1, we have

$$
\begin{aligned}
p_2 x_i &= p_1 x_{i+1}, \quad \forall i,\ i = 1, \ldots, k-1, \\
p_2 x_i &= p_3 x_{i-1}, \quad \forall i,\ i = k+1, \ldots, n.
\end{aligned}
$$

FIGURE 4.5. Optimal sublot structure when $p_2^2 > p_1 p_3$

Let $q_1 = p_1/p_2$ and $q_3 = p_3/p_2$. Substituting these in the above expressions, we obtain

$$\begin{aligned} x_i &= q_1 x_{i+1}, \quad \forall i,\ i = 1, \ldots, k-1, \\ &= q_3 x_{i-1}, \quad \forall i,\ i = k+1, \ldots, n, \\ x_i &= q_1^{k-i} x_k, \quad \forall i,\ i = 1, \ldots, k, \\ &= q_3^{i-k} x_k, \quad \forall i,\ i = k, \ldots, n. \end{aligned}$$

Using $\sum_{i=1}^{n} x_i = 1$, we get

$$\frac{1}{x_k} = \sum_{i=0}^{k-1} q_1^i + \sum_{i=1}^{n-k} q_3^i.$$

Hence, the sublot sizes can be expressed as follows:

$$\begin{aligned} x_i &= q_1^{k-i} x_k, \quad \forall i,\ i = 1, \ldots, k, \\ &= q_3^{i-k} x_k, \quad \forall i,\ i = k, \ldots, n. \end{aligned}$$

where

$$\begin{aligned} \frac{1}{x_k} &= \frac{q_1^k - 1}{q_1 - 1} + \frac{q_3^{n-k+1} - 1}{q_3 - 1} - 1, && \text{when } p_1 \neq p_2 \text{ and } p_2 \neq p_3, \\ &= k - 1 + \frac{q_3^{n-k+1} - 1}{q_3 - 1}, && \text{when } p_1 = p_2 \text{ and } p_2 \neq p_3, \\ &= \frac{q_1^k - 1}{q_1 - 1} + n - k, && \text{when } p_1 \neq p_2 \text{ and } p_2 \neq p_3. \end{aligned}$$

The makespan $M(x_k)$ for a crossover sublot k is given by

$$M(x_k) = \{p_1 x_1 + p_2 + p_3 x_n\}\, U.$$

Substituting $p_1 = q_1 p_2$, $p_3 = q_3 p_2$, $x_1 = q_1^{k-1} x_k$, and $x_n = q_3^{n-k} x_k$, we get

$$\begin{aligned} M(x_k) &= q_1 p_2 q_1^{k-1} x_k + p_2 + q_3 p_2 q_3^{n-k} x_k \\ &= \left\{ p_2 + p_2 x_k \left(q_1^k + q_3^{n-k+1} \right) \right\} U. \end{aligned}$$

The optimal crossover sublot k^* is the one that minimizes the makespan $M(x_k)$. Thus, we have

$$M(x_{k^*}) = \min_{1 \le k \le n} \left[\left\{ p_2 + p_2 x_k \left(q_1^k + q_3^{n-k+1} \right) \right\} U \right].$$

If more than one crossover sublot exists, then we choose k^* to be as small as possible. Lemma 4.2 gives the required properties to find the crossover sublot k^*.

Lemma 4.2 (Glass et al. [16]) *When* $(p_2)^2 > p_1 p_3$,

$$\begin{aligned} &M(x_k) > M(x_{k+1}), \quad \forall k,\ k = 1, \ldots, k^* - 1, \\ &M(x_{k^*}) \le M(x_{k^*+1}), \\ &M(x_k) < M(x_{k+1}), \quad \forall k,\ k = k^* + 1, \ldots, n - 1. \end{aligned}$$

Hence, to start with, we select an arbitrary sublot k, $k \in \{1, \ldots, n\}$, and find $\Delta k = M(x_{k+1}) - M(x_k)$. From Lemma 4.2, we know that if $\Delta k > 0$, then $k^* \le k$; if $\Delta k < 0$, then $k^* \ge k + 1$ and if $\Delta k = 0$, then $k^* = k$. Since $k^* \in \{1, 2, \ldots, n\}$, the search procedure requires $O(\log n)$ cpu time.

Next, we illustrate the above procedure through an example problem.

Example 4.2 Let $p_1 = 1$, $p_2 = 3$, and $p_3 = 4$; $U = 54$ and $n = 3$.

We have $q_1 = p_1/p_2 = 1/3$ and $q_3 = p_3/p_4 = 4/3$. First, we determine the crossover sublot k. Let $k = 2$. We have

$$\frac{1}{x_2} = \frac{\left(\frac{1}{3}\right)^2 - 1}{\left(\frac{1}{3}\right) - 1} + \frac{\left(\frac{4}{3}\right)^2 - 1}{\left(\frac{4}{3}\right) - 1} - 1 = \frac{8}{3}.$$

Therefore, $x_2 = 3/8$, and

$$M(x_2) = \left\{3 + 3\frac{3}{8}\left(\left(\frac{1}{3}\right)^2 + \left(\frac{4}{3}\right)^2\right)\right\} 54 = 276.8.$$

Also, consider $k + 1 = 3$. We have

$$\frac{1}{x_3} = \frac{\left(\frac{1}{3}\right)^2 - 1}{\left(\frac{1}{3}\right) - 1} + \frac{\left(\frac{4}{3}\right) - 1}{\left(\frac{4}{3}\right) - 1} - 1 = \frac{13}{9}.$$

Therefore, $x_3 = 9/13$, and

$$M(x_3) = \left\{3 + 3\frac{9}{13}\left(\left(\frac{1}{3}\right)^3 + \left(\frac{4}{3}\right)\right)\right\} 54 = 315.7.$$

Since $\Delta k = 315.7 - 276.8 > 0$, $k^* \le k - 1 = 1$. Next, consider $k = 1$. We have

$$\frac{1}{x_1} = \frac{\left(\frac{1}{3}\right) - 1}{\left(\frac{1}{3}\right) - 1} + \frac{\left(\frac{4}{3}\right)^3 - 1}{\left(\frac{4}{3}\right) - 1} - 1 = \frac{37}{9}.$$

Therefore, $x_1 = 9/37$, and

$$M(x_1) = \left\{3 + 3\frac{9}{37}\left(\frac{1}{3} + \left(\frac{4}{3}\right)^3\right)\right\} 54 = 268.5.$$

The optimal sublot sizes are

$$x_1^* = \frac{9}{37},$$
$$x_2^* = \frac{9}{37}\frac{4}{3},$$

and

$$x_3^* = \frac{9}{37}\left(\frac{4}{3}\right)^2.$$

Note that $x_1^* + x_2^* + x_3^* = 1$.

4.3 3/1/V/{NI,II}/{CV,DV}

Next, we consider the case of variable sublots for the above problem. When sublot sizes are variable and the no-idling constraint is enforced, the problem can be decomposed into two subproblems consisting of machines 1 and 2 and machines 2 and 3. For each pair of machines, the solution methodology for the 2/1/NI/CV problem can be used to obtain the continuous optimal sublot sizes.

When intermittent idling is permitted, we note that machine 1 operates continuously and machine 3 can be made to do so by right shifting the start times of the sublots on that machine. In the presence of intermittent idling and variable sublots, the optimal schedule must be a no-wait schedule. This follows by the fact that if any sublot i were to wait in between its processing on machines k and $(k+1)$, then its size could be increased resulting in the decrement of the makespan since the earlier sublots would be smaller, and hence, will get completed earlier.

If machine 1 and machine 3 are operating continuously, then a no-wait schedule is obtained if the time required to process any sublot i on machines 1 and 2 is equal to the time required to process the previous sublot on machines 2 and 3, i.e.,

$$s_{i+1}\left(p_1+p_2\right)=s_i\left(p_2+p_3\right).$$

Let

$$q=\frac{p_2+p_3}{p_1+p_2}. \tag{4.13}$$

We have

$$s_{i+1}=q s_i.$$

In terms of the size of the first sublot, the above can be written as

$$s_i=q^{i-1} s_1. \tag{4.14}$$

The value of s_1 can be calculated using the fact that $\sum_{i=1}^n s_i=U$. This gives

$$\begin{aligned} s_1 &= \left(\frac{1-q}{1-q^n}\right) U, \quad q \neq 1, \\ &= \frac{U}{n}, \qquad\qquad\quad q=1. \end{aligned} \tag{4.15}$$

This solution will be a no-wait schedule on machine 2 if this machine can process every sublot as soon as it is finished processing on machine 1, i.e.,

$$p_1 \sum_{u=1}^{i+1} s_u \geq p_2 \sum_{u=1}^{i} s_u+p_1 s_1 \quad \text{or} \quad p_1 \sum_{u=2}^{i+1} s_u \geq p_2 \sum_{u=1}^{i} s_u.$$

Since $s_u=q^{u-1} s_1$,

$$p_1 \sum_{u=2}^{i+1} q^{u-1} \geq p_2 \sum_{u=1}^{i} q^{u-1} \quad \text{or} \quad p_1 q \sum_{u=1}^{i} q^{u-1} \geq p_2 \sum_{u=1}^{i} q^{u-1},$$

i.e.,

$$p_1 q \geq p_2,$$

or

$$p_1 \left(\frac{p_2 + p_3}{p_1 + p_2} \right) \geq p_2,$$

or

$$p_2^2 - p_1 p_3 \leq 0.$$

Therefore, if this condition is not satisfied, then all three machines will operate continuously and, in that case, we can decompose the problem into two subproblems each comprising two machines, and consequently, leading to variable sublot sizes. These subproblems can be solved by using the two-machine procedures described earlier in Chap. 3 (for both continuous and discrete versions). However, if $p_2^2 - p_1 p_3 \leq 0$, then the optimal sublot sizes are consistent. The continuous sublot sizes are given by (4.14) and (4.15). To determine integer sublot sizes for this case, we proceed as follows. The latest start time for any sublot i on machine 3, to achieve makespan M, is given by $M - p_3 \left(U - \sum_{u=1}^{i-1} s_u \right)$. The second operation on machine 2 of sublot i must be completed on or before this time. Hence,

$$p_1 \sum_{u=1}^{i} s_u + p_2 s_i \leq M - p_3 \left(U - \sum_{u=1}^{i=1} s_u \right).$$

Let $S_i \left(= \sum_{u=1}^{i} s_u \right)$ represent the total number of items in sublots 1 to i. We have

$$p_1 S_i + p_2 (S_i - S_{i-1}) \leq M - p_3 (U - S_{i-1}),$$

or

$$S_i \leq \frac{p_2 S_{i-1} + M - p_3 (U - S_{i-1})}{p_1 + p_2},$$

or

$$S_i \leq \min \left\{ \frac{p_2 S_{i-1} + M - p_3 (U - S_{i-1})}{p_1 + p_2}, U \right\}.$$

This recursive formula can be used to calculate the sublot sizes. The procedure starts with $S_0 = 0$ and calculates the cumulative number of items as the largest integer permitted by the inequality. At the end, if $S_n < U$, then the makespan value from the continuous version needs to be incremented as follows:

$$\hat{M} = M + (p_1 + p_2) \min_{1 \leq i \leq n} \{1 - e_i\},$$

where

$$e_i = \min \left\{ \frac{p_2 S_{i-1} + M - p_3 (U - S_{i-1})}{p_1 + p_2}, U \right\} - S_i.$$

The algorithm is stated below in Fig. 4.6.

Example 4.3 Consider the data depicted in Table 4.4.

Step 1: Let $S_0=0$

Step 2: For $i=1,2,\ldots n$

Calculate $S_i = \lfloor \min\{M+p_2 \cdot S_{i-1}-p_3 \cdot (U-S_{i-1})/(p_1+p_2), U\} \rfloor$

S_i = Cumulative number of items in first i sublots

M = Optimal makespan from the continuous solution

U = Number of items in the batch

Step 3: If $S_n = U$

STOP

ELSE

Find new incremented value of $\hat{M}$ as

$\hat{M} = M + (p_1+p_2) \min (1-e_i)$

where $e_i = \min\{M+p_2\, S_{i-1}-p_3\,(U-S_{i-1})/p_1+p_2, U\}-S_{i-1}$

Go to STEP 2

FIGURE 4.6. Algorithm to obtain integer sublot sizes when $p_2^2 < p_1 p_3$

TABLE 4.4. Data for Example 4.3

Lot size (U)	20
Number of sublots (n)	4
Processing time on machine 1 (p_1)	3
Processing time on machine 2 (p_2)	3
Processing time on machine 3 (p_3)	4

TABLE 4.5. Continuous sublot sizes for Example 4.3

Parameter	Value
q	1.17
s_1	3.91
s_2	4.56
s_3	5.32
s_4	6.21
Makespan $\{p_1 U + (p_2 + p_3) s_n\}$	103.47

Since $p_2^2 - p_1 p_3 \leq 0$ holds for this example, as explained above, consistent sublots will be optimal. The value of q and s_1 can be calculated using (4.13) and (4.15), respectively. Using these in (4.14), we obtain the values of s_2, s_3, and s_4. These, along with the value of makespan, are shown in Table 4.5.

Next, we determine integer sublot sizes.

Integer Sublot Sizes

Step 1. Obtain continuous optimal sublot sizes and the corresponding makespan value (shown in Table 4.5).

Step 2. Referring to the algorithm presented in Fig. 4.6 determine cumulative sublot sizes S_1, S_2, S_3, and S_4 (shown below). Since $S_4(=16) < U$, we

Cumulative sublot sizes	e_i	$(1-e_i)$	Incremented makespan value
$S_1 = \lfloor \min(3.91, 20) \rfloor = 3$	0.91	0.09	$103.47 + (3+3) \times (0.09) = 104.01$
$S_2 = \lfloor \min(7.41, 20) \rfloor = 7$	0.41	0.59	
$S_3 = \lfloor \min(11.41, 20) \rfloor = 11$	0.41	0.59	
$S_4 = \lfloor \min(16.76, 20) \rfloor = 16$	0.24	0.76	

increment the value of M, which now becomes 104.01. The new cumulative sublot sizes are as follows:

Cumulative sublot sizes	e_i	$(1-e_i)$	Incremented makespan value
$S_1 = \lfloor \min(4, 20) \rfloor = 4$	0	1	
$S_2 = \lfloor \min(8.67, 20) \rfloor = 8$	0.67	0.33	$103.47 + (3+3) \times (0.33) = 105.99$
$S_3 = \lfloor \min(13.36, 20) \rfloor = 13$	0.36	0.64	
$S_4 = \lfloor \min(19.17, 20) \rfloor = 19$	0.17	0.83	

Still $S_4(=19) < U$. We further increment the value of M and obtain the following sublot sizes. Now since $S_4 = U$, we stop. The optimal sublot sizes are 4, 5, 5, and 6 with the corresponding makespan value of 106.

Cumulative sublot sizes	Sublot size
$S_1 = \lfloor \min(4.33, 20) \rfloor = 4$	4
$S_2 = \lfloor \min(9, 20) \rfloor = 9$	5
$S_3 = \lfloor \min(14.83, 20) \rfloor = 14$	5
$S_4 = \lfloor \min(20.67, 20) \rfloor = 20$	6

Again, when variable sublots and intermittent idling are permitted on the three machines, the solution procedure depends on the processing time on machine 2. If machine 2 dominates, then we can decompose the problem and obtain variable sublots using the two-machine procedures developed earlier in Chap. 3. However, if machine 2 is dominated, then consistent sublots are optimal and can be obtained by the algorithm described above.

4.4 3/1/C/{NI,II}/CV/{Lot-Detached Setup}

In the presence of setup times, the optimal continuous, consistent sublot sizes can once again be found by analyzing the structural properties of the network representation of this problem. But, first, let us see what impact does a detached setup have on the makespan. First of all, note that, if $t_1 > 0$, while $t_2 = t_3 = 0$, the entire schedule gets shifted to the right, and consequently, the makespan value increases by t_1. Assume that, initially, the starting times of each sublot on machines 2 and 3 have been delayed as much as possible without increasing the makespan. Consequently, it is easy to see that if, in addition, t_2 and t_3 are made large enough, they can further contribute to an increment in the makespan value. Hence, if M is the optimal makespan value in the presence of lot-detached setups on the machines and M_1 is the optimal makespan value obtained by using the following reduced setup values: $t_1' = t_1 - t_1$, $t_2' = \max(t_2 - t_1, 0)$, and $t_3' = \max(t_3 - t_1, 0)$, then $M = M_1 + t_1$, and the optimal sublot sizes that give M and M_1 are identical. Next, consider only the reduction of t_3 to zero; and let M_2 be the makespan value obtained under this case. Then, $M = \max(M_2, t_3 + p_3U)$, which follows by the fact that if g_3 is the starting time of the first sublot on machine 3 after having right shifted the sublots on machine 3 by the maximum amount without increasing the makespan, then $M = M_2$ in case $t_3 \leq g_3$, and $M = t_3 + p_3U$ in case $t_3 > g_3$. In either case, clearly, the optimal sublot sizes that give M_2 are also optimal for M. Now, combining the above two scenarios, if the setup times on the first and third machines are reduced to zero, and that on machine 2 to t_2', and the optimal makespan for this reduced problems be denoted by M_3, then

$$M = \max(M_3, \max\{t_3 - t_1, 0\} + p_3U) + t_1,$$

and the optimal sublots for M_3 are also optimal for M.

In view of the above discussion, the lot-detached setup problem on hand can be analyzed by considering an equivalent problem obtained by reducing the setup times on machines 1 and 3 to zero and that on machine 2 to $t_2 - t_1$. If $t_2 - t_1 > 0$, then the resulting problem is called a "reduced problem," and if $t_2 - t_1 \leq 0$ then it is called a "relaxed problem." This result is formally stated as follows.

Theorem 4.4 (Chen and Steiner [12]) *The three-machine lot streaming problem with setup times t_1, t_2, and t_3 on machines 1, 2, and 3, respectively, is equivalent to an alternative (called reduced) problem with no setup time on machine 1 and machine 3 and setup time on machine 2, $t_2' = t_2 - t_1$, with $t_2 > t_1$, or to a three-machine problem with no setup times (called relaxed) if $t_2' \leq 0$.*

Regarding the relationship between the relaxed and reduced problems, it can be shown (see [12] for a formal proof) that if $t_2' \leq p_1 x_1^{\text{Rel}}$, where x_1^{Rel} is the size of the first sublot in the relaxed version of the problem, the optimal sublot sizes for the relaxed problem are also optimal for the reduced problem. Hence, for subsequent discussion, we assume that $t_2' > p_1 x_1^{\text{Rel}}$. In the corresponding network representation, we add additional nodes at the beginning of each machine to represent the required setup on that machine based on the above discussion; this setup is zero on machines 1 and 3 and t_2' on machine 2. These nodes can be assumed to be associated with a dummy sublot, designated to be sublot 0.

We also have the following result.

Theorem 4.5 (Chen and Steiner [12]) *In an optimal solution of consistent sublots for the reduced problem, every sublot is critical. In addition, if $t_2' > p_1 x_1^{\text{Rel}}$, then the dummy sublot 0 is also critical.*

Similar to the approach taken earlier, we distinguish among the cases corresponding to $p_2^2 < p_1 p_3$, $p_2^2 = p_1 p_3$, and $p_2^2 > p_1 p_3$, analyze the critical path, and then, derive formulae for finding the optimal sublot sizes.

Case 1. $p_2^2 < p_1 p_3$

There can be two subcases corresponding to the value of t_2'. If t_2' is greater than a threshold value, then no segment on machine 1 can be critical, and hence, the problem reduces to a two-machine relaxed problem on machines 2 and 3, which can be solved by using the 2/1/{NI,II}/CV solution procedure. This threshold value is given as [12]

$$t_2' > p_1 - \left(\frac{p_2 \left[1 - \left(\frac{p_3}{p_2} \right)^{n-1} \right]}{1 - \left(\frac{p_3}{p_2} \right)^{n}} \right).$$

If, on the other hand, the value of t_2' does not exceed the threshold value, i.e., there is a critical segment on machine 1, then the critical paths have the properties summarized in Theorem 4.6.

Theorem 4.6 (Chen and Steiner [12]) *If $p_2^2 < p_1 p_3$, and there is a critical segment on machine 1, then there exists a $k \in \{1, 2, \ldots, n\}$ such that:*

- *Segment $(j, 1) - (j, 2)$ is not critical for all sublots j, $j = 1, \ldots, k-1$.*
- *Segment $(j, 2) - (j+1, 2)$ is not critical for all sublots j, $j = k, \ldots, n-1$.*
- *The segment $(k-1, 2) - (k, 2)$ may or may not be critical depending on actual data.*
- *Every other two node segment is critical except* $(0, 2) - (0, 3)$ *and* $(0, 3) - (1, 3)$.

These are shown in Fig. 4.7. It is easy to explain the solution specified by Theorem 4.6. Inherently, this solution will be like that shown in Fig. 4.3 except that, depending on the value of t_2', some sublots on machine 2 get shifted to the right, leading to a crossover sublot k, $k = 1, \ldots, n$, which now becomes the first sublot to be processed in a no-wait fashion on all three machines. As the sublots until sublot k are shifted to the right, there is no no-wait processing of the sublots until sublot k on machines 1 and 2, while they are processed in a no-wait fashion on machines 2 and 3. Also, depending on the shift, the segment $(k-1, 2) - (k, 2)$ may or may not be critical. The noncriticality of segments $(0, 2) - (0, 3)$ and $(0, 3) - (1, 3)$ follows because of a setup time of t_2' on machine 3.

Optimal Continuous Sublot Sizes

The optimal sublot sizes follow two different geometric progressions depending on the sublots appearing before or after the crossover sublot k. We derive the expressions for these sublot sizes next. From Fig. 4.7, we observe that the following are MCCs:

1. $(i, 3)$ and $(i + 1, 2)$, $\forall i, i = 1, \ldots, k - 2$. Therefore,

$$p_3 x_i = p_2 x_{i+1}.$$

Let $q_3 = (p_3/p_2)$. We have

$$x_{i+1} = q_3 x_i \quad \text{or} \quad x_i = q_3^{i-1} x_i, \quad \forall i, i = 1, \ldots, k - 1.$$

2. $(i, 3)$ and $(i + 1, 1)$, $\forall i, i = k, \ldots, n - 1$, Therefore,

$$(p_3 + p_2)\, x_i = (p_1 + p_2)\, x_{i+1}.$$

Let $q = (p_3 + p_2)/(p_1 + p_2)$. We have

$$x_{i+1} = q x_i \quad \text{or} \quad x_i = q^{i-k} x_k, \quad \forall i, i = k, \ldots, n.$$

Since $\sum_{i=1}^{n} x_i = 1$, we get

$$x_1 \sum_{i=1}^{k-1} q_3^{i-1} + x_k \sum_{i=1}^{n-k+1} q^{i-1} = 1.$$

Since $(0, 1) - (1, 1) - (k, 1) - (k, 2) - (k, 3)$ and $(0, 1) - (0, 2) - (1, 2) - (1, 3) - (k, 3)$ are critical segments, their lengths must be the same, i.e.,

$$p_1 x_1 \sum_{i=1}^{k-1} q_3^{i-1} + x_k (p_1 + p_2) = t_2' + x_1 p_2 + x_1 p_3 \sum_{i=1}^{k-1} q_3^{i-1}.$$

The values of x_1 and x_k can be determined by solving the above two equations to be as follows:

$$x_1 = \frac{D - t_2' B}{AD - BC}, \quad x_k = \frac{C - t_2' A}{BC - DA},$$

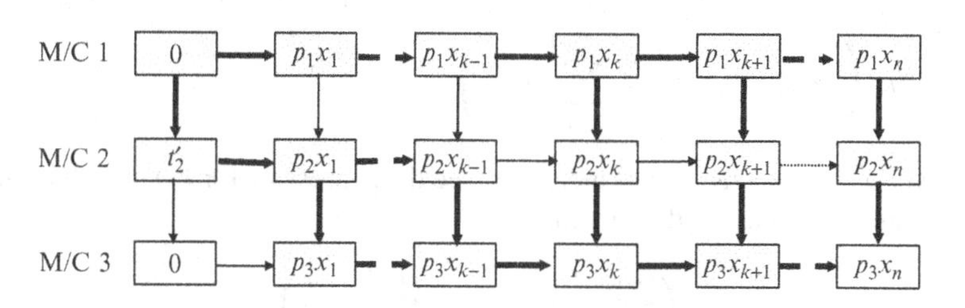

FIGURE 4.7. Optimal sublot structure when $p_2^2 < p_1 p_3$

where

$$A = \sum_{i=1}^{k-1} q_3^{i-1},$$
$$B = \sum_{i=1}^{n-k+1} q^{i-1},$$

$$C = (p_1 - p_3)\sum_{i=1}^{k-1} q_3^{i-1} - p_2,$$
$$D = p_1 + p_2,$$
$$q = \left(\frac{p_2 + p_3}{p_1 + p_2}\right),$$
$$q_3 = p_3/p_2.$$

And, consequently,

$$x_i = x_1 q_3^{i-1}, \quad \forall i = 2, \ldots, k-1,$$
$$x_i = x_k q^{1-k}, \quad \forall i = k+1, \ldots, n.$$

The makespan is given by

$$M = x_1 p_1 \sum_{i=1}^{k-1} q_3^{i-1} + x_k (p_1 + p_2) + x_k p_3 \sum_{i=k}^{n} q^{i-k}.$$

On the basis of the above discussion, the location of the crossover sublot k either remains the same or increases if t_2' is increased by a sufficient amount. Now, for a given crossover sublot k, if t_2' is not increased large enough, then $p_3 x_k(k) < p_2 x_{k-1}(k)$ (where $x_k(k)$ and $x_{k-1}(k)$ denote the kth and $(k-1)$th sublots corresponding to a given crossover sublot k), or, in other words, $x_k(k) < q_3 x_{k-1}(k) = q_3^{k-1} x_1(k)$, by using the expression for $x_i \forall i = 1, \ldots, k-1$ from above. Consequently, we will have

$$1 = x_1(k)\sum_{i=1}^{k-1} q_3^{i-1} + x_k(k)\sum_{i=1}^{n-k+1} q^{i-1} < x_1(k)\sum_{i=1}^{k-1} q_3^{i-1} + x_1(k) q_3^{k-1}\sum_{i=1}^{n-k+1} q^{i-1}$$
$$= x_1(k)\left(\sum_{i=1}^{k-1} q_3^{i-1} + q_3^{k-1}\sum_{i=1}^{n-k+1} q^{i-1}\right).$$

However, $x_1(k)$ decreases as t_2' increases (see the expression for x_1), which implies that t_2' can be incremented to obtain

$$x_1(k)\left(\sum_{i=1}^{k-1} q_3^{i-1} + q_3^{k-1}\sum_{i=1}^{n-k+1} q^{i-1}\right) = 1.$$

Therefore, the crossover sublot k over the interval $[1, n]$ can be determined as follows. We start with a value of $k_0 = \lceil n/2 \rceil$ and use the following relations:

$$x_1(k_0)\left[\left(\sum_{i=1}^{k_0-1} q_3^{i-1}\right) + q_3^{k_0-1}\left(\sum_{i=1}^{n-k_0+1} q^{i-1}\right)\right] = 1$$

and

$$\hat{t}_2(k_0) + x_1(k_0)\, p_2\left(1 + q_3 + \cdots + q_3^{k_0-2}\right) = x_1(k_0)\, p_1\left(1 + q_3 + \cdots + q_3^{k_0-1}\right),$$

where the second expression equates the critical paths on machines 2 and 1. If $\hat{t}_2(k_0) < t_2'$, then $k_0 < k$, and set $k_0 = [k_0 + (n - k_0)/2]$, else set $k_0 \geq k$, and $k_0 = \lceil k_0/2 \rceil$. This binary search will require $O(\log n)$ iterations.

Note that the threshold value of t_2', mentioned earlier, can be obtained by assuming the crossover sublot to be the sublot n, and thereby, equating critical segments $(0, 1), (1, 1), \ldots, (n, 1), (n, 2)$ and $(0, 1), (0, 2), (1, 2), \ldots, (n, 2)$.

Example 4.4 Consider an example where $p_1 = 4, p_2 = 1, p_3 = 2, t_1 = 2, t_2 = 4, t_3 = 3, U = 1$, and $n = 3$.

Since $(p_2)^2 < p_1 p_2$, Case 1 holds. We first check whether $t_2' (= t_2 - t_1) = 2$ is less than the threshold value

$$p_1 - \frac{p_2\left(1 - (p_3/p_2)^2\right)}{1 - (p_3/p_2)^3} = \frac{25}{7}.$$

Since this is not true, the problem cannot be reduced to a two-machine problem. We have $q = 3/5$ and $q_3 = 2$. We first find the crossover sublot k. Let $k_0 = 2$, we have

$$x_1(2)\left(\sum_{i=1}^{2-1} q_3^{i-1} + q_3^{2-1}\right)\left(\sum_{i=1}^{3-2+1} q^{i-1}\right) = 1,$$

which gives $x_1(2) = 5/21$.

By substituting this in the following expression,

$$\hat{t}_2(2) + x_1(2) p_2 = x_1(2) p_1 (1 + q_3),$$

we obtain $\hat{t}_2(2) = 55/21$. Since $\hat{t}_2(2) > t_2'$, we have $k \leq k_0 = 2$. Let $k_0 = \lceil 2/2 \rceil = 1$. Furthermore, $x_1(1)\left(\sum_{i=1}^{3-1+1} q^{i-1}\right) = 1$, which gives $x_1(1) = 25/49$. By substituting this in the following expression

$$\hat{t}_2(1) = x_1(1) p_1,$$

we obtain $\hat{t}_2(1) = 100/49$. Since $\hat{t}_2(1) > t_2'$, we have $k = 1$ to be the index of the crossover sublot in the optimal solution. Calculate the following intermediate constants by substituting $k = 1$:

$$A = \sum_{i=1}^{k-1} q_3^{i-1} = 0, \; B = \sum_{i=1}^{n-k+1} q^{i-1} = \frac{49}{25},$$

$$C = (p_1 - p_3) \sum_{i=1}^{k-1} q_3^{i-1} - p_2 = -1, \text{ and} D = p_1 - p_2 = 5.$$

Then, the size of the crossover sublot x_k (x_1 in this case) can be determined by substituting the above constants in the following expression

$$x_1 = \frac{C - t_2' A}{BC - DA} = \frac{25}{49}.$$

Moreover, $x_2 = x_1 q = 15/49$ and $x_3 = x_1 q^2 = 9/49$. The makespan is obtained by substituting the above variable values in the following expression:

$$\begin{aligned} M &= x_1 p_1 \sum_{i=1}^{k-1} q_3^{i-1} + x_k (p_1 + p_3) + x_k p_3 \sum_{i=k}^{n} q^{i-k} \\ &= \frac{321}{49}. \end{aligned}$$

Case 2. $p_2^2 = p_1 p_3$

Theorem 4.7 (Chen and Steiner [12]) *When* $p_2^2 = p_1 p_3$, *the optimal network has the following properties (see Fig. 4.8):*

- *Segment* $(0, 1) - (0, 2)$ *is critical.*
- *Segments* $(i, 2) - (i + 1, 2)$, $(j, 2) - (j, 3)$, *and* $(\ell, 3) - (\ell + 1, 3)$ *are critical,* $\forall i, i = 1, \ldots, n - 1, j = 1, \ldots, n,$ *and* $\ell = 1, \ldots, n - 1$.
- *All other two node segments are noncritical.*

Once again, this solution can be explained by referring to Fig. 4.4. The inclusion of setup time t_2' on machine 2 displaces the sublots on machine 2 maintaining a no-wait schedule on machines 2 and 3, while the sublots on machine 1 no longer constitute critical segments.

Optimal Continuous Sublot Sizes

Returning to Fig. 4.8, $(i + 1, 2)$ and $(i, 3)$ are MCCs. Therefore,

$$p_3 x_i = p_2 x_{i+1}.$$

Since $q_3 = (p_3/p_2)$, as defined earlier,

$$x_{i+1} = q_3 x_i, \quad \forall i = 1, \ldots, n - 1$$

or

$$x_i = q_3^{i-1} x_1, \quad \forall i = 1, \ldots, n.$$

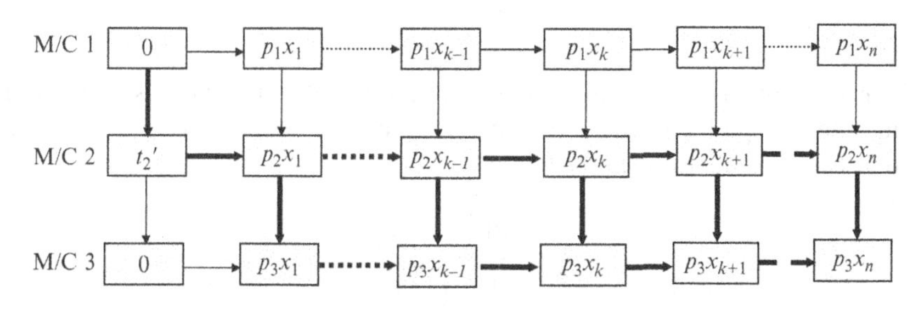

FIGURE 4.8. Optimal sublot structure when $p_2^2 = p_1 p_3$

As $\sum_{i=1}^{n} x_i = 1$, by substituting in it x_i from above and simplifying, we have

$$\begin{aligned} x_1 &= \frac{q-1}{q^n - 1}, \quad \text{if } p_2 \neq p_3, \\ &= \frac{1}{n}, \qquad\quad \text{if } p_2 = p_3, \end{aligned}$$

where $q = (p_3 + p_2)/(p_1 + p_2)$, as before. Note that, since $p_2^2 = p_1 p_3$, we have

$$\frac{p_2}{p_3} = \frac{p_1}{p_2},$$

$$\frac{p_2 + p_3}{p_3} = \frac{p_1 + p_2}{p_2},$$

or

$$\frac{p_3}{p_2} = \left(\frac{p_2 + p_3}{p_1 + p_2}\right),$$

i.e.,

$$q_3 = q.$$

The makespan is given by

$$\begin{aligned} M &= t_2' + p_2 \sum_{i=1}^{n} x_i + p_3 \\ &= t_2' + p_2 x_1 \left(\sum_{i=1}^{n} q^{i-1} \right) + p_3 \\ &= t_2' + \frac{p_2(1-q)}{1-q^n} + p_3, \text{ if } p_2 \neq p_3 \\ &= t_2' + \frac{p_2}{n} + p_3, \text{ if } p_2 = p_3. \end{aligned}$$

Case 3. $p_2^2 > p_1 p_3$

The analysis of this case is similar to that of Case 1, in that if t_2' exceeds a threshold value, then the problem reduces to a two-machine problem involving machines 2 and 3. This threshold value is as follows:

$$t_2' > \frac{p_1\left(1 - \frac{p_3}{p_2}\right)}{1 - \left(\frac{p_3}{p_2}\right)^n}.$$

If this threshold value is not exceeded, then the optimal solution has the following properties.

Theorem 4.8 (Chen and Steiner [12]) *If $p_2^2 > p_1 p_3$ and there is a critical segment on machine 1, then there exist sublots k and j, $k \in \{1, \ldots, n\}$, $j \in \{k, k+1\}$ such that:*

- *Segment $(i, 1) - (i, 2)$ is not critical for all sublots $i, i = k+1, \ldots, n$.*
- *Segment $(i, 2) - (i, 3)$ or $(i, 3) - (i+1, 3)$ are not critical for all sublots $i, i = 0, \ldots, j-1$.*
- *The segments $(k, 2) - (k, 3)$ and $(k, 3) - (k+1, 3)$ may or may not be critical depending on actual data.*
- *Every other two node segment is critical.*

This is depicted in Fig. 4.9 (for $j = k$).

Note that k denotes the last sublot for which the segment $(k, 1) - (k, 2)$ is critical, and j denotes the first sublot for which the segment $(j, 2) - (j, 3)$ is critical. Now, j cannot be less than k because, following the definition of critical corners, it would contradict the assumption that $p_2^2 > p_1 p_3$. Also, j cannot be greater than $k+1$ because, then, it would imply that sublot $k+1$ is not critical, which is not possible because, by Theorem 4.5, every sublot must be critical. Therefore, j can only take the value of k or $k+1$. Also, when $k = n$, j must be equal to n since the total number of sublots is n.

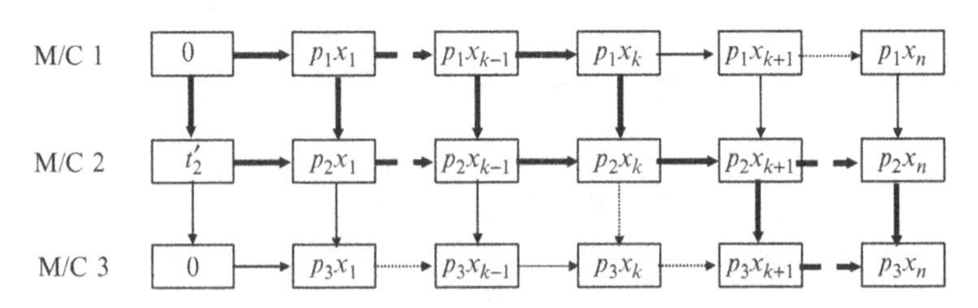

FIGURE 4.9. Optimal sublot structure when $p_2^2 > p_1 p_3$

Once again, we can explain a threshold value of t'_2 and the nature of the critical segments specified by Theorem 4.8 by referring to Fig. 4.5. Given that all segments corresponding to machine 2 are critical, a positive value of t'_2, while shifting the sublots on machine 2 to the right, maintains the criticality of all of these segments but decreases the location of crossover sublot k from that for the no setup case. Thus, if t'_2 is sufficiently large, then none of the segments on machine 1 will remain critical, and the optimal sublot sizes can be obtained by solving a two-machine problem involving machines 2 and 3. On the other hand, in case the value of t'_2 is not large enough, then some critical segments on machine 1 do exist and the nature of the solution is like that shown in Fig. 4.9 (which in a way is similar to that in Fig. 4.5) in which all the segments on machine 2 are critical, sublots from 1 to k are processed in a no-wait fashion on machines 1 and 2. While in Fig. 4.5, the sublots k to n are processed in a no-wait fashion on machines 2 and 3, in this case, either sublots k to n or sublots $k+1$ to n are processed in a no-wait fashion on machines 2 and 3, depending on the data.

Continuous Optimal Sublot Sizes

From Fig. 4.9, $(i, 2)$ and $(i+1, 1)$ are MCCs for $i = 1, \ldots, k-1$. Therefore,

$$p_2 x_i = p_1 x_{i+1}, \quad i = 1, \ldots, k-1.$$

If $q_2 = (p_2/p_1)$, we have

$$x_i = q_2^{i-1} x_1, \quad \forall i = 1, \ldots, k.$$

Similarly, $(i+1, 2)$ and $(i, 3)$ are MCCs for $k+1, \ldots, n-1$, which give

$$p_2 x_{i+1} = p_3 x_i,$$

or, if $q_3 = (p_3/p_2)$,

$$x_i = q_3^{i-(k+1)} x_{k+1}, \quad \forall i = k+1, \ldots, n-1.$$

Since $\sum_{i=1}^{n} x_i = 1$, we have

$$x_1 \left(1 + q_2 + \cdots + q_2^{k-1}\right) + x_{k+1} \left(1 + q_3 + \cdots + q_3^{n-(k+1)}\right) = 1$$

or,

$$x_1 \left(\frac{1 - q_2^k}{1 - q_2}\right) + x_{k+1} \left(\frac{1 - q_3^{n-k}}{1 - q_3}\right) = 1.$$

Since segments $(0, 1) - (0, 2) - (1, 2)$ and $(0, 1) - (1, 1) - (1, 2)$ are critical,

$$t'_2 = x_1 p_1.$$

By solving the above two equations for x_{k+1}, we obtain

$$
\begin{aligned}
x_i &= x_1 q_2^{i-1}, \ \forall i = 1, \ldots, k, \\
x_{k+1} &= \frac{(1-q_2)\, p_1 - \left(1-q_2^k\right) t_2'}{\left(1-q_3^{n-k}\right) p_1} \left(\frac{1-q_3}{1-q_2}\right), \quad \text{if } q_2 \neq 1, q_3 \neq 1, \\
&= \frac{p_1 - t_2' \sum_{i=0}^{k-1} q_2^i}{p_1 \sum_{i=0}^{n-(k+1)} q_3^i}, \quad \text{otherwise}, \\
x_i &= x_{k+1} q_3^{i-(k+1)}, \ \forall i = k+1, \ldots, n.
\end{aligned}
$$

where q_2 and q_3 are as defined above.

The makespan is given by

$$
\begin{aligned}
M &= p_1 x_1 \sum_{i=0}^{k-1} q_2^i + p_2 x_1 q_2^{k-1} + x_{k+1} \left(p_2 + p_3 \sum_{i=0}^{n-(k+1)} q_3^i \right) \\
&= t_2' \sum_{i=0}^{k-1} q_2^i + \frac{t_2' p_2 q_2^{k-1}}{p_1} + x_{k+1} \left(p_2 + p_3 \sum_{i=0}^{n-(k+1)} q_3^i \right),
\end{aligned}
$$

by substituting $x_1 = t_2'/p_1$ from above. To find the value of k, we first find the value k^* corresponding to the crossover sublot when $t_2' = 0$, i.e., when no setups are present (see Sect. 4.2). As the location of the crossover sublot decreases with increment in t_2', a binary search is then performed over the interval $[1, k^*]$, with a trial value of $k_0 = \lceil k^*/2 \rceil$ and using the following equations:

$$
\hat{t}_2(k_0) = p_1 x_1(k_0) \quad \text{and} \quad \sum_{i=1}^{n} x_i(k_0) = 1,
$$

where

$$
\begin{aligned}
x_i(k_0) &= x_1(k_0)\, q_2^{i-1}, \quad \forall i = 1, \ldots, k_0, \\
x_i(k_0) &= x_{k_0}(k_0)\, q_3^{i-k_0}, \quad \forall i = k_0, \ldots, n.
\end{aligned}
$$

x_k can be obtained from the expression given above for x_{k+1} by replacing $(k+1)$ by k_0.

If $\hat{t}_2(k_0) > t_2'$, then $k_0 \leq k$ and we set $k_0 = \lceil k_0 + (n-k_0)/2 \rceil$; else $k_0 > k$ and we set $k_0 = \lceil k_0/2 \rceil$.

Example 4.5 To illustrate the procedure for Case 3, consider another example. Let $p_1 = 4$, $p_2 = 5$, $p_3 = 2$, $t_1 = 2$, $t_2 = 4$, $t_3 = 3$, $U = 1$, and $n = 3$.

Note that $(p_2)^2 > p_1 p_2$. We first check whether $t_2' (= t_2 - t_1) = 2$ is less than the threshold value. We have

$$
\frac{p_1 (1 - p_3/p_2)}{1 - (p_3/p_2)^3} = \frac{100}{39}.
$$

Since this is not true, the problem cannot be reduced to a two-machine problem. We have $q_1 = 4/5$, $q_2 = 5/4$, and $q_3 = 2/5$.

To start with, we first find the crossover sublot k^*, corresponding to the case $t_2' = 0$ (please refer to Case 3 in Sect. 4.2). Let $k = 2$. Then,

$$\frac{1}{x_2} = \frac{q_1^2 - 1}{q_1 - 1} + \frac{q_3^{3-2+1} - 1}{q_3 - 1} - 1 = \frac{11}{5}.$$

After simplification, we have $x_2 = 5/11$, and

$$M(x_2) = p_2 + p_2 x_2 \left(q_1^2 + q_3^{3-2+1}\right) = \frac{75}{11}.$$

Next, try $k = 3$. We have

$$\frac{1}{x_3} = \frac{q_1^3 - 1}{q_1 - 1} + \frac{q_3^{3-3+1} - 1}{q_3 - 1} - 1 = \frac{61}{25}.$$

This gives $x_3 = 25/61$, and

$$M(x_3) = p_2 + p_2 x_3 \left(q_1^3 + q_3^{3-3+1}\right) = \frac{419}{61}.$$

Since $M(x_3) > M(x_2)$, $k^* \leq 2$. We try $k = 1$. We have

$$\frac{1}{x_1} = \frac{q_1^1 - 1}{q_1 - 1} + \frac{q_3^{3-1+1} - 1}{q_3 - 1} - 1 = \frac{39}{25},$$

which gives $x_1 = 25/39$, and

$$M(x_1) = p_2 + p_2 x_1 \left(q_1^1 + q_3^{3-1+1}\right) = \frac{101}{39}.$$

Since $M(x_2) < M(x_1)$, $k^* \geq 2$. Hence, for the case of $t_2' = 0$, we have $k^* = 2$. With this value of k^*, the search for the desired crossover sublot is performed over the interval $[1, k^*]$. Try $k = \lceil k^*/2 \rceil = 1$. Solve $x_1(1) + q_3 x_1(1) + q_3^2 x_1(1) = 1$, i.e., $x_1(1) + \frac{2}{5} x_1(1) + \frac{2}{5}\frac{2}{5} x_1(1) = 1$, which gives $x_1(1) = 25/39$. Since $t_2'(1) = p_1 x_1(1) > t_2'$, try $k = 2$. Solve $x_1(2) + q_2 x_1(2) + q_3 q_2 x_1(2) = 1$, i.e., $x_1(2) + \frac{5}{4} x_1(2) + \frac{2}{5}\frac{5}{4} x_1(2) = 1$, which gives $x_1(2) = 4/11$. Since $t_2'(2) = p_1 x_1(2) < t_2'$, this indicates that the index of the crossover sublot is $k = 1$. We use the following expressions to calculate the sublot sizes:

$x_1 = t_2'/p_1 = 1/2$, $x_2 + x_2 q_2 = 1/2 \Rightarrow x_2 = 5/14$, and $x_3 = 1/7$. The makespan value, $M = 65/7$.

4.5 3/1/C/II/CV/{Lot-Attached Setup}

We now consider the case of lot-attached setups. The sublot sizes are consistent and can take on continuous values. Furthermore, intermittent idling is also permitted on the machines.

The approach based on the network representation of the underlying problem, as discussed in the earlier sections of this chapter, can be used for the lot-attached setups as well. However, it is not as straightforward. But, the optimal sublot sizes can still be obtained in $O(n)$ time.

The lot-attached setup prohibits start of the lot on other machines while the first machine is still being setup. Hence, if we define a *reduced* problem in which only the setup time on machine 1 is reduced to zero while those on the other machines remain unchanged, then

$$C_{\max} = t_1 + C^1_{\max},$$

where $C_{\max}$ is the optimal makespan for the 3/1/C/CV/II/Lot-Attached Setup problem, and $C^1_{\max}$ the optimal makespan for the reduced problem.

It follows that the sublot sizes that are optimal for the reduced problem are also optimal for the original problem. Hence, it is sufficient to solve the reduced problem, which is identical to the original problem, except that $t_1 = 0$. The corresponding network representation (see Fig. 4.10) consists of nodes for each sublot on every machine. Node (1, 1) has a weight of p_1x_1, node $(1, k)$ has a weight of $t_k + p_kx_1, k = 1, 2, 3$, and node (i, k) has a weight of $p_kx_i, k = 1, 2, 3$, $i = 1, \ldots, n$.

Since the setups are attached, it is only appropriate to assume that there is at least one item on a machine before a setup can begin. Since the sublot sizes are consistent, it is enough to assume that $x_1 \geq \Delta$ where $\Delta = 1/U$. Accordingly, the solution procedure consists of solving the reduced problem separately under three different cases:

1. $x_1 = \Delta$, with the resulting makespan designated by $M_1(x)$.
2. $x_1 > \Delta$ and no critical segment on machine 1, with the resulting makespan designated by $M_2(x)$.
3. $x_1 > \Delta$ and a critical segment on machine 1, with the resulting makespan designated by $M_3(x)$.

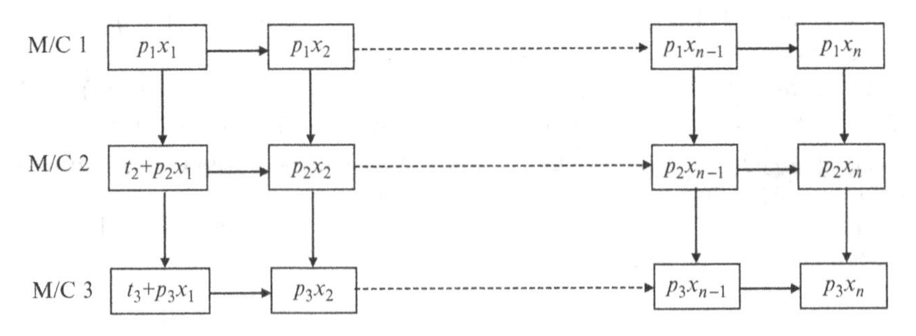

FIGURE 4.10. Network representation for the reduced 3/1/C/II/CV/Lot-Attached Setup problem

The optimal solution will be $M(x) = \min\{M_1(x), M_2(x), M_3(x)\}$. Case 1 is separated out because it reduces the problem to the one involving lot-detached setups. We analyze each of these cases next.

Case 1. $x_1 = \Delta$

When x_1 is fixed at Δ, then machine 2 can begin its setup at Δp_1. Consequently, machine 3 can, then, begin its setup at $t_2 + \Delta(p_1 + p_2)$. This implies that the problem can be viewed as the one with detached setups of zero, Δp_1 and $t_2 + \Delta(p_1 + p_2)$ on machines 1, 2, and 3, respectively, and a lot of size $U(1 - \Delta)$ with $(n - 1)$ sublots (the first sublot is of size Δ). Consequently, the procedures developed for the detached setups (see Sect. 4.4) can be applied here. Note that the detached setup on machine 2 is as follows:

$$t_2' = \Delta p_1 + t_2 + \Delta p_2 = t_2 + \Delta (p_1 + p_2),$$

and on machine 3, it is

$$t_3' = t_2 + \Delta (p_1 + p_2) + t_3 + \Delta p_3 = t_2 + t_3 + \Delta (p_1 + p_2 + p_3).$$

Consider the case of $x_1 > \Delta$ next. It can be shown [11] that there exists an optimal schedule in which the sublot sizes on the first and the second machines and on the second and the third machines are the same, i.e., $x_{i,1} = x_{i,2}$ and $x_{i,2} = x_{i,3} \forall i, i = 1, \ldots, n$. Furthermore, when $x_1 > \Delta$ in an optimal solution with consistent sublot sizes for the reduced problem on three machines, then every sublot is critical. These results imply that when $x_1 > \Delta$ in an optimal solution for the reduced problem on three machines, then there exists an optimal solution in which all sublot sizes are consistent and positive.

Case 2. $x_1 > \Delta$ and no critical segment on machine 1

The structure of the critical paths for this case is shown in Fig. 4.11. As before, the heavy lines indicate critical segments and the light lines indicate noncritical segments.

Theorem 4.9 formally states the structure of critical paths in the optimal network representation.

Theorem 4.9 (Chen and Steiner [10]) *If the optimal solution has no critical segment on machine 1 and $x_1 > \Delta$, then:*

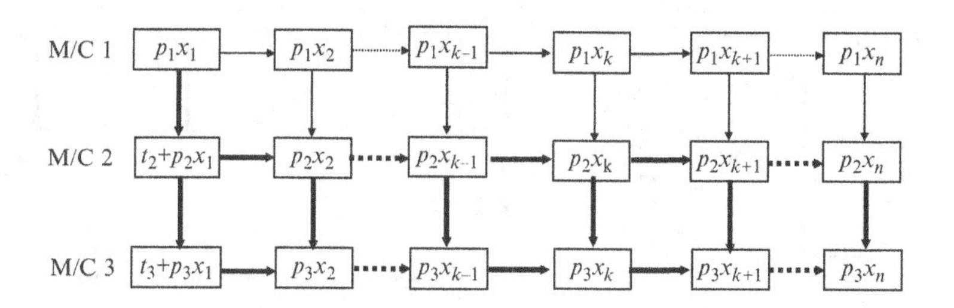

FIGURE 4.11. Optimal sublot structure for Case 2

1. $(i, 2) - (i, 3)$ *is critical for* $i = 1, \ldots, n$.
2. $(i, 2) - (i + 1, 2)$ *is critical for* $i = 1, \ldots, n - 1$.
3. $(i, 3) - (i + 1, 3)$ *is critical for* $i = 1, \ldots, n - 1$.

The criticality of segment $(1, 1) - (1, 2)$ is obvious as the processing of the first sublot on machine 2 does not begin unless it is transferred to that machine. Given that none of the segments on machine 1 is critical and we desire to minimize makespan, it is clear that segments on machine 2 will be critical and the schedule between machines 2 and 3 will be a no-wait schedule, thereby, implying

$$p_3 x_i = p_2 x_{i+1}, \forall i = 2, \ldots, n-1 \quad \text{and} \quad t_3 + p_3 x_1 = p_2 x_2.$$

All of these lead to the solution specified in Theorem 4.9 above.

Optimal Sublot Sizes

From the above theorem, both $(i, 2) - (i + 1, 2) - (i + 1, 3)$ and $(i, 2) - (i, 3) - (i + 1, 3)$ are critical for $i = 1, \ldots, n - 1$. Hence,

$$t_3 + p_3 x_1 = p_2 x_2 \quad \text{and} \quad p_3 x_1 = p_2 x_{i+1}, \quad \forall i = 2, \ldots, n-1.$$

These equations, in conjunction with $\sum_{i=1}^{n} x_i = 1$, can be used to determine the unique sublot sizes in $O(n)$ time.

Case 3. $x_1 > \Delta$ and a critical segment on machine 1
Like the discussion in Sect. 4.4, in this case, we need to distinguish between the situations depending on $p_2^2 < p_1 p_3$, $p_2^2 = p_1 p_3$, and $p_2^2 > p_1 p_3$.

Case 3a. $p_2^2 < p_1 p_3$

If $p_2^2 < p_1 p_3, x_1 > \Delta$ and there is a critical horizontal segment on machine 1 in an optimal solution x, then the structure of the critical paths is as shown in Fig. 4.12 and is stated formally in Theorem 4.10.

Theorem 4.10 (Chen and Steiner [10]) *If* $p_2^2 < p_1 p_3, x_1 > \Delta$ *and there exists a critical segment on machine 1 in an optimal solution* x*, then there exists a crossover sublot* $k \in \{2, \ldots, n\}$ *such that:*

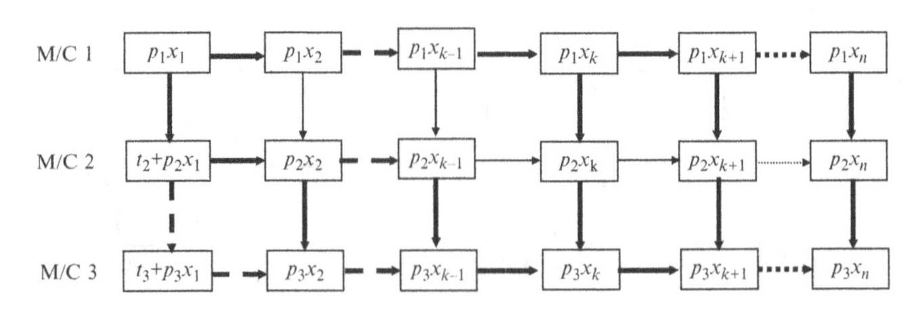

FIGURE 4.12. Optimal sublot structure for Case 3a: $p_2^2 < p_1 p_3$

1. *Segment $(i, 1) - (i, 2)$ is not critical for $i = 2, \ldots, k - 1$;*
2. *Segment $(i, 2) - (i + 1, 2)$ is not critical for $i = k, \ldots, n - 1$;*
3. *Segments $(1, 2) - (1, 3)$, $(1, 3) - (2, 3)$, and $(k - 1, 2) - (k, 2)$ are critical or not depending on the problem instance*
4. *Every other two-node segment is critical.*

Interestingly, the structure of the optimal solution for this case is identical to that for the case of $p_2^2 < p_1 p_3$ under detached setup (see Theorem 4.6 and Fig. 4.7), except for that corresponding to node (0, 3) in Fig. 4.7 and node (1, 3) in Fig. 4.12. Therefore, it can be explained following a similar argument as follows. Inherently, both of these situations are identical to the case of no setup (see Fig. 4.3). The lot-attached setup can be viewed to shift the sublots on machine 2 to the right from those in Fig. 4.3, thereby, leading to a crossover sublot k (the first sublot, now, that is processed in a no-wait fashion on all three machines). The processing time of node (0, 3) in Fig. 4.7 is zero while that of node (1, 3) in Fig. 4.12 is $t_3 + p_3 x_1$. As a result, depending on the values of p_3 and t_3, either one or both of the segments $(1, 2) - (1, 3)$ and $(1, 3) - (2, 3)$ may be critical. The same is true about the criticality of segment $(k - 1, 2) - (k, 2)$.

Optimal Sublot Sizes

For a specific value of k, we observe that the following are MCCs:

1. $(i, 3)$ and $(i + 1, 2)$, $\forall i = 2, \ldots, k - 3$
2. $(i, 3)$ and $(i + 1, 1)$, $\forall i = k + 1, \ldots, n - 1$

Therefore,

$$p_3 x_i = p_2 x_{i+1}, \quad \forall i = 2, \ldots, k - 3$$

and

$$(p_2 + p_3) x_i = (p_1 + p_2) x_{i+1}, \quad \forall i = k, \ldots, n - 1.$$

This implies,

$$x_i = x_2 q_2^{i-2}, \quad \forall i = 2, \ldots, k - 1 \tag{4.16}$$

and

$$x_i = x_k q^{i-k}, \quad \forall i = k, \ldots, n, \tag{4.17}$$

where $q_2 = p_3/p_2$ and $q = (p_2 + p_3)/(p_2 + p_1)$.

Substituting these into $\sum_{i=1}^{n} x_i = 1$, we have

$$x_1 + x_2 \left(1 + q_2 + \cdots + q_2^{k-3}\right) + x_k \left(1 + q + \cdots + q^{n-k}\right) = 1. \tag{4.18}$$

Since both paths $(1, 1) \cdots (k, 1) - (k, 2) - (k, 3)$ and $(1, 1) - (1, 2) - (2, 2) - (2, 3) \cdots (k, 3)$ are critical, they should have the same length, i.e.,

$$A p_1 x_2 + (p_1 + p_2) x_k = t_2 + p_2 x_1 + p_2 x_2 + A p_3 x_2, \tag{4.19}$$

where $A = 1 + q_2 + \cdots + q_2^{k-3}$.

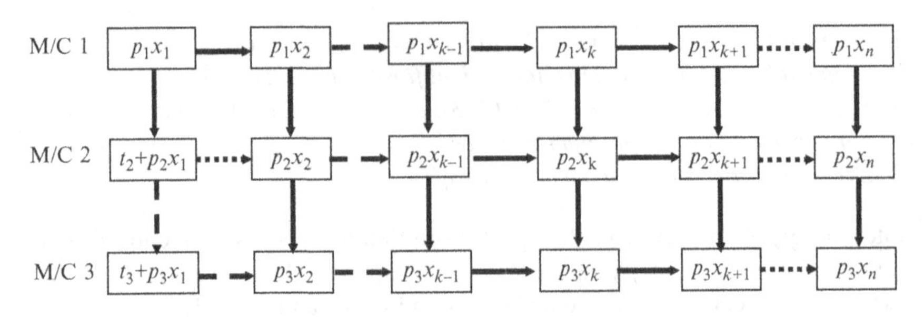

FIGURE 4.13. Optimal sublot structure for Case 3b: $p_2^2 = p_1p_3$

Using (4.18) and (4.19), it can be shown [11] that $x_k = \min\{C_1, C_2\}$, where

$$C_1 = \frac{t_2 + p_2}{\left(\dfrac{Bp_2 + p_1 + p_2 - (p_2 + A(p_3 - p_2 - p_1))}{q_2^{k-2}}\right)},$$

$$C_2 = \frac{p_2 + A\,(p_3 - p_2 - p_1) + (t_2 + p_2)\left(A + \dfrac{p_2}{p_3}\right) + t_3\left(\dfrac{p_2 + A(p_3 - p_2 - p_1)}{p_3}\right)}{\left(A + \dfrac{p_2}{p_3}\right)(Bp_2 + p_2 + p_1) - B(p_2 + A(p_3 - p_2 - p_1))},$$

and

$$B = 1 + q + \cdots + q^{n-k}.$$

Since the path $(1,1) - (1,2) \cdots (1,n) - (2,n) - (3,n)$ is critical, we can write the makespan expression as follows:

$$M(x) = p_1\,(x_1 + \cdots + x_n) + p_2x_n + p_3x_n = p_1 + (p_2 + p_3)\,x_n = p_1 + (p_2 + p_3)\,q^{n-k}x_k.$$

The optimal value of k can, then, be determined by searching over all possible values of $k (k = 1, \ldots, n)$ computing x_k from above and selecting the one that minimizes

$$M(x) = p_1 + (p_2 + p_3)\,q^{n-k}x_k.$$

This will require $O(n)$ time. Observe that once x_k is known, we can use (4.18) and (4.19) to obtain x_1 and x_2; $x_i, \forall i = k, \ldots, n$, can be obtained from (4.17) and $x_i, \forall i = 2, \ldots, k-1$, can be obtained from (4.16).

Case 3b. $p_2^2 = p_1p_3$

The structure of critical paths in this case is shown in Fig. 4.13 and is stated formally in Theorem 4.10.

Theorem 4.11 (Chen and Steiner [10]) *If $p_2^2 = p_1p_3, x_1 > \Delta$ and there exists a critical horizontal segment on machine 1 in an optimal solution x, then the optimal structure of sublot sizes can be described as follows:*

1. *Segments $(1,2) - (1,3)$ and $(1,3) - (2,3)$ are critical if $t_3 > (p_3/p_2)t_2$, segment $(1,2) - (2,2)$ is critical if $t_3 < (p_3/p_2)t_2$, and all of these segments are critical if $t_3 = (p_3/p_2)t_2$.*

2. *All other two-node segments are critical.*

Note that if $t_3 = (p_3/p_2)t_2$, which is true when $t_3 = t_2 = 0$, the above situation reduces to the case of no setup (see Fig. 4.4) and the schedule is compact and all the paths in the network will be critical. In case, segment $(1, 2) - (2, 2)$ is critical but segments $(1, 2) - (1, 3)$ and $(1, 3) - (2, 3)$ are not, then we have $x_1 p_2 + t_2 = x_2 p_1$, as $(1, 2)$ and $(2, 1)$ are MCCs. Furthermore,

$$\begin{aligned} t_3 + p_3 x_1 &< p_2 x_2 \\ &= \frac{p_2}{p_1} x_2 p_1 \\ &= \frac{p_2}{p_1} (x_2 p_2 + t_2), \end{aligned}$$

which, after further reduction and using $p_2^2 = p_1 p_3$, results in $t_3 < (p_3/p_2)t_2$. However, if $t_3 > (p_3/p_2)t_2$, then a similar reasoning will imply that segments $(1, 2) - (1, 3)$ and $(1, 3) - (2, 3)$ are critical while segment $(1, 2) - (2, 2)$ is not.

Optimal Sublot Sizes

From Theorem 4.11, $(i+1, 1)$ and $(i, 2)$ are MCCs for $i = 2, \ldots, n-1$. Therefore, $p_2 x_i = p_1 x_{i+1}$, which implies:

$$x_{i+1} = q^{i-2} x_2, \quad \forall i = 2, \ldots, n-1, \tag{4.20}$$

where

$$q = \frac{p_2 + p_3}{p_2 + p_1} = \frac{p_2}{p_1}.$$

If $t_3 < (p_3/p_2)t_2$, then from Theorem 4.11, segment $(1, 2) - (2, 2)$ is critical and $(1, 2)$ and $(2, 1)$ are MCCs. Hence,

$$t_2 + p_2 x_1 = p_1 x_2. \tag{4.21}$$

Using (4.20) and (4.21) in conjunction with $\sum_{i=1}^{n} x_i = 1$, we can determine all the sublot sizes in $O(n)$ time.

Similarly, if $t_3 \geq (p_3/p_2)t_2$, then from Theorem 4.11, segments $(1, 2) - (1, 3)$ and $(1, 3) - (2, 3)$ are critical and $(1, 3)$ and $(2, 1)$ are MCCs. Hence,

$$t_3 + t_2 + (p_2 + p_3) x_1 = (p_2 + p_1) x_2. \tag{4.22}$$

Using (4.20) and (4.22) in conjunction with $\sum_{i=1}^{n} x_i = 1$, we can once again determine all the sublot sizes in $O(n)$ time.

Case 3c. $p_2^2 > p_1 p_3$

The structure of critical paths in this case is shown in Fig. 4.14 and is stated formally in Theorem 4.12.

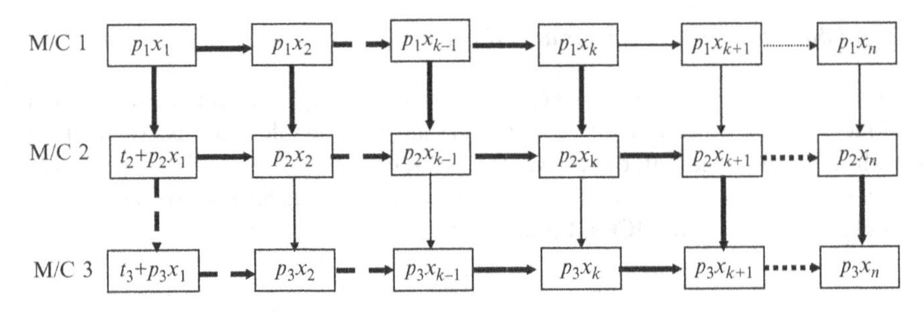

FIGURE 4.14. Optimal sublot structure for Case 3c: $p_2^2 > p_1p_3$

Theorem 4.12 (Chen and Steiner [10]) *If $p_2^2 > p_1p_3, x_1 > \Delta$ and there exists a critical horizontal segment on machine 1 in an optimal solution, then there exist sublots k (the crossover sublot) and j, with $k \in \{1, \ldots, n\}$, $j \in \{k, k+1\}$, such that:*

1. *Segment $(i, 1) - (i, 2)$ is not critical for $i = k+1, \ldots, n$.*
2. *Segment $(i, 2) - (i, 3)$ is not critical for $i = 2, \ldots, j-1$.*
3. *Segments $(1, 2) - (1, 3) \cdots (k+1, 3)$ and $(k, 2) - (k, 3) - (k+1, 3)$ are critical or not depending on the data.*
4. *Every other two-node segment is critical.*

Note the similarity of this solution with the no-setup solution (Fig. 4.5) and detached-setup solution (Fig. 4.9). The criticality of segments as specified by (3) above is impacted by setup time t_3.

Optimal Sublot Sizes

For a specific value of k, we observe that the following are MCCs:

1. $(i+1, 1)$ and $(i, 2)$, $\forall i = 2, \ldots, k-1$
2. $(i+1, 2)$ and $(i, 3)$, $\forall i = k+1, \ldots, n-1$

Therefore,

$$p_2x_i = p_1x_{i+1}, \quad \forall i = 2, \ldots, k-1$$

and

$$p_3x_i = p_2x_{i+1}, \quad \forall i = k+1, \ldots, n-1.$$

This implies,

$$x_i = x_2q_1^{i-2}, \quad \forall i = 2, \ldots, k \tag{4.23}$$

and

$$x_{i+1} = x_{k+1}q_2^{i-k}, \quad \forall i = k+1, \ldots, n-1, \tag{4.24}$$

where $q_1 = p_2/p_1$ and $q_2 = p_3/p_2$.

Substituting these into $\sum_{i=1}^{n} x_i = 1$, we have

$$x_1 + x_2A_1 + B_1x_{k+1} = 1, \tag{4.25}$$

where

$$A_1 = \left(1 + q_1 + \cdots + q_1^{k-2}\right) \quad \text{and} \quad B_1 = \left(1 + q_2 + \cdots + q_2^{n-k-1}\right).$$

Since (1, 2) and (2, 1) are MCCs,

$$t_2 + p_2 x_1 = p_1 x_2. \tag{4.26}$$

It can be shown [10] that $x_{k+1} = \min\{F_1, F_2\}$, where

$$F_1 = \frac{p_3 q_1^{k-2}\left(t_2 + p_2\right)}{p_1 p_2 + A_1 p_2^2 + p_3 p_2 B_1 q_1^{k-2}}$$

and

$$F_2 = \frac{t_3\left(p_1 + A_1 p_2\right) + p_1 p_3 - A_1 t_2 + A_1\left(p_2 - p_3\right)\left(t_2 + p_2\right)}{p_1 p_2 + A_1 p_2^2 - A_1 B_1 p_2^2 + A_1 B_1 p_1 p_2 + p_1 p_3 B_1}.$$

The index $k + 1$ can be determined in $O(n)$ time by searching over all possible values and selecting the one that minimizes

$$M(x) = D + E x_{k+1},$$

where

$$D = t_2 + p_2 + \frac{p_1^2 - A_1 t_2 p_1}{p_1 + A_1 p_2}$$

and

$$E = p_3 q_2^{n-k-1} - \frac{p_1^2 B_1}{p_1 + A_1 p_2}.$$

Observe that once x_{k+1} is known, we can solve (4.25) and (4.26) to obtain x_1 and x_2. Also, $x_i, \forall i = k + 1, \ldots, n - 1$, can be obtained from (4.24) and $x_i, \forall i = 2, \ldots, k$, can be obtained from (4.23).

Example 4.6 To illustrate the above procedure, consider the following example. Let $p_1 = 5, p_2 = 8, p_3 = 13, t_1 = 0, t_2 = 4, t_3 = 3, U = 1, n = 3$, and $\Delta = 1/22$.

We will explore all the three cases discussed above.

Case 1. If $x_1 = \Delta$, the problem becomes a three-machine problem with two sublots and detached setup times given by

$$t_2' = t_2 + (p_1 + p_2)\,\Delta = \frac{101}{22},$$
$$t_3' = t_2 + t_3 + (p_1 + p_2 + p_3)\,\Delta = \frac{90}{11}.$$

Using the method in Sect. 4.4, we obtain the following sublot sizes and the makespan value:

$$x_1 = \frac{1}{22}, \quad x_2 = \frac{4}{11}, \quad x_3 = \frac{13}{22}, \quad \text{and the makespan, } M = \frac{453}{22}.$$

Case 2. $x_1 = \Delta$ and no critical segment on machine 1. We use the following three equations to determine the sublot sizes:

$$\begin{aligned} t_3 + p_3 x_1 &= p_2 x_2, \\ p_3 x_2 &= p_2 x_3, \\ x_1 + x_2 + x_3 &= 1. \end{aligned}$$

After substituting the values of p_1, p_2, and p_3, we have

$$\begin{aligned} 3 + 13x_1 &= 8x_2, \\ 13x_2 &= 8x_3, \\ x_1 + x_2 + x_3 &= 1. \end{aligned}$$

This gives $x_1 = -0.0146 < 0 < \Delta$. Hence, this case is not possible.
Case 3. $x_1 = \Delta$, and there is a critical segment on machine 1. There are three possible subcases depending on $p_2^2 < p_1 p_3$, $p_2^2 = p_1 p_3$, or $p_2^2 > p_1 p_3$. However, based on the given data, we have $p_2^2 < p_1 p_3$. We calculate

$$q = \frac{p_2 + p_3}{p_2 + p_1} = \frac{21}{13} \quad \text{and} \quad q_2 = \frac{p_3}{p_2} = \frac{13}{8}.$$

First, try crossover sublot $k = 2$. We have

$$\begin{aligned} A &= 0, \\ B &= 1 + q = \frac{34}{13}, \\ C_1 &= \frac{t_2 + p_2}{Bp_2 + p_1 + p_2 - \dfrac{p_2}{q_2^{2-2}}} = \frac{4+8}{\dfrac{34}{13}8 + 5} = \frac{156}{337}, \end{aligned}$$

and

$$C_2 = \frac{p_2 + (t_2 + p_2)\left(\dfrac{p_2}{p_3}\right) + t_3\left(\dfrac{p_2}{p_3}\right)}{\left(\dfrac{p_2}{p_3}\right)(Bp_2 + p_1 + p_2) - Bp_2} = \frac{8 + 12\dfrac{8}{13} + 3\dfrac{8}{13}}{\dfrac{8}{13}\left(\dfrac{34}{23}8 + 5 + 8\right) - \dfrac{34}{13}8} = -234.$$

Therefore, $x_2 = \min(C_1, C_2) = -234 < 0$, which means that this case does not exist.

Next, try $k = 3$. We have the intermediate constants:

$$\begin{aligned} A &= 1, \\ B &= 1, \\ C_1 &= \frac{t_2 + p_2}{Bp_2 + p_1 + p_2 - \dfrac{p_2 + A(p_3 - p_2 - p_1)}{q_2^{3-2}}} = \frac{4+8}{8 + 5 + 8 - \dfrac{8/13}{8}} = \frac{156}{209}, \end{aligned}$$

and

$$C_2 = \frac{p_2 + A(p_3 - p_2 - p_1) + (t_2 + p_2)\left(A + \frac{p_2}{p_3}\right) + t_3\left(\frac{p_2 + A(p_3 - p_2 - p_1)}{p_3}\right)}{\left(A + \frac{p_2}{p_3}\right)(Bp_2 + p_1 + p_2) - B(p_2 + A(p_3 - p_2 - p_1))}$$
$$= \frac{8 + 12\left(1 + \frac{8}{13}\right) + 3\left(\frac{8}{13}\right)}{\left(1 + \frac{8}{13}\right)(8 + 5 + 8) - 8} = \frac{380}{337}.$$

This gives

$$x_3 = \min\left\{\frac{156}{209}, \frac{380}{337}\right\} = \frac{156}{209}.$$

We obtain x_1 and x_2 by substituting x_3 in the following equations:

$$x_1 + x_2 + x_3 = 1,$$
$$Ap_1x_2 + (p_1 + p_2)x_3 = t_2 + p_2(x_1 + x_2) + Ap_3x_2.$$

We have

$$x_1 + x_2 + \frac{156}{209} = 1,$$
$$5x_2 = 4 + 8(x_1 + x_2),$$

which gives $x_1 = 1.2057$, $x_2 = -0.9521$. Therefore, this case also does not exist. Since, we have reached the given number of sublots, stop. The only feasible case for this problem is Case 1. The optimal sublot sizes are $x_1 = 1/22$, $x_2 = 4/11$, $x_3 = 13/22$, and the makespan, $M = 453/22$.

For ease of reference, we can summarize the above procedure as follows:

Step 1. Set $k = 2$.

Step 2. Calculate the size of crossover sublot using the value of $x_k = \min\{C_1, C_2\}$. If $x_k \leq 0$, set $k = k + 1$, and repeat *Step 2*. Otherwise, go to *Step 3*.

Step 3. Substitute the value of x_k into (4.18) and (4.19) to obtain the values of x_1 and x_2. If $x_1 \leq \Delta$ or $x_2 \leq 0$, then set $k = k + 1$ and go to *Step 2*. Otherwise, calculate $M(x_k)$. If $k = n$, go to *Step 4*, otherwise, set $k = k + 1$ and go to *Step 2*.

Step 4. $M^*(x) = \min\{M(x_k)\}$.

4.6 Chapter Summary

Clearly, the optimal sublot sizes, both continuous and discrete, for a single lot, three-machine lot streaming problem can be determined by using the underlying mathematical programming model. However, some of its special cases can be easily analyzed by exploiting the inherent structural properties. For the case of consistent and continuous sublot sizes and no setups, the problem can be solved by analyzing the network representation of the processing of the sublots on the machines. In this network representation, the nodes represent the sublots, and the directed arcs between the nodes capture the sequence in which the sublots travel from one machine to another as well as the order of their processing on the machines. The critical path of this network gives the minimum makespan value. The nature of the critical path depends on whether or not the second machine is dominant. In case $(p_2)^2 < p_1 p_3$, i.e., the second machine is dominated, then machine 2 does not lie on the critical path and the critical path passes through machines 1 and 3. The optimal schedule is a no-wait schedule, and optimal sublot sizes are geometric in the ratio $(p_2 + p_3) : (p_1 + p_2)$. In case $p_2 = p_1 p_3$, then all paths are critical and the schedule is compact. The sublot sizes are, once again, geometric. In case $p_2 > p_1 p_3$, the critical path is such that for a certain k, $1 \leq k \leq n$, all sublots on machines 1 and 2 before sublot k lie on the critical path, sublot k lies on the critical path and all sublots on machines 2 and 3 after sublot k lie on the critical path. The optimal sublot sizes can be determined accordingly.

When sublot sizes are variable and no idling among the sublots is permitted, optimal sublot sizes can be obtained by decomposing the problem into two subproblems. One of these consists of machines 1 and 2 and the other consists of machines 2 and 3. Both of these can be solved by using the two-machine procedures discussed in Chap. 3 (for both continuous and discrete versions). In case the sublot sizes are variable and intermittent idling among the sublots is permitted, then the optimal solution depends on the fact of the second machine being dominant or not. In case, the second machine is dominant, the problem can, once again, be split into 2 two-machine subproblems consisting of machines 1 and 2 and machines 2 and 3, and accordingly solved by the procedures presented in Chap. 2. However, in case the second machine is dominated, then consistent sublots are optimal and sublot sizes are geometric.

The situation of consistent sublot sizes, but in the presence of lot-detached/attached setups, can also be analyzed by using the network representations of the underlying problem. For the lot-detached case, the problem can be reduced to an equivalent problem having the setup times on machines 1 and 3 to be zero and that on machine 2 to be $t_2 - t_1$. In case $t_2 - t_1 > 0$, then the resulting problem is called a "reduced problem," and if $t_2 - t_1 \leq 0$ then it is called a "relaxed problem." In either case, critical paths and optimal sublot sizes can be determined as before for the situations of $p_2^2 < p_1 p_3$, $p_2^2 = p_1 p_3$, and $p_2^2 > p_1 p_3$. For the lot-attached case, it is assumed that there is at least one item on a machine before a setup is begun. Since the sublot sizes are consistent, it is enough to assume that $x_1 > \Delta$, where $\Delta = 1/U$. Accordingly, there are the following three cases that need to be considered (1) $x_1 = \Delta$, (2) $x_1 > \Delta$ and no critical segment on machine 1, and (3) $x_1 > \Delta$ and a critical segment on machine 1. The critical paths and sublot sizes are, then, determined for each case.

5
m-Machine Lot Streaming Models

5.1 A Brief Overview

The m-machine lot streaming problem is more difficult than its two- and three-machine counterparts. Even though structural properties have been identified for some versions of this problem, yet it is not uncommon to find heuristic approaches that have been proposed for its solution.

One of the first efforts in this regard is the work of Baker and Pyke [4], who developed a two-sublot-based heuristic procedure for the streaming of a single lot to determine continuous and consistent sublot sizes so as to minimize makespan. The bottleneck machine plays a central role in this procedure. For the same problem, Glass and Potts [17] have used a network representation of the underlying problem for analysis, which is an extension of their work on the three-machine lot streaming problem (see Sect. 4.2), and they propose a relaxation algorithm for the determination of continuous, consistent, and optimal sublot sizes. Their algorithm also relies on the concept of dominant machines that helps in reducing problem size. The discrete version of this problem is addressed by Chen and Steiner [12]. Kalir and Sarin [23] address the same problem in the presence of sublot-attached setups and equal sublot sizes. Their problem essentially reduces to the determination of optimal number of sublots, n^*, since $s^* = (U/n^*)$. They develop an optimum seeking polynomial time algorithm that searches along the feasible region, which is convex and is defined by makespan functions corresponding to dominant machines.

Once we increase the number of lots to more than one, the issue of sequencing the lots also needs to be addressed besides determining the number of sublots of a lot and sublot sizes. This sequencing issue, while assuming known equal-size sublots for each lot, is addressed by Kalir and Sarin [22]. A heuristic procedure has been proposed, which attempts to maximize the time buffer prior to the bottleneck machine, thereby, minimizing potential bottleneck idleness, while also looking ahead to sequencing the lots with large remaining processing times early in the schedule.

The lot streaming problem in no-wait flow shops, in the presence of detached setups and no-intermingling constraint, has been addressed by Kumar et al. [26].

For the single batch problem, a linear programming formulation can be used to obtain continuous optimal sublot sizes, while integer sublot sizes can be obtained from the continuous optimal sublot sizes by using heuristic procedures. For the multiple batch case, the lot sequencing and sublot sizing problems are not independent of each other as in the two-machine case. A procedure in this regard is to sequence the lots by solving a traveling salesman problem (TSP) once the integer-sized sublots for each lot have been obtained (by using a procedure developed for the single batch case). A genetic algorithm (GA) for the simultaneous determination of sublot sizes and the sequence in which to process the lots has been developed by Kumar et al. [26]. However, it has been observed that, although the solution quality of GA-based algorithm is acceptable, its computational requirement is high.

Unlike the approaches above, where the objective is to find sublot sizes in order to minimize the makespan, the problem of minimizing the mean absolute deviation of lot completion times from due dates has also been addressed by Yoon and Ventura [39, 40]. For a given sequence of lots, linear programming formulations have been developed to obtain the optimal sublot sizes for each lot when buffers in between the successive machines have infinite or finite capacity, sublots are equal or consistent and when the flow shop is no-wait. The initial sequence is obtained by applying any of the following: EDD, smallest slack time on last machine, smallest overall slack time (OSL), or smallest overall weighted slack time. These sequences are improved upon by any of the following neighborhood search mechanisms: adjacent pairwise interchange of lots, nonadjacent pairwise interchange of lots (NAPI), and selecting two lots and moving a lot in front/back of the other lot. Experimental results show that the smallest OSL rule, in conjunction with the nonadjacent lot interchange heuristic, generates the best solutions. For the case of infinite buffer capacity and equal sublot sizes, a hybrid genetic algorithm (HGA) has been developed to work, in conjunction with OSL/NAPI, to obtain a better sequence.

The problem of minimizing the sum of the weighted sublot completion times has been addressed by Topaloglu et al. [34]. Two algorithms have been proposed for the m-machine single-lot problem involving consistent sublot sizes with only two sublots on each machine. The feasible region is convex and helps in designing these algorithms. One algorithm searches for the optimal solution by moving from one candidate point to an adjacent one, while the other finds the optimal solution by performing a bisection search-based procedure on the candidate optimal points.

The m-machine single-lot problem with consistent and equal sublot sizes to minimize a weighted sum of the makespan, (sublot) mean flow time, average WIP, sublot-attached setup, and transfer time has been addressed by Kalir et al. [25]. The problem essentially reduces to finding the optimal number of sublots n^*. The unified cost function has been shown to be a segmental strictly convex function. The proposed algorithm searches for the minimizing solution over each segment of the function. The optimum solution is identified if it occurs at an intersection point, else an approximate solution is determined based on a closed-form formula for n^*.

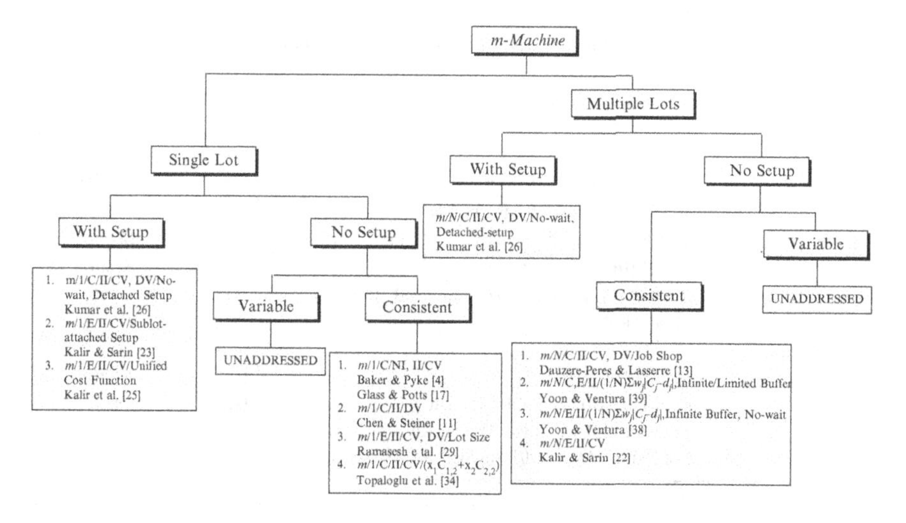

FIGURE 5.1. Overview of m-machine lot streaming models

A related problem that has been addressed by Ramasesh et al. [29] is to determine the optimal lot size, i.e., number of items in a lot used for processing an item having a constant, deterministic demand over an infinite time horizon. The objective is to minimize total cost that is made up of sublot transfer times and waiting or delay times, which are independent of lot size and setup time on each machine. The machines may have equal/unequal production rates but sublot sizes are equal and the number of sublots is assumed known. The problem essentially amounts to the determination of optimal sublot size.

An iterative heuristic, consisting of a sublot sizing and lot sequencing procedure for solving the n-job m-machine lot streaming problem in a job shop environment with no intermingling of sublots, has also been developed by Dauzere-Peres and Lasserre [13]. The intermingling of sublots is avoided by including precedence restrictions between sublots belonging to the same lot. The continuous sublot sizes are determined by using a linear programming formulation, and then, integer-sized sublots are obtained by using a rounding procedure. The sequencing procedure treats each sublot as a separate job and sequences them by using a modified shifting bottleneck procedure (SBP) presented in Dauzere-Peres and Lasserre [13, 14], which is an improved version of the original SBP of Adams et al. [1], so as to minimize the makespan.

A schematic representation of work reported on the m-machine lot streaming models is depicted in Fig. 5.1. Table 5.1 shows the problem status for the various m-machine single and multiple batch lot streaming problems. Table 5.2 gives a brief summary of the major results and/or solution methodologies used to solve these problems.

Next, we discuss each of the problems listed in Table 5.2.

TABLE 5.1. Problem status of m-machine lot streaming problems

No.	Problem	Problem type	Status
A1. Single lot			
1.	m/1/C/II/CV	$\Im^L(s)$	Section 5.2
2.	m/1/C/II/DV	$\Im^L(s)$	Section 5.3
3.	m/1/E/II/CV/Sublot-Attached Setup	$\Im^L(s)$	Section 5.4
4.	$m/1/C/II/CV/\sum_{i=1}^{2} x_i\ C_{im}$	$\Im^L(s)$	Section 5.6
5.	m/1/E/II/CV/Unified Cost Function	$\Im^L(s)$	Section 5.7
6.	m/1/V/NI/DV/Sublot Transfer and Removal Times, Release Times	$\Im^L(s)$	Open
A2. Multiple lots			
1.	m/N/E/II/CV	$\Im^L(s, S)$	Section 5.5
2.	m/N/CV/NI/DV/{Sublot Intermingling, Sublot Transfer and Removal Times, Release Times, Lot/Sublot-Attached/Detached Setup Times}	$\Im^L(s, S)$	Open

TABLE 5.2. Summary of major results/solution methodology for the m-machine lot streaming problems

No.	Problem	Major results/solution methodology	Complexity	
			$\Im^L(s)$	$\Im^L(S)$
1.	m/1/C/II/CV [4, 17]	The first algorithm analyzes the two-sublot case, which relies on the presence of a bottleneck machine and extends the result to the m-machine case. The second algorithm, based on a network representation, consists of a relaxation algorithm which identifies the dominant machines. Then, the critical path structure of the optimal solution is characterized, which is used to derive optimal sublot sizes.	$O\left(m + m'n^{m'-2}\right)$	NA
2.	m/1/C/II/DV [11]	The approximation algorithms use the continuous optimal solution to obtain integer-sized sublots. The first algorithm finds a sublot index where the sum of the differences between the rounded-up and continuous optimal sublot sizes equals the sum of the difference between the continuous optimal and rounded-down sublot sizes. All sublots before and including this sublot are rounded up and the remaining are rounded down. The second algorithm distributes the fractional parts in a more balanced manner.	$O(n)$, after the continuous optimal solution is obtained	NA
3.	m/1/E/II/CV/ Sublot-Attached Setup [23]	The problem reduces to finding the optimal number of sublots n^*, since $s^* = (U/n^*)$. The corresponding algorithm sequentially searches along the feasible region, which is shown to be convex.	$O(m)$	NA

4.	m/N/E/II/CV [22]	The sequencing heuristic attempts to maximize the time buffer prior to the bottleneck machine, thereby minimizing potential bottleneck idleness, while also looking ahead to sequence lots with large remaining processing times early in the schedule.	–	NA
5.	m/1/C/II/CV/ $\sum_{i=1}^{2} x_i C_{im}$ [34]	The feasible region of the problem under consideration is convex. The first algorithm searches for the optimum solution by moving from one candidate point to an adjacent one. The second algorithm finds the optimal solution by performing a bisection search-based procedure on the candidate optimal points.	$O(m^2)$	NA
6.	m/1/E/II/CV/ Unified Cost Function [25]	The problem reduces to finding the optimal number of sublots n^*, since $s^* = (U/n^*)$. The unified cost function $Z(n)$ consists of a weighted sum of the makespan, (sublot) mean flow time, average WIP, sublot-attached setup, and transfer time. $Z(n)$ is a segmental strictly convex function and the corresponding algorithm searches for the minimizing solution over each segment of $Z(n)$. It finds the optimum solution if it occurs at an intersection point; else, it finds an approximate solution based on a closed-form formula for n^*.	$O(m^2) + O(m)$	NA

5.2 $m/1/C/II/CV$

5.2.1 *Equal Sublot Heuristic*

The simplest situation for this problem would be the one in which all sublots are equal in size, i.e., $s_i = U/n, \forall i = 1, \ldots, n$. As an illustration, consider the following example.

Example 5.1 Consider a system consisting of four machines having processing times 4, 2, 6, and 8 on machines 1, 2, 3, and 4, respectively. A single lot consists of 14 items and the number of sublots is 5.

Since all sublots are of the same size, it can be obtained as $14/5 = 2.8$. The resulting makespan is 145.6.

The equal sublot heuristic ignores information about the processing times of the machines. Furthermore, the ratio of consecutive sublots is 1, which is restrictive since unequal sublot sizes might result in a lower makespan.

5.2.2 *Derived Two-Machine Problem Heuristic*

This heuristic is similar to the method of Campbell et al. [6] for a flow shop problem in which an m-machine problem is converted to several pseudo-two-machine problems with the processing time of a job on machine 1 to be the sum of its processing times on machines 1 to k, and that on the other machine to be equal to the sum of its processing times on machines $m - k + 1$ to m, where $k = 1, 2, \ldots, m - 1$. These pseudo-two-machine problems are then solved to obtain optimal makespan values, and the best among them is the desired solution. Accordingly, for the problem on hand, this heuristic creates a derived two-machine problem with processing times $P(1, k) \left(= \sum_{i=1}^{k} p_i\right)$ on machine 1 and $P(m - k + 1, m) \left(= \sum_{i=m-k+1}^{m} p_i\right)$ on machine 2. The value of k is varied from 1 to the largest integer less than or equal to $m/2$, i.e., $k = \lfloor m/2 \rfloor$. Once the processing times for the derived two-machine problem have been computed, the ratio $P(m - k + 1, m)/P(1, k)$ is calculated, and the sublot sizes are found such that the consecutive sublots are in the above ratio. The motivation for this allocation is that, in the two-machine lot streaming problem, the optimal sublot sizes are in the ratio $p_2 : p_1$. The algorithm is stated below.

Algorithm for the Derived Two-Machine Problem Heuristic

Step 1. Initialize by setting $k = 1$.
Step 2. Is $k \leq \lfloor m/2 \rfloor$?
If yes, then go to Step 3.
Else, go to Step 7.
Step 3. Create a derived two-machine problem and calculate the processing times on machine 1 and machine 2 as $P(1, k)$ and $P(m-k+1, m)$, respectively.
Step 4. Calculate the ratio for the sublot sizes as

$$Q = \frac{P(m - k + 1, m)}{P(1, k)}.$$

Step 5. Calculate the sublot sizes as

$$s_i = Q^{i-1} s_1,$$

where

$$s_1 = \left(\frac{1 - Q}{1 - Q^n}\right) U, \quad Q < 1$$
$$= \left(\frac{Q - 1}{Q^n - 1}\right) U, \quad Q > 1.$$

Step 6. Increment k, i.e., $k = k + 1$, and go to Step 2.
Step 7. The minimum makespan is given by $\min_{1 \leq k \leq \lfloor m/2 \rfloor} C_{\max}(k)$.

Example 5.2 Consider the data of Example 5.1. The algorithm is illustrated below.

Step 1. Let $k = 1$.
Step 2. Since $k < 2$, we go to Step 3.
Step 3. The derived problem has processing times $P(1, 1) = 4$ and $P(4, 4) = 8$.

Step 4. The ratio Q is 2.

Step 5. The sublot sizes are {0.45, 0.90, 1.80, 3.60, and 7.20} and the makespan is 117.42. Note that the sublot sizes add to 13.95 (instead of 14) due to rounding.

Step 6. The value of k is incremented to 2 and the above steps are repeated.

The derived problem now has processing times of 6 and 14. The ratio Q is 2.33 and the sublot sizes are {0.27, 0.64, 1.49, 3.48, and 8.12} and the makespan is 115.24. The value of k is incremented to 3 but the condition in Step 2 fails and the algorithm terminates. Hence, the final sublot sizes are {0.27, 0.64, 1.49, 3.48, and 8.12} and the makespan is 115.24. Note that this makespan value is significantly better than the one obtained using the equal sublot heuristic.

5.2.3 *Two-Sublot Heuristic*

In this heuristic, we first determine an optimal solution in the presence of only two sublots. To extend the method to the case when the number of sublots is greater than two, the ratio of consecutive sublot sizes is set to that obtained in the optimal solution for the two-sublot problem.

Let $P(k, \ell)$ denote the sum of the processing times per item on machines k to ℓ, i.e.,

$$P(k, \ell) = \sum_{u=k}^{l} p_u, \quad k = 1, \ldots, m, \quad l = k, \ldots, m.$$

Let the total amount of work be allocated between two sublots x_1 and x_2. To simplify the notation, we use a single variable x, where $x = x_1$ and $(1-x) = x_2$.

The optimal makespan for the m-machine two-sublot problem can be obtained as follows:

$$C_{\max} = \max_k \{(xP(1,k) + (1-x)P(k,m))\,U\}.$$

The machine for which the maximum value of makespan is obtained is referred to as the *critical machine* and both the sublots are processed without any idle time between them on this machine.

Characterizing the Optimal Value of *x*

We can rewrite the above makespan expression as follows:

$$C_{\max} = \max_k \{(x[P(1,k) - P(k,m)] + P(k,m))\,U\}. \tag{5.1}$$

Assume that machine k is critical. Then, for $\ell \neq k$, the following must hold:

$$\{x\,[P(1,l) - P(l,m)] + P(l,m)\} \leq \{x\,[P(1,k) - P(k,m)] + P(k,m)\}.$$

After collecting terms involving x, we have

$$x\,\{P(1,l) - P(l,m) - P(1,k) + P(k,m)\} \leq P(k,m) - P(l,m). \tag{5.2}$$

For $\ell < k$, the above expression simplifies to

$$x\{P(l+1,k)+P(l,k-1)\} \geq P(l,k-1),$$

or,

$$x \geq \frac{P(l,k-1)}{P(l+1,k)+P(l,k-1)}. \tag{5.3}$$

Similarly, for $\ell > k$, the expression simplifies to

$$x\{P(k+1,l)+P(k,l-1)\} \leq P(k,l-1),$$

or,

$$x \leq \frac{P(k,l-1)}{P(k+1,l)+P(k,l-1)}. \tag{5.4}$$

Expressions (5.3) and (5.4) must hold simultaneously for machine k to be critical. Therefore,

$$x_{\mathrm{L}} = \max_{l<k}\left\{\frac{P(l,k-1)}{P(l+1,k)+P(l,k-1)}\right\}$$

and

$$x_{\mathrm{U}} = \min_{l>k}\left\{\frac{P(k,l-1)}{P(k+1,l)+P(k,l-1)}\right\}.$$

In case $x_{\mathrm{L}} > x_{\mathrm{U}}$, then machine k cannot be critical. The optimal value of x in (5.1) depends on its coefficients in (5.1). If the coefficient is positive, then the makespan will be minimized by choosing the smallest value of x, i.e., $x = x_{\mathrm{L}}$. On the other hand, if the coefficient is negative, the makespan will be minimized by choosing x to be as large as possible, i.e., $x = x_{\mathrm{U}}$. Hence, if $P(1,k) - P(k,m) \geq 0$, then $x = x_{\mathrm{L}}$ and if $P(1,k) - P(k,m) < 0$, then $x = x_{\mathrm{U}}$.

The algorithm for calculating the sublot sizes for minimizing the makespan starts with the first machine (i.e., $k = 1$) and calculates x_{L} and x_{U}. If $x_{\mathrm{L}} > x_{\mathrm{U}}$ we move on to the next machine; else, we calculate the coefficient $P(1,k) - P(k,m)$ of x in (5.1). If the coefficient is greater than or equal to zero, then $x = x_{\mathrm{L}}$; else, $x = x_{\mathrm{U}}$. The makespan can be calculated by using (5.1). This procedure is repeated for all the machines, and the sublot sizes corresponding to the machine with the least makespan are optimal. The algorithm is stated below.

Algorithm for Optimal Sublot Sizes for the Two-Sublot, m-Machine Problem

Step 1. Let the first machine be critical, i.e., $k = 1$.

Step 2. If $k = m$, go to Step 4.

Else, calculate the values for x_{L} and x_{U} as below:

$$x_{\mathrm{L}} = \max_{l<k}\left\{\frac{P(l,k-1)}{P(l+1,k)+P(l,k-1)}\right\}$$

and

$$x_{\mathrm{U}} = \min_{l>k}\left\{\frac{P(k,l-1)}{P(k+1,l)+P(k,l-1)}\right\}.$$

Step 3.

If $x_L > x_U$, increment k, i.e., $k = k + 1$, and go to Step 2.
If $x_L \leq x_U$, calculate the coefficient as $\theta_k (= P(1,k) - P(k,m))$ of x.
If $\theta_k \geq 0, x = x_L$; else, $x = x_U$.
Calculate the makespan as

$$C_{\max}(k) = \{(x[P(1,k) - P(k,m)] + P(k,m))\, U\}.$$

Increment k and go to Step 2.

Step 4. Calculate x_L for $k = m, \theta_m = P(1,m) - P(m,m), x = x_L$, and $C_{\max}(m)$ for this x. Include this in the determination of $C^*_{\min}$ along with the other $C_{\max}(k)$ for which $x_{Lk} \leq x_{Uk}$. Output the minimal makespan and sublot sizes, i.e.,

$$C^*_{\min} = \min\{C_{\max}(k) : x_{Lk} \leq x_{Uk} \text{ for } k = 1, \ldots, m-1 \text{ and } C_{\max}(m)\}.$$

Example 5.3 Consider a system comprising of four machines with processing times per item of 4, 2, 6, and 4 on machines 1, 2, 3, and 4, respectively, and a single lot with 14 items.

The calculations are summarized below:

k	x_L	x_U	Skip	θ_k	x	$C_{\max}/U$
1	–	$\min\{2/3, 1/2, 3/7\} = 3/7$	N	−12	3/7	10.8571
2	$\max\{2/3\} = 2/3$	$\min\{1/4, 4/9\} = 1/4$	Y	–	–	–
3	$\max\{3/7, 1/4\} = 3/7$	$\min\{3/5\} = 3/5$	N	2	3/7	10.8571
4	$\max\{1/2, 4/9, 3/5\} = 3/5$	–	N	12	3/5	11.2

Hence, the optimal allocation of work is 3/7 and 4/7 to the first and second sublots, respectively, resulting in sublot sizes of 6 and 8. The corresponding makespan is 152.

Characteristics of the Optimal Solution

1. The equality in (5.2) implies that there exists a second critical machine on which no inserted idle time exists; call these critical machines c_1 and c_2.
2. The second sublot is processed continuously on machines c_1 to c_2. The time interval between the completions of the first sublot on machines c_1 and c_2 is given by $xP(c_1+1, c_2)$. For the second sublot, the time interval between the start times on machines c_1 and c_2 is $(1-x)P(c_1, c_2-1)$. For the above property to hold, these must be equal, i.e.,

$$xP(c_1+1, c_2) = (1-x)P(c_1, c_2-1).$$

3. In view of (2), the ratio $x/(1-x)$ of the sublot sizes is as follows:

$$\frac{x}{1-x} = \frac{P(c_1, c_2-1)}{P(c_1+1, c_2)}.$$

4. By expression (5.1), either machine c_1 or c_2 has the maximum processing time.

The last characteristic of the optimal solution can be utilized to simplify the solution procedure since we now need to find only one more critical machine. Further, if the coefficient for the first critical machine is positive, then $x = x_{\mathrm{L}}$. In order to find x_{L}, we only need to consider the machines with an index lower than the first critical machine. On the other hand, if the coefficient of the first critical machine is negative, then $x = x_{\mathrm{U}}$ and we need to consider all machines with higher indexes. The second critical machine is identified by the index giving x_{L} or x_{U}. The modified algorithm is given below.

Modified Algorithm for the Two-Sublot, m-Machine Problem

Step 1. Identify the machine k with the largest processing time per item. This is the first critical machine.

Step 2. Calculate its coefficient as

$$\theta_k = P(1, k-1) - P(k+1, m).$$

Step 3. If $\theta_k \geq 0$, calculate x' as follows:

$$x' = \max_{l<k} \left\{ \frac{P(l, k-1)}{P(l+1, k)} \right\}.$$

If $\theta_k < 0$, calculate x' as follows:

$$x' = \min_{l>k} \left\{ \frac{P(k, l-1)}{P(k+1, l)} \right\}.$$

The machine for which the maximum/minimum value of x' is obtained is the second critical machine.

Step 4. The optimal sublot sizes are given by

$$x = \left(\frac{x'}{1+x'} \right)$$

and

$$1 - x = \left(\frac{1}{1+x'} \right).$$

The minimum makespan is $\{(\theta_k x + P(k, m))U\}$.

Example 5.4 Consider the data as above in Example 5.3.

Step 1. Machine 3 has the largest processing time, and hence, is the first critical machine.

Step 2. Its coefficient is calculated as $\theta_3 = \{P(1, 2) - P(4, 4)\} = 2$.

Step 3. Since $\theta 3 > 0$,

$$x' = \max_{l<3} \left\{ \frac{P(l, 2)}{P(l+1, 3)} \right\} = \max\{3/4, 1/3\} = 3/4.$$

Since the maximum value is achieved for $\ell = 1$ (machine 1), it is the second critical machine.

Step 4. The optimal sublot sizes are 6 and 8 and the optimal makespan is 152.

Let E denote the relative error and be defined as below

$$E = \frac{C^{\mathrm{H}}_{\max} - C^{*}_{\max}}{C^{*}_{\max}},$$

where $C^{\mathrm{H}}_{\max}$ is the makespan obtained by the heuristic procedure, and $C^{*}_{\max}$ the optimal makespan obtained by the linear programming solution.

Some experimentation on the use of these heuristics has been presented by Baker and Pyke [4]. It is shown that the two-sublot heuristic performs better than the equal sublot heuristic and the derived two-machine heuristic. Moreover, the derived two-machine heuristic turns out to be not as effective in this case as it is in the traditional flow shop environment.

All of the above heuristics obtain sublot sizes from a single ratio; however, the optimal sublot sizes need not follow a single ratio. The procedure below relaxes this assumption.

5.2.4 *Different Sublot Ratio Algorithm*

This algorithm builds a schedule for the m-machine problem by using the optimal solution to the two-sublot, m-machine problem as an initial solution. At each iteration, it finds the optimal size of an additional sublot i, referred to as the *last sublot,* while maintaining the relative sizes of the previous sublots.

We first discuss a procedure for sizing the last sublot, and then, illustrate the working of the overall algorithm through an example.

Sizing the Last Sublot

We have U items that are to be allocated to the first $(n-1)$ sublots and the nth sublot. Let $C_{i,k}$ depict the completion time of the ith sublot on machine k when $U = 1$. The completion time of each sublot and the makespan are directly proportional to the total number of items. Thus, if the size of the nth sublot is x, then the $(n-1)$ sublots are constituted from the remaining lot size of $(1-x)$. Consequently, the completion time of the ith sublot on machine k reduces to $(1-x)C_{ik}$. For any machine k, the makespan M of the schedule must satisfy the following relation:

$$M \geq ((1-x)C_{n-1k} + xP(k,m))\,U.$$

This can be written as

$$M \geq \left(C_{n-1k} + x\left[P(k,m) - C_{n-1k}\right]\right) U.$$

The maximum among the values on the right side will determine the makespan, i.e.,

$$M = \max_{1 \leq k \leq m} \left\{(C_{n-1k} + x\,(P(k,m) - C_{n-1k}))\,U\right\}.$$

The optimal size of the nth sublot would be the minimum of all possible values of x, i.e.,

$$M = \min_{0 \leq x < 1} \left\{ \max_{1 \leq k \leq m} \left(C_{n-1k} + x \left[P(k,m) - C_{n-1k} \right] \right) U \right\}. \tag{5.5}$$

The term $C_{n-1,k} + x[P(k,m) - C_{n-1,k}]$ represents the equation of a straight line with intercept $C_{n-1,k}$ and slope $P(k,m) - C_{n-1,k}$. The intersection point of lines corresponding to any two machines can be obtained as

$$C_{n-1k} + x \left[P(k,m) - C_{n-1k} \right] = C_{n-1\ell} + x \left[P(\ell,m) - C_{n-1\ell} \right].$$

After simplification,

$$x = \frac{C_{n-1l} - C_{n-1k}}{P(k,m) - P(l,m) + C_{n-1l} - C_{n-1k}}.$$

In (5.5), the maximum over all machines lies on the upper envelope formed by the intersection of m lines. The minimum point on this upper envelope corresponds to the minimum value of x. The algorithm first applies the two-sublot, m-machine algorithm to obtain the completion times of the second sublot on each machine. It then calculates the intersection points of lines corresponding to a given machine k and machine $\ell, \ell = (k+1)$ to m. The maximum of these points $x^k_{\max}$ may or may not lie on the outer envelope. It will lie on the outer envelop if $x^k_{\max} < x^{k-1}_{\max}$, and if so, we include it in X^{OE}, the set containing the points lying on the outer envelope. The minimum makespan can be obtained by identifying the first $x^k_{\max}$, such that $M\left(x^k_{\max}\right) < M\left(x^{k+1}_{\max}\right)$. The algorithm is stated below.

Algorithm for Sizing the Last Sublot

Step 1. Use the two-sublot, m-machine algorithm to obtain the optimal solution for the two-sublot problem.

Step 2. For $k = 1$ to $(m-1)$

For $\ell = (k+1)$ to m

Calculate the intersection points as below

$$x^{k,l} = \frac{C_{n-1l} - C_{n-1k}}{P(k,m) - P(l,m) + C_{n-1l} - C_{n-1k}},$$

where

$$C_{i-1\ell} = C_{2\ell}$$

and

$$C_{i-1k} = C_{2k}.$$

END

Find the maximum value of $x^{k,l}$ as

$$x_{\max}^{k} = \max_{k+1 \le l \le m} \left\{ x^{k,l} \right\}.$$

END

Step 3. Identify points lying on the outer envelope as

$$X^{\mathrm{OE}} = \left\{ x_{\max}^{k} : x_{\max}^{k} < x_{\max}^{k-1}, \forall k = 1, \ldots, (m-1) \right\},$$

where $x_{\max}^{0} = 1$.

Step 4. Calculate the minimum makespan as

$$M^{*}\left(x_{\max}^{k}\right) = \min_{1 \le k \le m-1} \left\{ \left(C_{i-1k} + x_{\max}^{k} \left[P(k,m) - C_{i-1k} \right] \right) U : x_{\max}^{k} \in X^{\mathrm{OE}} \right\}.$$

The above algorithm can be embedded in an iterative procedure for the m-machine, n-sublot version of the problem, which is presented below.

Different Sublot Ratio Algorithm

Step 1. Obtain the two-sublot solution by the application of the "two-sublot, m-machine algorithm."

Step 2. For $i = 3$ to n

Apply the "last sublot sizing algorithm" to find the size of sublot i, while maintaining the relative sizes of the previous $(i-1)$ sublots.

END

Next, we illustrate the working of both the algorithms discussed above.

Example 5.5 Let the processing times on machines 1, 2, 3, and 4 be 4, 2, 6, and 4, respectively, and there are 14 items in a lot and the number of sublots is 5.

Step 1. Application of the m-machine two-sublot algorithm results in the allocation of work of 3/7 and 4/7 to the first two sublots and sublot sizes of 6 and 8, as noted earlier (in Example 5.3). The completion times of the second sublot on each machine $C_{n-1,k}$ (obtained by using (5.1) for $m = 1, 2, 3,$ and 4) and $P(k, m)$, the total processing time on machine k to m, are given below:

	Machine 1	Machine 2	Machine 3	Machine 4
$P(k,m)$	16	12	10	4
$C_{n-1,k}$	4	5.143	8.571	10.857

Step 2. We now use the *last sublot sizing algorithm* to size sublots 3, 4, and 5.

Sizing of the Third Sublot

The calculations are summarized below:

	$x^{k\ell}$				$\max\{x^{k\ell}\}$	X^{OE}	Makespan
	M1	M2	M3	M4			
M1	–	0.222	0.423	0.364	0.423	Yes	9.184
M2	–	–	0.632	0.417	0.632	No	–
M3	–	–	–	0.276	0.276	Yes	8.965

Accordingly, we allocate $(1 - 0.276) = 0.724$ to the first two sublots with relative sizes of 3/7 and 4/7, and 0.276 to the third sublot. Thus, the proportion of work allocated is 0.311 $(= 0.724 \times 3/7)$, 0.413 $(= 0.724 \times 4/7)$, and 0.276 resulting in sublot sizes of 4.354, 5.782, and 3.864.

Sizing of the Fourth Sublot

The completion times of the third sublot on each machine and $P(k, m)$ are given below:

	Machine 1	Machine 2	Machine 3	Machine 4
$P(k, m)$	16	12	10	4
$C_{n-1,k}$	4	4.552	7.866	8.97

The calculations for the size of the fourth sublot are summarized below:

	$x^{k\ell}$				$\max\{x^{k\ell}\}$	X^{OE}	Makespan
	M1	M2	M3	M4			
M1	–	0.121	0.392	0.293	0.392	Yes	8.704
M2	–	–	0.624	0.356	0.624	No	–
M3	–	–	–	0.155	0.155	Yes	8.197

We allocate $(1 - 0.155) = 0.845$ to the first three sublots with relative sizes of 0.311, 0.413, and 0.276, and 0.155 to the fourth sublot. Thus, the proportion of work allocated is 0.263, 0.349, 0.233, and 0.155.

Sizing of the Fifth Sublot

The completion times of the fourth sublot on each machine and $P(k, m)$ are given below:

	Machine 1	Machine 2	Machine 3	Machine 4
$P(k, m)$	16	12	10	4
$C_{n-1,k}$	4	4.31	7.578	8.2

The calculations for the size of the fifth sublot are summarized below:

	$x^{k\ell}$				$\max\{x^{k\ell}\}$	X^{OE}	Makespan
	M1	M2	M3	M4			
M1	–	0.072	0.374	0.259	0.374	Yes	8.488
M2	–	–	0.620	0.327	0.620	No	–
M3	–	–	–	0.094	0.094	Yes	7.806

This results of the allocation of $(1 - 0.094) = 0.906$ to the first four sublots with relative sizes of 0.263, 0.349, 0.233, and 0.155, and 0.094 to the fifth sublot. Thus, the proportion of work allocated is 0.238, 0.316, 0.14, 0.211, and 0.094.

This algorithm has been shown to perform quite well [4]. However, the performance of this algorithm can be improved further in the presence of multiple critical machines. As we indicated earlier, for the example problem on hand, the solution of the two-sublot problem consists of two critical machines. Since the critical machines affect the makespan considerably, we could apply the above algorithm to a subset of machines. Let k and ℓ denote the indices of the critical machines and $b = \max(k, \ell)$. The modified algorithm applies the last sublot sizing algorithm from machine 1 to b at every stage except at the last stage, when it is applied to all the machines. This algorithm requires fewer computations since the number of intersection points to be calculated decreases from $n(n-1)/2$ to $b(b-1)/2$. A still better performance, at the expense of computational time, can be obtained by applying the modified heuristic described above to both the original and reverse problems and choosing the better of the two schedules.

5.2.5 *Two-Stage, Network Representation-Based Algorithm*

We now discuss a solution procedure to find optimal sublot sizes for the above problem, based on a network representation of the problem. It is an extension of the network representation used for the three-machine lot streaming problem (see Sect. 4.2). Also, it exploits the concept of machine dominance that is well known in the traditional flow shop literature [27]. In particular, in that context, a machine i is said to dominate machine $i+1$ if

$$\min_{j=1,\ldots,n} p_{ij} \geq \max_{j=1,\ldots,n} p_{i+1j},$$

where, recall that, p_{ij} represents the processing time of job j on machine i. Machine i is termed dominant and machine $i+1$ the dominated machines. The key consequence of this fact is that the dominated machine $i+1$ does not contribute to the makespan. In general, if machines $i+1, i+2, \ldots, i+\ell$ are dominated machines, then all of these machines do not contribute to the makespan, and, as a result, could be substituted by a single pseudo-machine on which the processing time of a job is equal to the sum of its processing times on machines $i+1, \ldots, i+\ell$. In the lot streaming context on hand, if $m = 3$, machine 2 is said to be dominated if $p_2 \leq p_1 p_3$ (subscript j is suppressed for being items of the same lot), and machine 2 does not contribute to the makespan. We observed this fact for the three-machine lot streaming problem (see Sect. 4.2). Interestingly, just as in the case of the traditional flow shop scheduling problem, this dominance can be extended to the m-machine lot streaming problem as well. Consequently, it helps in reducing the problem size and the corresponding network representation that we need to analyze in order to determine optimal sublot sizes. Thus, this method consists of two stages. In the first stage, a relaxation algorithm is used to identify the dominant machines. The second stage is an iterative process in which, at each iteration, the problem is solved to optimality taking into account only the dominant machines. The sublot sizes thus obtained are optimal for the original problem as well. We describe this method next.

Machine Dominance. Any machine ℓ is said to be dominated by machine k and machine $m(k \leq \ell \leq m)$ if the following condition holds:

$$\left(\sum_{u=k+1}^{\ell} p_u\right)\left(\sum_{v=\ell}^{m-1} p_v\right) \leq \left(\sum_{u=k}^{\ell-1} p_u\right)\left(\sum_{v=\ell+1}^{m} p_v\right).$$

Note that if $k+1=\ell$ and $m=\ell+1$, then the above condition reduces to $p_\ell^2 \leq p_k p_m$. In case, we have $m=3$, i.e., a three-machine flow shop problem, then the above condition becomes $p_2^2 \leq p_1 p_3$, which is the same dominance condition that was mentioned in Chap. 4 (see Sect. 4.2) for the three-machine flow shop problem. So, we can view the above dominance condition as a generalization of the dominance condition of the three-machine problem. We denote this as $k \succ \ell \prec m$. If the above condition is violated, then machine ℓ is not dominated by machines k and m and we denote it as $k \nsucc \ell \nprec m$. The above definition implies that machines k and m are dominant. The following result relates the above definition to the length of a critical path in the corresponding network representation.

Theorem 5.1 (Glass and Potts [17]) *If machine ℓ is dominated by machines k and $m(k < \ell < m)$, then for any sublot sizes of sublots i and $j, i \leq j, i, j \in \{1, 2, \ldots, n\}$, the length of the path $(i,k)-(j,k)-(j,\ell)-(j,m)$ or $(i,k)-(i,\ell)-(i,m)-(j,m)$ is at least as long as that of the path $(i,k)-(i,l)-(j,\ell)-(j,m)$.*

An implication of this theorem is that there will always exist a critical path which will not contain any arc $(i,\ell)-(i+1,\ell), i=1,\ldots,n-1$, when machine ℓ is dominated by machines $(\ell-1)$ and $(\ell+1)$. The dominance property is transitive [17]. That is, if machine $\ell+1$ is dominated by machines ℓ and $\ell+2$, and machine $\ell+2$ is dominated by machines $\ell+1$ and $\ell+3$, then both machines $\ell+1$ and $\ell+2$ are dominated by machines ℓ and $\ell+3$. This permits us to collapse all dominated machines between two dominance machines into a single (delay) machine on which the processing time of a sublot is equal to the sum of its processing times on all dominated machines. This is accomplished by using a *relaxation algorithm*, which converts the original flow shop problem Γ into a relaxed problem $\mathfrak{R}$. This relaxed problem consists of $m^{\mathfrak{R}}$ machines, each having a processing time per item of $p_{k^{\mathfrak{R}}}, k^{\mathfrak{R}} = 1, \ldots, m^{\mathfrak{R}}$. Between any two consecutive machines $k^{\Re}$ and $k^{\Re}+1$ in $\Re$, there exists a delay machine with processing time of $d_{k^{\Re}}$, $k^{\Re}=1,\ldots,m^{\Re}-1$, which represents the minimum time that must elapse between the completion of any sublot on machine $k^{\Re}$ and start of its processing on machine $k^{\Re}+1$. Each machine $k^{\Re}(k^{\Re}=1,\ldots,m^{\Re})$ in $\Re$ is associated with a machine $\varphi_{k^{\Re}}$ in Γ, where $1=\varphi_{1^{\Re}}<\cdots<\varphi_{m^{\Re}}=m$. We refer to machines $\varphi_{1^{\Re}},\ldots,\varphi_{m^{\Re}}$ as capacitated machines. Thus, $p_{k^{\Re}}=p_{\varphi_{k^{\Re}}}$ $(k^{\Re}=1,\ldots,m^{\Re})$ and the processing times $d_{k^{\Re}}$ of the delay machines, as noted earlier, are given by $\sum_{\nu=\varphi_{k^{\Re}}+1}^{\varphi_{k^{\Re}+1}-1}p_{\nu}$ $(k^{\Re}=1,\ldots,m^{\Re}-1)$, which correspond to the processing times between machines $\varphi_{k^{\Re}}$ and $\varphi_{k^{\Re}+1}$. If $\varphi_{k^{\Re}}+1=\varphi_{k^{\Re}+1}$, then we insert a dummy machine in the original problem Γ (between machine $\varphi_{k^{\Re}}$ and $\varphi_{k^{\Re}+1}$) with zero processing time.

Thus, the relaxed problem consists of capacitated (or dominant) machines and delay machines. The capacitated machines are identical to machines in the original problem Γ while the delay machines (appearing between any pair of capacitated machines) represent an aggregation of machines in the original problem.

The relaxation algorithm presented below is basically an iterative procedure that identifies the capacitated machines. It begins by creating an initial relaxed problem in which all machines are assumed to be capacitated machines. The machine dominance condition is checked to determine a machine that is dominated by two neighboring nondelay machines. If a dominant machine is found, it is removed from further consideration and its processing time is added to the delay time. This procedure is repeated until no further dominant machine can be identified.

Relaxation Algorithm for Identifying Capacitated Machines in an m-Machine Flow Shop

Step 1. Construct an initial relaxed problem in which the number of capacitated machines is equal to the number of machines in the original problem and delay machines, with zero processing times, are inserted between every pair of adjacent machines.

$$\begin{aligned}
&m^{\Re}=m,\\
&\varphi_{k^{\Re}}=k,\quad \forall k=1,\ldots,m,\\
&d_{k^{\Re}}=0,\quad \forall k^{\Re}=1,\ldots,m^{\Re}-1,\\
&p_{\varphi_{k^{\Re}}}=p_k,\quad \forall k=1,\ldots,m,\\
&k^{\Re}=2.
\end{aligned}$$

Step 2. Check for dominance by verifying the condition below:

$$\left(d_{k^{\Re}-1}+p_{k^{\Re}}\right)\left(p_{k^{\Re}}+d_{k^{\Re}}\right)\leq\left(p_{k^{\Re}-1}+d_{k^{\Re}-1}\right)\left(d_{k^{\Re}}+p_{k^{\Re}+1}\right).$$

Step 3. If the condition is satisfied, i.e., machine $k^{\Re}$ is dominated, then combine its processing time $p_{k^{\Re}}$ and the processing time of the associated delay machine $d_{k^{\Re}}$ with that of the previous delay machine $d_{k^{\Re}-1}$

$$
\begin{aligned}
&d_{k^{\Re}-1} = d_{k^{\Re}-1} + p_{k^{\Re}} + d_{k^{\Re}},\\
&\varphi_{\ell^{\Re}} = \varphi_{\ell^{\Re}+1}, \quad \forall \ell^{\Re} = k^{\Re}, \ldots, m^{\Re} - 1,\\
&d_{\ell^{\Re}} = d_{\ell^{\Re}+1}, \quad \forall \ell^{\Re} = k^{\Re}, \ldots, m^{\Re} - 2,\\
&p_{\ell^{\Re}} = p_{\ell^{\Re}+1}, \quad \forall \ell^{\Re} = k^{\Re}, \ldots, m^{\Re} - 1,\\
&m^{\Re} = m^{\Re} - 1,\\
&k^{\Re} = \max\{1, k^{\Re} - 2\}.
\end{aligned}
$$

Else
Go to Step 4.

Step 4. If $k^{\Re} = m^{\Re} - 1$, then STOP.
Else, increment $k^{\Re}$, i.e., $k^{\Re} = k^{\Re} + 1$, and go to Step 2.

Example 5.6 Consider a five-machine flow shop with processing time per item of 4, 2, 6, 4, and 5, respectively. The number of sublots is 5 and there are 20 items in the lot. The working of the algorithm is illustrated below.

Step 1. The initialization step sets $m^{\Re} = 5$, $p_{\varphi_{1^{\Re}}} = 4$, $d_{1^{\Re}} = 0$, $p_{\varphi_{2^{\Re}}} = 2$, $d_{2^{\Re}} = 0$, $p_{\varphi_{3^{\Re}}} = 6$, $d_{3^{\Re}} = 0$, $p_{\varphi_{4^{\Re}}} = 4$, $d_{4^{\Re}} = 0$, $p_{\varphi_{5^{\Re}}} = 5$, and $k^{\Re} = 2$.

Step 2. Check for machine dominance, i.e.,

$$(0+2)(2+0) \leq (4+0)(0+6),$$

or,

$$4 \leq 24.$$

Step 3. Since machine $2^{\Re}$ is dominated, we combine its processing time and that of the associated delay machine with the delay time of the previous delay machine to obtain $d_{1^{\Re}} = 0+2+0 = 2$. Renumber the remaining machines as below:

$$
\begin{aligned}
\varphi_{2^{\Re}} &= \varphi_{3^{\Re}},\\
\varphi_{3^{\Re}} &= \varphi_{4^{\Re}},\\
\varphi_{4^{\Re}} &= \varphi_{5^{\Re}},\\
d_{2^{\Re}} &= d_{3^{\Re}},\\
d_{3^{\Re}} &= d_{4^{\Re}},\\
m^{\Re} &= 4,
\end{aligned}
$$

and

$$k^{\Re} = 1.$$

Step 4. Increment $k^{\Re}$ to 2.

We now check for dominance again. Since $48 \not\leq 24$, machine $2^{\Re}$ (machine 3 in Γ) is not dominated. We increment $k^{\Re}$ to 3 and check for dominance. Since $16 \leq 30$, machine $3^{\Re}$ (machine 4 in Γ) is dominated by machine $2^{\Re}$ (machine 3 in Γ) and machine $4^{\Re}$ (machine 5 in Γ), and we merge both its processing time and that of the associated delay machine with that of the previous delay machine to obtain $d_{2^{\Re}} = 4$. We also set $\varphi_{3^{\Re}} = \varphi_{4^{\Re}}$, $m^{\Re} = 3$, and $k^{\Re} = 1$. We increment $k^{\Re}$ to 2 and check for dominance. Since $80 \not\leq 54$, machine $2^{\Re}$ (machine 3 in Γ) is not dominated by machine 1 and machine $3^{\Re}$ (machine 5 in Γ). Since $k^{\Re} = m^{\Re} - 1$, the algorithm terminates. The relationship between the original problem Γ and the relaxed problem $\Re$ for the above example is shown in Fig. 5.2.

It can be shown [17] that, in the original problem Γ, any delay machine is dominated by an adjacent pair of capacitated machines and that all capacitated machines are dominant machines. Also, there exists a critical path in Γ, which does not contain any arc of a delay machine created by the relaxation algorithm. Finally, it can be shown that the optimal sublot sizes for $\Re$ are also optimal for Γ and they result in identical makespan for the two problems.

The optimal sublot sizes for the relaxed problem have a distinctive block structure (like that shown in Fig. 3.3). The corresponding network representation consists of blocks $B(k^{\Re}, h_{k^{\Re}}; k^{\Re} + 1, h_{k^{\Re}+1})$, $1 \leq k^{\Re} \leq m^{\Re} - 1$, where $h_{k^{\Re}}$ are integers such that $1 = h_{1^{\Re}} \leq \cdots \leq h_{m^{\Re}} = n$. The ratio of sublot sizes between any two consecutive integers is as below

$$\frac{x_i}{x_{i+1}} = \frac{p_{k^{\Re}} + d_{k^{\Re}}}{p_{k^{\Re}+1} + d_{k^{\Re}}}, \quad \forall i = h_{k^{\Re}}, \ldots, h_{k^{\Re}+1}, \quad \forall k^{\Re} = 1, \ldots, m^{\Re} - 1.$$

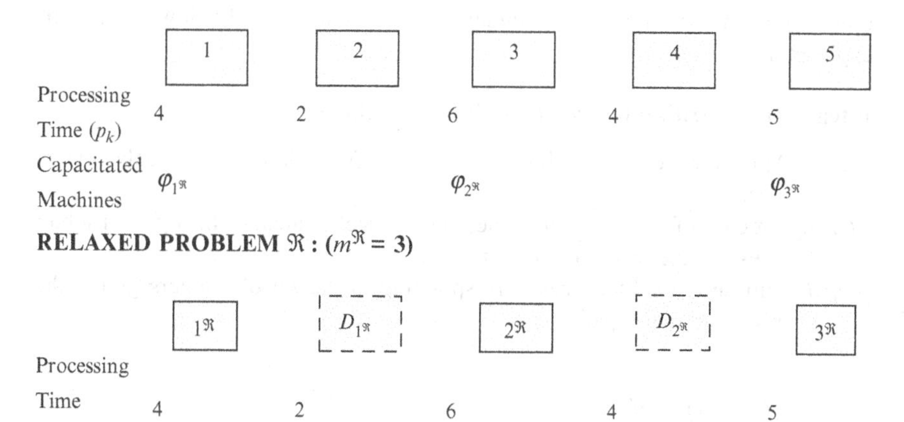

FIGURE 5.2. The original problem, and the relaxed problem created by the relaxation algorithm

Therefore,

$$x_i = \rho_{k^{\Re}} x_{i+1}, \quad \forall i = h_{k^{\Re}}, \ldots, h_{k^{\Re}+1} - 1,$$
$$\rho_{k^{\Re}} = \frac{p_{k^{\Re}} + d_{k^{\Re}}}{p_{k^{\Re}+1} + d_{k^{\Re}}}, \quad \forall k^{\Re} = 1, \ldots, m^{\Re} - 1.$$

Therefore,

$$x_{h_{k^{\Re}}} = \rho_{k^{\Re}}^{h_{k^{\Re}+1} - h_{k^{\Re}}} x_{h_{k^{\Re}+1}}, \quad \forall k^{\Re} = 1, \ldots, m^{\Re} - 1.$$

Since

$$\sum_{i=1}^{n} x_i = 1,$$

$$\sum_{k^{\Re}=2}^{m^{\Re}} \sum_{i=h_{k^{\Re}-1}}^{h_{k^{\Re}}-1} x_1 + x_n = 1,$$

or,

$$\sum_{k^{\Re}=2}^{m^{\Re}} \sum_{i=0}^{h_{k^{\Re}} - h_{k^{\Re}-1}} \rho_{k^{\Re}-1}^{i} x_{h_{k^{\Re}}} + x_n = 1.$$

The corresponding makespan can be calculated as

$$C_{\max} = \sum_{k^{\Re}=2}^{m^{\Re}} \left(p_{k^{\Re}-1} \sum_{i=0}^{h_{k^{\Re}} - h_{k^{\Re}-1}} \rho_{k^{\Re}-1}^{i} + d_{k^{\Re}-1} \right) x_{h_{k^{\Re}}} + p_{m^{\Re}} x_n.$$

Hence, the sublot sizes can be calculated for a given set of integers $h_{1^{\Re}}, \ldots, h_{m^{\Re}}$. For any problem, there are a maximum of $n^{m^{\Re}-2}$ sets of integers. The optimal makespan corresponds to the set, which gives the minimum makespan. The integrated algorithm for the problem under consideration is stated below and is self-explanatory.

Integrated Algorithm for the *m*/1/C/II/CV Problem

Step 1. Apply the "relaxation algorithm" to identify the dominant and delay machines.

Step 2. For each of the $n^{m^{\Re}-2}$ feasible set of integers, calculate the optimal sublot sizes and the resulting makespan.

Step 3. Choose the sublot sizes corresponding to the set of integers giving the minimum makespan.

5.3 *m*/1/C/II/DV

We now discuss the discrete version of the above problem and present two approximation procedures, which obtain integer sublot sizes by rounding the optimal continuous sublot sizes. These procedures differ in the manner they allocate the fractional part of the continuous sublot sizes.

5.3.1 *Approximate Rounding Procedure*

Step 1. Solve the continuous version of the problem and obtain the continuous optimal sublot sizes.

Step 2. Calculate the difference υ between the lot size and the rounded-down continuous optimal sublot sizes as follows

$$\upsilon = U - \sum_{i=1}^{N} \left\lfloor s_i^{\mathrm{C}} \right\rfloor,$$

where s_i^{C} is the continuous optimal size of sublot. The ν so computed gives the number of continuous-size sublots which when rounded up, the sum of the differences of their rounded up and actual values is equal to the sum of the differences of the rounded-down values of the remaining continuous-sized sublots from their actual values.

Step 3. For the first υ noninteger sublots in the continuous optimal solution, calculate the integer sizes by rounding up the continuous values. For sublots $(\upsilon + 1)$ to n, the integer sizes are given by rounding down the continuous values, i.e.,

$$s_i^{\mathrm{INT}} = \left\lceil s_i^{\mathrm{C}} \right\rceil, \quad i = 1, \ldots, \upsilon,$$

$$s_i^{\mathrm{INT}} = \left\lfloor s_i^{\mathrm{C}} \right\rfloor, \quad i = (\upsilon + 1), \ldots, n.$$

Example 5.7 Assume that the optimal continuous sizes are

$$s^{\mathrm{C}} = \{10.5, 10, 12.4, 19.6, 5.6, 1.9\}.$$

Therefore,

$$\upsilon = 60 - (10 + 10 + 12 + 19 + 5 + 1) = 3.$$

Hence, we round up the first three noninteger sublot sizes from their continuous and optimal values, and round down the remaining continuous and optimal sublot sizes. In this case, since the second sublot is integer, we do not consider it. Therefore,

$$s^{\mathrm{INT}} = \{11, 10, 13, 20, 5, 1\}.$$

It can be shown [11] that the makespan of this approximation is within $\left\{\sum_{k=2}^{m} p_k + (p_{\max} - p_{\min})(n/2)\right\}$ of the optimal makespan.

5.3.2 *Balanced Re-distribution Procedure*

Step 1. Solve the continuous version of the problem to obtain (continuous) optimal sublot sizes.

Step 2. Calculate the fractional parts as below

$$\Delta_i = s_i^{\mathrm{C}} - \left\lfloor s_i^{\mathrm{C}} \right\rfloor, \quad i = 1, \ldots, n.$$

Step 3. If $\Delta_i = 0$, for any $i = 1, \ldots, n$, i.e., the optimal continuous sublot size is integer, then $s_i^{\mathrm{INT}} = s_i^{\mathrm{C}}$. Else, go to Step 4.

Step 4a. Find the first noninteger sublot and denote it by i.
Step 4b. Find a sublot $k \geq i$, such that the sum of the fractional parts exceeds 1 when sublot $(k+1)$ is included, i.e.,

$$\sum_{j=i}^{k} \Delta_j \leq 1 < \sum_{j=i}^{k+1} \Delta_j.$$

Step 4c. Round up sublot i, and round down sublots $(i+1), \ldots, k$, i.e.,

$$s_i^{\text{INT}} = \left\lceil s_i^{\text{C}} \right\rceil,$$
$$s_j^{\text{INT}} = \left\lfloor s_j^{\text{C}} \right\rfloor, \quad j = (i+1), \ldots, k.$$

Step 4d. Adjust the size of sublot $(k+1)$ to take into account the consequence of the net change in Step 4c above, as follows:

$$s_{k+1}^{\text{C}} = s_{k+1}^{\text{C}} - \left\{ (1 - \Delta_i) - \sum_{j=i+1}^{k} \Delta_j \right\}.$$

Step 5. If all the sublots are integers, Stop; else, go to Step 4a.

Example 5.8 Consider the above example where $s^{\text{C}} = \{10.5, 10, 12.4, 19.6, 5.6, 1.9\}$.

Step 2. The fractional parts are $\Delta = \{0.5, 0, 0.4, 0.6, 0.6, 0.9\}$.
Step 3. Since the second sublot is integer, we let $s_2^{\text{INT}} = 10$.
Step 4a. The first noninteger sublot is $i = 1$.
Step 4b. $k = 3$, since $0.9 \leq 1 < 1.5$.
Step 4c. $s_1^{\text{INT}} = 11$, $s_2^{\text{INT}} = 10$, and $s_3^{\text{INT}} = 12$.
Step 4d. $s_4^{\text{C}} = s_4^{\text{C}} - \{(1 - 0.5) - (\Delta_2 + \Delta_3)\} = 19.5$.

The resultant sublot sizes are $s^{\text{C}} = \{11, 10, 12, 19.5, 5.6, 1.9\}$.

Now, $i = 4$ and $k = 4$, since $0.5 \leq 1 < 1.1$. Hence, $s_4^{\text{INT}} = 20$ and $s_5 = 5.1$. The sublot sizes now become $s^{\text{C}} = \{11, 10, 12, 20, 5.1, 1.9\}$. Next, $i = 5, k = 6$, since $1 \leq 1$. Hence, $s_5^{\text{INT}} = 6$ and $s_6^{\text{INT}} = 1$. Hence, the integer sublot sizes are $\{11, 10, 12, 20, 6, 1\}$.

The makespan of the balanced redistribution procedure can be shown [9] to lie within $\sum_{i=1}^{m-1} p_i$ of the optimal makespan.

5.4 m/1/E/II/CV/Sublot-Attached Setup

We now discuss the problem of minimizing the makespan in an m-machine flow shop, consisting of a single lot and sublot-attached setups. We assume equal and continuous sublot sizes to enable development of an analytical framework. Equal sublot sizes can be obtained as

$$s_i^* = \frac{U}{n^*}, \quad \forall i = 1, \ldots, n^*.$$

Thus, the problem under consideration, essentially, reduces to the determination of optimal number of sublots n^*. We first present an expression for the makespan when setups are sublot-attached, and then, develop some structural properties, which provide the necessary background for the sublot sizing algorithm.

As an illustration, consider a three-machine flow shop with a single lot consisting of $12(U)$ items, with $3(n)$ equal-size sublots of size 4. The processing times (p_k) are 2, 3, and 2, and sublot-attached setup times (τ_k) are 3, 2, and 1 units on machines 1, 2, and 3, respectively. The Gantt chart for this schedule is as shown in Fig. 5.3. Note that the $\max_{1 \le k \le 3} \{(U/n)p_k + \tau_k\}$ for this problem is obtained for $k = 2$.

The makespan expression for the problem on hand is as follows:

$$C_{\max}(n) = \left\{ \frac{U}{n} \sum_{\ell=1}^{m} p_\ell + \sum_{\ell=1}^{m} \tau_\ell \right\} + (n-1) \max_{1 \le k \le m} \left\{ \frac{U}{n} p_k + \tau_k \right\}. \tag{5.6}$$

Since the first two terms are independent of the machine index, the above expression can be written as

$$C_{\max}(n) = \max_{1 \le k \le m} \left\{ \frac{U}{n} \sum_{\ell=1}^{m} p_\ell + \sum_{\ell=1}^{m} \tau_\ell + (n-1) \left(\frac{U}{n} p_k + \tau_k \right) \right\}.$$

Let $C_{\max_k}(n)$ denote the expression within the brackets, which is the workload on machine k. Then, $C_{\max}(n)$ can be written as

$$C_{\max}(n) = \max_{1 \le k \le m} \left\{ C_{\max_k}(n) \right\},$$

where

$$C_{\max_k}(n) = \frac{U}{n} \sum_{\ell=1}^{m} p_\ell + \sum_{\ell=1}^{m} \tau_\ell + (n-1) \left(\frac{U}{n} p_k + \tau_k \right).$$

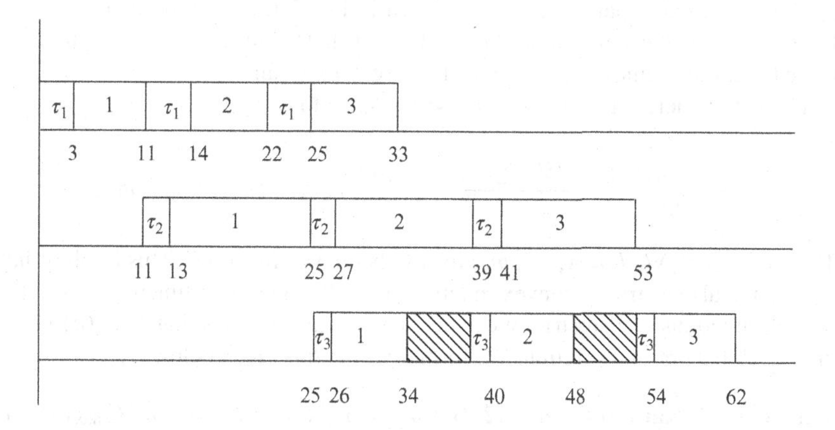

FIGURE 5.3. Lot streaming with sublot-attached setups and equal sublot sizes

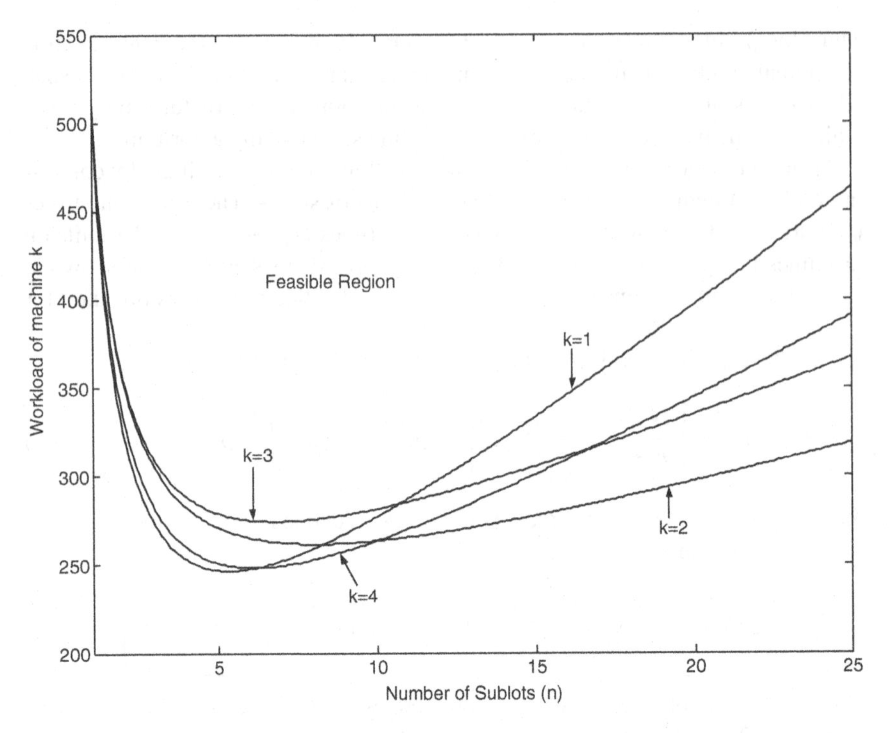

FIGURE 5.4. Typical makespan functions for $m/1/E/CV/\{$Sublot-Attached Setups$\}$

For the special case of $n = 1$,

$$C_{\max} = C_{\max_k} = \frac{U}{n}\sum_{\ell=1}^{m} p_\ell + \sum_{\ell=1}^{m} \tau_\ell, \quad \forall k = 1, \ldots, m.$$

A plot of the makespan functions is shown in Fig. 5.4. As can be seen from the figure, $C_{\max_k}$ is the same for all $k(k = 1, \ldots, m)$. The upper envelope formed by these functions defines $C_{\max}(n)$, which we seek to minimize.

The second derivative of $C_{\max_k}(n)$ with respect to n is given as

$$\frac{\mathrm{d}^2 C_{\max_k}(n)}{\mathrm{d}n^2} = \frac{2U\left(\sum_{\ell=1}^{m} p_\ell - p_k\right)}{n^3} > 0, \quad \forall k = 1, \ldots, m.$$

Thus, $C_{\max_k}(n), \forall k, k = 1, \ldots, m$, are strictly convex functions. This implies that $C_{\max}(n)$ is also a strictly convex function since the maximum function of strictly convex functions is also strictly convex. This further implies that $C_{\max}(n)$ has a unique global optimal solution. We formally state this result below.

Theorem 5.2 (Kalir and Sarin [21]) *$C_{\max_k}(n), \forall k = 1, \ldots, m$, and $C_{\max}(n)$ are strictly convex functions.*

For any $k = 1, \ldots, m$, let n_k^* denote the optimal number of sublots corresponding to the minimum of $C_{\max_k}(n)$, i.e., n_k^* is the solution to the equation $\mathrm{d}C_{\max_k}(n)/\mathrm{d}n = 0$. This is given as

$$n_k^* = \sqrt{\frac{U\left(\sum_{\ell=1}^{m} p_\ell - p_k\right)}{\tau_k}}.$$

Although Fig. 5.4 shows a typical feasible region for the problem under consideration, we cannot rule out the possibility of this region being defined entirely by a single machine, say ℓ. In that case, the minimum of machine ℓ, n_ℓ^* coincides with the global optimum of $C_{\max}(n)$, i.e., ℓ is the maximizing index of $\left\{\frac{U}{n_\ell^*} p_k + \tau_k\right\}_{k=1}^{m}$. If no such ℓ exists, the optimum must occur at an intersection point, $n_{\ell k}$, for which $C_{\max_\ell}(n_{\ell k}) = C_{\max_k}(n_{\ell k})$, i.e., the optimum solution coincides with the intersection point of two $C_{\max_k}(n)$ functions. This is formally stated below.

Theorem 5.3 (Kalir and Sarin [23]) *For any $\ell = 1, \ldots, m$, let n_ℓ^* be the solution to the equation $\mathrm{d}C_{\max_\ell}(n)/\mathrm{d}n = 0$. If there exists such a ℓ, such that ℓ is the maximizing index of $\left\{\frac{U}{n_\ell^*} p_k + \tau_k\right\}_{k=1}^{m}$, then n_ℓ^* is the global optimum of $C_{\max}(n)$. Otherwise, if no such ℓ exists, the optimum must occur at an intersection point, $n_{\ell k}$, for which $C_{\max_\ell}(n_{\ell k}) = C_{\max_k}(n_{\ell k})$.*

The intersection point can be calculated by equating $C_{\max_\ell}(n_{\ell k})$ and $C_{\max_k}(n_{\ell k})$. This gives

$$n_{\ell k} = \frac{U(p_\ell - p_k)}{\tau_k - \tau_\ell}.$$

Since the feasible region is defined by a dominant machine between any two consecutive intersection points, the efficiency of an algorithm, which searches for the optimum along this boundary, will increase if the number of intersection points, i.e., the number of machines, is reduced. We formalize this idea next.

For any pair of machines (k, ℓ), if $p_k \geq p_\ell$ and $\tau_k \geq \tau_\ell$, then it follows that

$$\frac{U}{n} p_k + \tau_k \geq \frac{U}{n} p_\ell + \tau_\ell.$$

This further implies that $C_{\max_k}(n) \geq C_{\max_\ell}(n)$, since the other terms in the expression for the workload of a machine are independent of the machine index. Hence, machine ℓ is dominated by another machine. If the above condition holds, we refer to this result as the *dominance property*, which is formally stated below.

Theorem 5.4 (Kalir and Sarin [23]) *For any pair of machines (k, ℓ), if $p_k \geq p_\ell$ and $\tau_k \geq \tau_\ell$, then machine ℓ need not be considered for optimality.*

Let S and $\bar{S}$ denote the original and reduced set of dominating machines after the application of Theorem 5.4. Then, it follows that $\bar{S}$ consists of machines such that for any machine pair (k, ℓ) either $p_k > p_\ell$ and $\tau_k < \tau_\ell$ or $p_k < p_\ell$ and $\tau_k > \tau_\ell$

must hold, since otherwise, they would have been eliminated by the *dominance property*.

Recall that the workload of machine k is given by

$$C_{\max_k}(n) = \frac{U}{n}\sum_{\ell=1}^{m} p_\ell + \sum_{\ell=1}^{m} \tau_\ell + (n-1)\left(\frac{U}{n}p_k + \tau_k\right).$$

Observe that $\lim_{n\to\infty} C_{\max_k}(n) = \tau_k n$. Hence, for sufficiently large n, the function with higher setup dominates the other. Thus, it must also be the case for all $n > n_{k\ell}$.

Hence, if $\tau_k > \tau_\ell$, then $C_{\max_k}(n) > C_{\max_\ell}(n), \forall n > n_{k\ell}$. As shown earlier, both $C_{\max_k}(n)$ and $C_{\max_\ell}(n)$ are strictly convex functions which intersect at only two points, i.e., $n = 1$ and $n = n_{k\ell}$ since there is no other solution to $C_{\max_k}(n) = C_{\max_\ell}(n)$. It follows from the strict convexity of the functions that for $1 \le n \le n_{k\ell}$ the inequality flips and we have $C_{\max_k}(n) < C_{\max_\ell}(n)$. This result is stated below.

Theorem 5.5 (Kalir and Sarin [23]) *If for any pair of machines* $(k, \ell) \in \bar{S}$, $\tau_k < \tau_\ell$, *then* $C_{\max_k}(n) > C_{\max_\ell}(n)$, $1 \le n < n_{k\ell}$.

Note that, inherently, we assume that $n_{k\ell} > 1$. In case $n_{k\ell} < 1$, machine k will dominate machine ℓ if one of the following holds: $\tau_k > \tau_\ell$, or $\tau_k = \tau_\ell$ and $p_k > p_\ell$, and machine ℓ can be dropped from further consideration.

A consequence of the above theorem is that the machine with the least setup in $\bar{S}$ will define $C_{\max}(n)$ until its first intersection point, which can be calculated as

$$n_{\ell k} = \frac{U(p_\ell - p_k)}{\tau_k - \tau_\ell},$$

where

$$\ell = \arg\min_{i\in\bar{S}} \{\tau_i\} \quad \text{and} \quad k = \arg\min_{i\in\bar{S}} \left\{\frac{p_\ell - p_i}{\tau_i - \tau_\ell}\right\}.$$

On the basis of the above analysis, we now present an optimum seeking algorithm for the m/1/E/II/CV/Sublot-Attached Setup problem. The algorithm searches for the optimal solution along the $C_{\max}(n)$ function. Starting with $n = 1$, it checks for optimality along the first dominant $C_{\max_k}(n)$ function, its first intersection point, the second dominant $C_{\max_k}(n)$ function, and its intersection point, and so on. It can easily be shown that the order of complexity of this algorithm is $O(m)$, a low polynomial order, which guarantees obtaining the optimal solution quickly and efficiently.

Sublot Sizing Algorithm for *m*/1/E/II/CV/Sublot-Attached Setup

Let $C_{\max}$ be denoted by Z.

Step 1. Set segment start point $n_s = 1$; $Z_s = Z(1)$. Also, set the current best solution $n^* = 1$; $Z^* = Z(1)$.
Apply the dominance property ($\tau_k \ge \tau_\ell$ and $p_k \ge p_1$), to eliminate some of the candidate machines. Denote by $\bar{S}$ the set of dominant machines.
Locate the first bottleneck machine: $j = \arg\min_{k\in\bar{S}} \{\tau_k\}$.
For the sake of convenience, we designate $Z(n_j)$ as Z_j in the sequel.

Step 2. Compute n_j^* by using the following expression:

$$n_j^* = \sqrt{\frac{U\left(\sum_{\ell=1}^{m} p_\ell - p_j\right)}{\tau_j}}.$$

Case 1. $|\bar{S}| = 1$ (we are at the last segment)

Set segment end point $n_e = Q$; $Z_e = Z(Q)$. Compare Z_e and Z^*. If $Z^* > Z_e$, then set $Z^* = Z_e$ and $n^* = n_e^*$.

If $n_s \le n_j^* \le n_e$, compare Z_j^* and Z^*. If $Z^* > Z_j^*$, then set $Z^* = Z_j^*$ and $n^* = n_j^*$. Otherwise, the current best solution is optimal. Stop.

Case 2. $|\bar{S}| > 1$ (we have more than one segment to search)

Let

$$\ell = \arg\min_{k \in \bar{S}} \left\{ \frac{p_k - p_j}{\tau_j - \tau_k} \right\},$$

To find the next intersection point, calculate $n_{j\ell} = U(p_j - p_\ell)/(\tau_\ell - \tau_j)$.

$$\text{Set } n_e = n_{j\ell}; \quad Z_e = Z_{j\ell}.$$

If $n_e > n_s$, then continue. Otherwise go to Step 3.

If $Z^* > Z_e$, then set $Z^* = Z_e$ and $n^* = n_e$.

Compare n_j^* with both n_s and n_e. There are three possible cases:

(a) $n_j^* < n_s$: the current n^* is still the best.

(b) $n_s \le n_j^* < n_e$: if $Z^* > Z_j^*$, then set $Z^* = Z_j^*$ and $n^* = n_j^*$. Otherwise, the current n^* is still the best.

(c) $n_j^* \ge n_e$: the end point is already checked, and hence, the current n^* is still the best.

After cases (a), (b), or (c) whichever is encountered, set $n_s = n_e$.

Step 3. Remove j from $\bar{S}$.

Set $j \leftarrow \ell$.

Go to Step 2.

The complexity of this algorithm is $O(m)$ computations and $O(m^2)$ comparisons.

Example 5.9 Consider a four-machine flow shop with a single lot consisting of 100 items. The processing (setup) times are 2(12), 3(5), 1(10), and 2(10) on machines 1, 2, 3, and 4, respectively.

Step 1. Machines 3 and 4 are dominated by machine 1. Thus $\bar{S} = \{1, 2\}$, with $\{2\}$ having the least setup.

Step 2. We get $n_2^* = 10$; $n_{12} = 14.28$.

Since $n_2^* < n_{12}\left(n_2^* > 1\right)$, it is optimal, and the makespan is $Z_2\left(n_2^*\right) = 432$.

To demonstrate what happens if the optimal solution occurs at an intersection point, suppose that the setup time of the first machine is changed to $\tau_1 = 20$ instead of 12. In that case, $n_{12} = 6.67$ and we continue to Step 3 as follows.

Step 3. Set $n_1 = 6.67$
Set $j = \{1\}$.
Remove $\{2\}$ from $\bar{S}$.
Go to Step 2

Step 2. We get $n_1^* = 5.47$. Since $n_1^* < n_1$, it is infeasible. Also since $\bar{S}$ is empty, the last intersection point $n_{12} = 6.67$ is optimal.

Note that determination of the optimal integer solution is easy, since the functions are strictly convex, and hence, the integer solution must be either the rounded-down value or the rounded-up value of the optimal continuous solution.

5.5 m/N/E/II/CV

We now consider an m-machine flow shop with N lots to be processed. The optimal solution to this problem would consist of specifying a sequence in which to process the lots as well as the sublot sizes for each lot. The problem on hand, thus, involves both the lot sequencing and sublot sizing issues. However, here, we restrict ourselves to the lot sequencing problem by assuming that the sublot sizes are known a priori. Moreover, we assume that the sublot sizes of all lots are equal, i.e., $s_i = s, \forall i = 1, \ldots, N$, all sublots of a lot are processed consecutively, i.e., there is no intermingling of the sublots, and no setup is involved.

The proposed heuristic, referred to as the bottleneck minimal idleness (BMI) heuristic, relies on identifying and using the bottleneck machine. In particular, BMI heuristic attempts to maximize the time buffer prior to the bottleneck machine, defined as $\arg\left\{\max_{1\leq k<m}\left\{\sum_{i=1}^{N} U_i p_{ik}\right\}\right\}$ where, recall that, U_i is the size of lot i, thereby minimizing potential bottleneck idleness, while also looking ahead to sequence lots with large remaining processing times early in the schedule. It is motivated by the fact that, in the lot streaming sequencing problem, when sufficiently small sublot sizes are used, minimizing idle time on the bottleneck machine must necessarily result in a near optimal schedule. The result below, referred to as Bottleneck Dominance Theorem, plays a key role in the BMI heuristic. It identifies a property of the problem on hand, which is used to generate schedules that minimize the bottleneck idle time under lot streaming.

Theorem 5.6 (Bottleneck Dominance Theorem) (Kalir and Sarin [22]) *For any lot i, if the difference $p_{i,\mathrm{BN}} - \max_{1\leq k<\mathrm{BN}}\{p_{ik}\}$ is nonnegative, then under lot streaming, there would be no idle time created between the sublots of lot i on the bottleneck machine.*

We define a lot i as "bottleneck dominant" if it satisfies the bottleneck dominant property and as "bottleneck dominated" otherwise. We now address the question of sequencing the bottleneck dominant lots. Our main objective is to construct a sequence in which the idle time between the lots on the bottleneck machine is minimal. To accomplish this task, an attempt is made to maintain as large a difference as possible between the completion times of a lot on the bottleneck machine (BN) and on the machine immediately preceding it (BN-1), while constructing the sequence, since the larger this time difference is, the less likely it will be for the next lot to create idle time on the bottleneck machine. In order to maximize the buffer between BN and BN-1, a lexicographic rule is used. This rule sequences lots in the decreasing order of the closeness of their secondary bottleneck machines to the primary bottleneck machine, i.e., the lots which are closest to the primary bottleneck are scheduled first. The secondary bottleneck refers to a machine with the largest unit processing from among the machines upstream of the bottleneck machine. By utilizing this approach, some of the idle time that might have been created on the machines closer to the bottleneck machine is absorbed because it overlaps with the processing of the previous lots. We illustrate the working of the lexicographic rule via an example.

Example 5.10 Consider a five-machine four-lot flow shop system, with per unit processing times and lot sizes as given in Table 5.3.

Note that machine 5 is the bottleneck machine and all lots are bottleneck dominant. For simplicity, we will consider unit-size sublots.

The lexicographic rule works as follows. First, the secondary bottleneck machines of all the lots are identified. These are machines 4, 3, 3, and 2 for lots 1, 2, 3, and 4, respectively. Then, the lot closest to the primary bottleneck, i.e., lot 1, is placed first in the sequence. Next, there is a tie between lots 2 and 3 since both lots have machine 3 as their secondary bottleneck. To break the tie, we compare their subsequent bottleneck machines. For lot 2 it is machine 2 and for lot 3 it is machine 1. Hence, we sequence lot 2 before lot 3. To complete the sequence, lot 4 is placed last in the sequence. The resulting sequence is {1–2–3–4}. The corresponding schedule is shown in Fig. 5.5.

This schedule achieves a time buffer, i.e., difference between completion times of lots on BN and BN-1, of $90 - 56 = 34$ time units. Note that the first two sublots of lot 2 take only 2 time units each to complete on machine 4, as opposed to 3 time units each taken by the third and fourth sublots. In other words, the idle time is absorbed for the first two sublots. This availability was created by scheduling a lot (lot 1 in this case) whose secondary bottleneck is closer to the BN, thereby

TABLE 5.3. Data for illustration of lexicographic rule of BMI

Lot	Machine 1	Machine 2	Machine 3	Machine 4	Machine 5	U_i
1	1	2	3	4	5	4
2	1	3	4	2	5	4
3	2	1	4	3	5	4
4	3	3	2	1	5	4

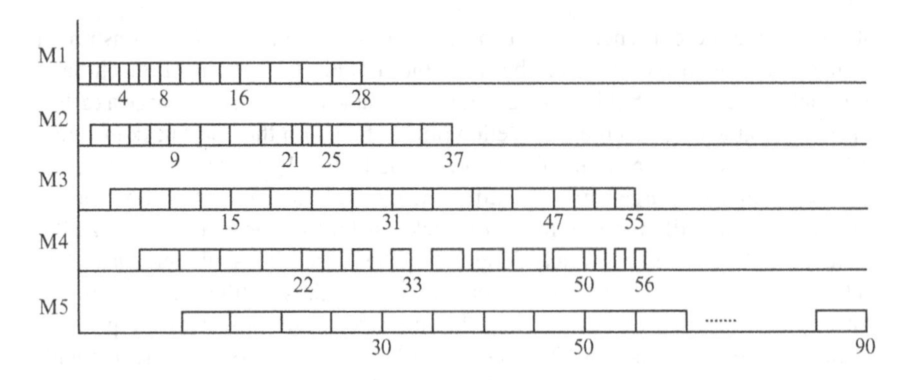

FIGURE 5.5. Schedule generated by the lexicographic rule

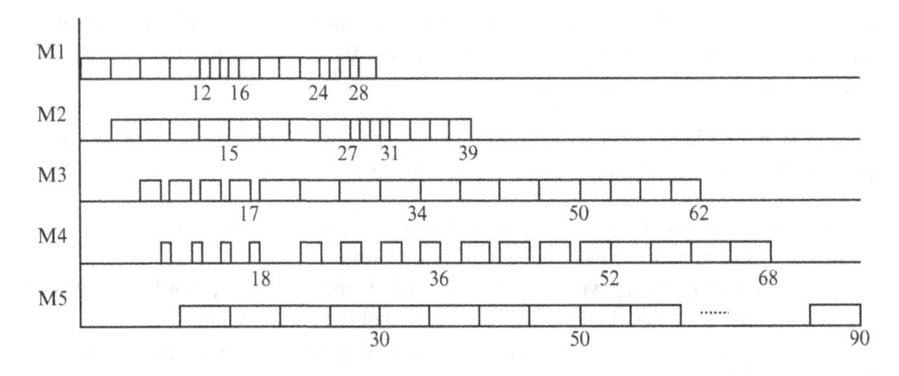

FIGURE 5.6. Schedule generated by an arbitrary sequence

creating a queue of work upstream of the machine. Similarly, observe that idle time is also avoided for the first two sublots of lot 4 on machine 4. To appreciate the effect of this rule, consider an arbitrary sequence {4–2–3–1}, constructed by interchanging lots 1 and 4. The resulting schedule is shown in Fig. 5.6.

In this schedule, the idle time in between the sublots and lots on machine 4 has increased from 10 in the original schedule to 20. Consequently, the completion time on machine 4 for this sequence has increased to 68, resulting in a reduced time buffer of $90 - 68 = 22$ time units.

Having sequenced the bottleneck dominant lots, we now address the question of sequencing the bottleneck dominated lots. To keep consistent with our strategy, we would like to sequence at least some of these lots in their entirety according to the lexicographic rule, even though this may decrease the time buffer as long as it does not create bottleneck idleness. However, in many cases, we might not be able to do this for all bottleneck dominated lots. If the bottleneck machine does not have enough time buffer to ensure zero bottleneck idleness, this may result in a downstream machine becoming the bottleneck thereafter. In that event, it would have been best if the lots with the larger remaining processing time after the bottleneck machine would have been scheduled earlier ("pushed forward") in the sequence. This would have enabled absorption of these large processing times early in the schedule, while better utilizing the future downstream bottleneck machine. In addition, this would also have resulted in lesser workload for the downstream bottleneck machine once it has become the bottleneck.

To implement this "pushing forward" of the lots with large processing times on downstream machines, we use a "tail rule" in the BMI heuristic. This rule is implemented as follows. For each lot, the largest unit processing time from amongst the machines after the bottleneck machine is identified. This is defined as the "tail" of the lot. The lots are then ordered in decreasing order of their "tails," i.e., a lot with a larger tail is sequenced before a lot with a smaller tail.

Finally, we address the issue pertaining to the dominance of the bottleneck machine. In cases of strong dominance (when the processing time of the bottleneck machine is far greater than that of the other machines), it may suffice to consider only the bottleneck machine in the implementation of the heuristic. However, if the machines are relatively balanced (i.e., no strong dominance of any machine exists), then it is necessary to consider several machines, instead of a single machine, as potential primary bottlenecks. Toward that end, we use a statistical measure to determine a set of machines to be considered as potential primary bottleneck machines. This set consists of machines whose total processing time is a certain distance away from the average total processing time. Let T_k denote the total processing time of lots on machine k. The average total processing time and its standard deviation are

$$\bar{T} = \frac{\sum_{k=1}^{m} T_k}{m},$$

$$\sigma_{\bar{T}} = \frac{S}{\sqrt{m}},$$

where

$$S = \sqrt{\frac{\sum_{k=1}^{m} \left(T_k - \bar{T}\right)^2}{m-1}}.$$

The desired set of machines consists of all k, such that T_k satisfies $T_k \geq \left(\bar{T} + u\sigma_{\bar{T}}\right)$, where u is some predetermined value.

The BMI heuristic is given below.

Bottleneck Minimal Idleness Heuristic for *m/N/E/II/CV*

Step 1. Let S be the set of lots: $S \equiv \{1, 2, \ldots, N\}$. Compute

$$T_k = \sum_{i=1}^{N} U_i p_{ik}, \forall k, \quad \bar{T} = \frac{\sum_{k=1}^{m} T_k}{m}, \quad \sigma_{\bar{T}} = \frac{S}{\sqrt{m}},$$

where

$$S = \sqrt{\frac{\sum_{k=1}^{m} \left(T_k - \bar{T}\right)^2}{m-1}}.$$

Determine the ordered set of candidate machines: $D = \{k : T_k \geq \left(\bar{T} + u\sigma_{\bar{T}}\right)\}$.

Step 2. Let k be the index of the first machine in set D. Set $D = D - \{k\}$.
Step 3. Let $\text{BN} = k$. If $k \neq m$, compute the "tail" for each lot as follows:

$$\text{tail}(i) = \max_{\text{BN}+1 \leq r \leq m} \{p_{ir}\}, \quad \forall i = 1, \ldots, N.$$

Step 4. Let S_1 be the set of bottleneck dominant lots and S_2 be the set of bottleneck dominated lots, i.e.,

$$S_1 = \left\{ i : p_{i,\text{BN}} - \max_{1 \leq r < \text{BN}} \{p_{ir}\} \geq 0 \right\},$$

$$S_2 = \left\{ i : p_{i,\text{BN}} - \max_{1 \leq r < \text{BN}} \{p_{ir}\} < 0 \right\}.$$

Step 5. For each lot in S_1 and S_2, determine its secondary bottleneck machines starting with machine BN-1, throughout the upstream machines, until machine 1.
Step 6. Determine an ordered set S by arranging the lots in the decreasing order of the closeness of the secondary bottleneck machine to BN. Break ties in accordance with the following rules, in decreasing order of preference:

- The closeness of the subsequent secondary bottleneck machines to BN
- The value of tail(i) (larger value first)
- Arbitrarily (smaller sequential number)

Step 7. Construct a two-dimensional array with the ordering as one of its dimensions and the set type as its other dimension ("1" for S_1 and "2" for S_2).
Step 8. Construct the sequence as follows:
Step 8.1. Schedule the first "1"-type lot in the array first. Push back any "2"-type lots that precede it in the array to follow the "1"-type lot in the array.
Step 8.2. If the next lot in the array is "1"-type, schedule it. Else, if the next lot is "2"-type and if scheduling this lot (in its entirety) does not result in bottleneck idleness, schedule the "2"-type lot next. Else, if the "2"-type lot does result in bottleneck idleness, push it back in the array to follow the next "1"-type lot in the array. If there are no more "1"-type lots, go to Step 9.
Step 8.3. Repeat Step 8.2 until all the "1"-type lots are scheduled.
Step 9. If all "2"-type lots have been scheduled, then go to Step 10. Otherwise add the remaining "2"-type lots in nonincreasing order of their tail(i).
Step 10. Select the best sequence among those generated so far: k, S, $C_{\max}(S)$.
Step 11. If $D = \phi$, then STOP; otherwise, go to Step 2.
Step 12. If $|D| > 1$, select the best sequence among those generated.

In Step 1, the set D of potential primary bottleneck machines is identified. In Step 2, the first machine in set D is considered as the primary bottleneck machine and the heuristic begins. In Step 3, the tail of each lot is calculated and stored for future use. The set of bottleneck dominant and bottleneck dominated lots are identified in Step 4. In Step 5, the recursive vector of secondary bottleneck machines for each lot is identified, while in Step 6, it is used to arrange the lots via the lexicographic rule. Then in Step 7, an array is constructed which contains the sequence of lots according to the lexicographic rule and the lot type ("1" for bottleneck dominant and "2" for bottleneck dominated). Step 8 contains the heart of the heuristic as it is the schedule construction step. It constructs the schedule in accordance with the lexicographic rule with the exception of pushing back "2"-type lots after the next "1"-type lot in case they create bottleneck idleness when taken in their current location. This essentially achieves the same objective of pushing "2"-type lots forward when they are considered as a separate set which follows the set of "1"-type lots. Then, in Step 9, after all "1"-type lots have been scheduled, the remaining "2"-type lots are scheduled in decreasing order of their tails. The entire process is then repeated for all machines in set D.

Example 5.11 Consider a five-machine four-lot flow shop with per unit processing times and lot sizes as shown in Table 5.4.

Application of Steps 1 and 2 of the heuristic results in identification of machine 3 as the primary bottleneck machine. Next, the sequence {1–3–4–2} is generated based on the lexicographic rule and the "tail" rule. Lot 1 is first because it had machine 2 as its secondary bottleneck machine while the others have machine 1 as their secondary bottleneck machine. Lot 3 precedes lots 4 and 2 because it has a larger tail (6 as compared to 5 and 1). Similarly, lot 4 precedes lot 2 because it has a larger tail between them. Lots 1, 2, and 3 are bottleneck dominant lots whereas lot 4 is a bottleneck dominated lot. This completes Steps 3–7. The schedule is constructed in Step 8 as follows. Lot 1 is scheduled first since it is a "1"-type lot and is next in the sequence. Lot 3 is scheduled next for the same reason. Lot 4, which is next in the sequence, is a "2"-type lot, and hence, must be tried before being scheduled, to make sure that it does not create any bottleneck idleness. This is indeed the case, and hence, lot 4 is scheduled next in the sequence. Finally, lot 2 is scheduled to complete the sequence. Step 9 is skipped in this case and the algorithm either terminates or repeats the procedure with a different primary bottleneck machine. The final sequence generated is {1–3–4–2}, which is indeed the unique optimal solution (verified using total enumeration).

Next, consider the case in which the sublot sizes of different lots are different. This problem is an extension of the two-machine problem discussed in Sect. 3.14. Recall that for the two-machine case, starting from unit-size sublots, the procedure iterates by increasing sublot sizes of lots and resequencing the lots using Johnson'salgorithm. For increments in their sublot sizes, the lots are considered

TABLE 5.4. Data for the illustration of BMI

Lot	Machine 1	Machine 2	Machine 3	Machine 4	Machine 5	U_i
1	2	3	7	10	10	8
2	5	3	5	1	1	11
3	6	4	10	6	4	12
4	10	1	7	1	5	14

in the decreasing order of the idle time caused by a lot. The same procedure can be applied for the m-machine case as well except that for the re-sequencing of lots after a change in lot sizes, we can use the BMI heuristic (see Sect. 5.5). In case the sublot-attached setup time is also present, the sublot processing time of a sublot can be determined to be $\tau_{ij} + Lp_{ij}$. The complexity of this algorithm is $O\left(NmU_{\max}^2\right)$.

So far, in this chapter, we have considered the objective of minimizing the makespan. However, next, we deviate from this assumption and consider the objective of minimizing the total sublot completion time in Sect. 5.6 and a more general unified cost function in Sect. 5.7, albeit for $U = 1$.

5.6 $m/1/C/II/CV/\sum_{i=1}^{2} x_i C_{im}$

Assume the number of sublots to be restricted to two on each machine, and the sublots sizes to be continuous and consistent. As before, let x_1 and $x_2 = (1 - x_1)$ be the proportion of work allocated to the first and second sublots, respectively. Let $C_{i,k}$ denote the completion time of the ith sublot on machine k and p_k be the time process a lot per item on machine k.

The mathematical formulation for this problem is as follows:

$$\text{Minimize}: \quad (x_1 C_{1m} + x_2 C_{2m})$$

$$\text{Subject to}:$$

$$\begin{aligned}
&C_{11} \geq x_1 p_1, \\
&C_{2k} \geq C_{1k} + x_2 p_k, \quad k = 1, \ldots, m, \\
&C_{ik+1} \geq C_{ik} + x_i p_{k+1}, \quad i = 1, 2, k = 1, \ldots, (m-1), \\
&x_1 + x_2 = 1, \\
&C_{ik}, x_1, x_2 > 0.
\end{aligned}$$

Let the objective function be denoted as

$$F(x_1, x_2) = (x_1 C_{1m} + x_2 C_{2m}).$$

The completion times of the sublots can be written as

$$C_{1m} = x_1 \sum_{k=1}^{m} p_k$$

and

$$C_{2m} = \max_{1 \leq k \leq m} \left\{ x_1 \sum_{l=1}^{k} p_l + x_2 \sum_{l=k}^{m} p_l \right\}.$$

Making the above substitutions along with $x_2 = 1 - x_1$, in the expression for flow time, we have

$$F(x_1) = x_1^2 \sum_{k=1}^{m} p_k + (1 - x_1) \max_{1 \leq k \leq m} \left\{ x_1 \sum_{l=1}^{k} p_l + (1 - x_1) \sum_{l=k}^{m} p_l \right\}.$$

After simplification and rearrangement, the above expression reduces to

$$F(x_1) = \max_{1 \le k \le m} \left\{ x_1^2 \left(\left(2 \sum_{l=k}^{m} p_l \right) - p_k \right) + x_1 \left(\sum_{l=1}^{k} p_l - 2 \sum_{l=k}^{m} p_l \right) + \sum_{l=k}^{m} p_l \right\}.$$

Let

$$a_k = \left(2 \sum_{l=k}^{m} p_l \right) - p_k, \quad b_k = \left(\sum_{l=1}^{k} p_l - 2 \sum_{l=k}^{m} p_l \right), \quad \text{and} \quad c_k = \sum_{l=k}^{m} p_l.$$

We have

$$F(x_1) = \max_{1 \le k \le m} \left\{ a_k x_1^2 + b_k x_1 + c_k \right\}.$$

Note that the maximum operator inherently takes care of the above constraints on completion times. Hence, an equivalent formulation can be written as

$$\begin{aligned} &\text{Minimize}: \quad F(x_1) \\ &\text{Subject to}: \\ &\qquad F(x_1) \ge a_k x_1^2 + b_k x_1 + c_k, \quad \forall k = 1, \ldots, m, \end{aligned}$$

where

$$a_k = \left(2 \sum_{l=k}^{m} p_l \right) - p_k,$$

$$b_k = \left(\sum_{l=1}^{k} p_l - 2 \sum_{l=k}^{m} p_l \right),$$

and

$$c_k = \sum_{l=k}^{m} p_l.$$

Let $y_k = a_k x_1^2 + b_k x_1 + c_k$. Since $\mathrm{d}^2 y_k / \mathrm{d}x_1^2 = 2a_k > 0$, y_k is a convex function. The area defined by each constraint is the epigraph of this convex function, and is a convex set. Since the intersection of convex sets is convex, the feasible region of the problem constitutes a convex set. Also, the objective function is linear. Therefore, the optimum solution will occur at the boundary of the feasible region. The feasible region, which consists of candidate points for the optimum solution, is made up of intersection points and individual minimums of each constraint. Since each constraint is quadratic, any pair of constraints has two intersection points, thereby, giving $2\binom{m}{2} = m(m-1)$ intersection points and m individual minimum points. The feasible region is illustrated in Fig. 5.7. It is convex and its boundary is piecewise in which any edge (lying between two consecutive intersection points) corresponds to one or more binding constraints over an interval on the x_1 axis.

We now discuss some properties of the constraints and the feasible region.

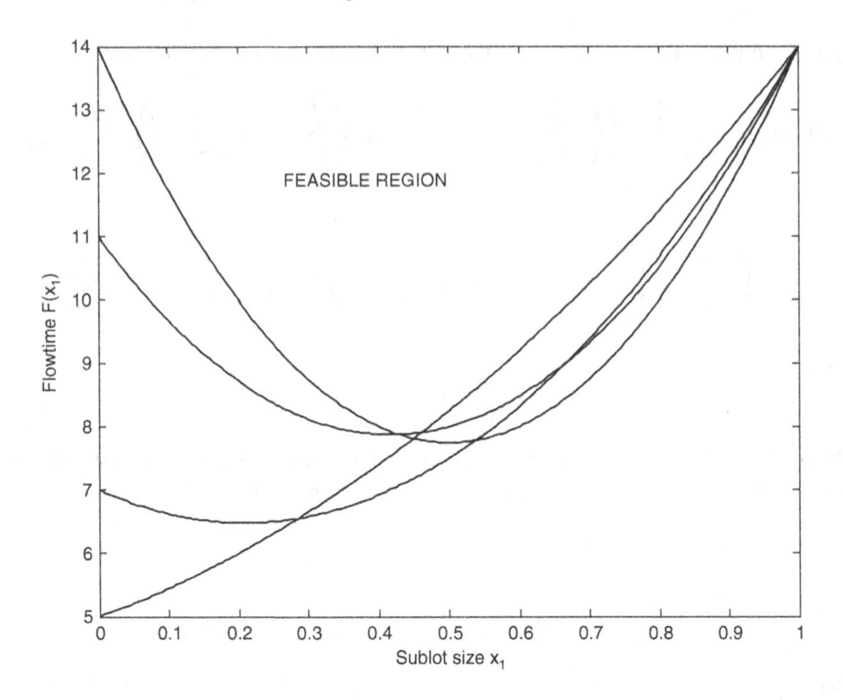

FIGURE 5.7. Feasible region for the $m/1/CV/C/\sum_{i=1}^{2} x_i C_{i,m}$ problem

Property 1 [Topaloglu et al. [34]]. *The intersection points of any two constraints, k and $\ell(k < \ell)$, can be obtained by solving the following quadratic equation*

$$a_{k\ell}x_1^2 + b_{k\ell}x_1 + c_{k\ell} = 0,$$

where

$$a_{k\ell} = a_k - a_\ell = \left(2 \sum_{i=k+1}^{\ell-1} p_i\right) + p_k + p_\ell,$$

$$b_{k\ell} = b_k - b_\ell = -3 \sum_{i=k+1}^{\ell-1} p_i - 2p_k - p_\ell,$$

and

$$c_{k\ell} = c_k - c_\ell = \sum_{i=k+1}^{\ell-1} p_i + p_k.$$

Let $w = \sum_{i=k+1}^{\ell-1} p_i$, then

$$a_{k\ell} = 2w + p_k + p_\ell,$$
$$b_{k\ell} = -3w - 2p_k - p_\ell,$$

and

$$c_{k\ell} = w + p_k.$$

The roots of this quadratic equation are given by

$$\frac{-b_{k\ell} \pm \sqrt{b_{k\ell}^2 - 4a_{k\ell}c_{k\ell}}}{2a_{k\ell}}.$$

The determinant $\Delta = b_{k\ell}^2 - 4a_{k\ell}c_{k\ell}$ can be simplified and written as $(w + p_\ell)^2$. The first root, $x_{1,1}^{k,\ell}$, is given by

$$\frac{3w + 2p_k + p_\ell - (w + p_\ell)}{4w + 2p_k + 2p_\ell},$$

which can be rearranged and written as follows:

$$x_{1,1}^{k,\ell} = \frac{c_k - c_\ell}{a_k - a_\ell}.$$

The second root is given by

$$x_{1,2}^{k,\ell} = \frac{3w + 2p_k + p_\ell + (w + p_\ell)}{4w + 2p_k + 2p_\ell} = 1.$$

Hence, for any two constraints k and ℓ, the intersection points are given by $\frac{c_k - c_\ell}{a_k - a_\ell}$ and 1.

Property 2 [Topaloglu et al. [34]]. *The intercept of any constraint k on the $F(x_1)$ axis is given by c_k. This implies that the intercept decreases as k increases.*

Property 3 [Topaloglu et al. [34]]. *The minimum of any constraint k can be obtained as*

$$x_1^{k*} = \frac{-b_k}{2a_k} = \frac{-\left(\sum_{i=1}^{k} p_i - 2\sum_{i=k}^{m} p_i\right)}{4\sum_{i=k}^{m} p_i - 2p_i},$$

which can be simplified and written as

$$x_1^{k*} = \frac{1}{2} - \frac{\sum_{i=1}^{k-1} p_i}{2\left(2\sum_{i=k+1}^{m} p_i + p_k\right)}.$$

This implies that as k increases, the minimum of each constraint shifts to the left on the x_1 axis, and is never greater than 1/2.

Property 4 [Topaloglu et al. [34]]. *Since the minimum of each constraint occurs when $x_1 \le 1/2$, the slope of the tangent of each constraint over the interval (1/2, 1] is positive. This implies that, in the interval (1/2, 1], there are no intersection points, since at an intersection point the slopes of the two tangents must have opposite signs. Hence x_1^*, the optimal proportion of work allocated to the first sublot, must be less than or equal to 1/2.*

Property 5 [Topaloglu et al. [34]]. *The slope of constraint k at* $x_{12}^{k} = 1$ *is given by* $s^k(1)$*, where* $s^k(1) = 2a_k + b_k = \sum_{i=1}^{m} p_i + \sum_{i=k+1}^{m} p_i$*, which decreases as k increases.*

These properties are summarized below:

1. Constraints k and $(k+1)$ intersect at two points, namely,

$$x_{11}^{kk+1} = \frac{c_k - c_{k+1}}{a_k - a_{k+1}} \quad \text{and} \quad x_{12}^{kk+1} = 1.$$

2. The intercept $F^{k+1}(x_1)$ of the constraint $k+1$ on the $F(x_1)$ axis is lower than $F^k(x_1)$, the intercept of constraint k, i.e.,

$$F^{k+1}(x_1) < F^k(x_1).$$

3. The minimum of constraint $k+1$ is lower than that of constraint k, i.e.,

$$x_1^{(k+1)^*} < x_1^{k^*},$$

where

$$x_1^{k^*} = \frac{1}{2} - \frac{\sum_{i=1}^{k-1} p_i}{2\left(2\sum_{i=k+1}^{m} p_i + p_k\right)}.$$

4. The slope $s^{k+1}(1)$ of constraint $k+1$ at $x_1 = 1$ is lower than $s^k(1)$, the slope of constraint k at $x_1 = 1$, i.e.,

$$s^{k+1}(1) < s^k(1).$$

A consequence of Property 1 is that we can now only check the nontrivial intersection point of any two constraints k and $k+1$ given by x_{11}^{kk+1}. This reduces the candidate intersection points for the optimum by half, thus giving $m(m+1)/2$ points.

On the basis of the above development, there are the following two algorithms that can be used to obtain the optimal sublot sizes.

5.6.1 *Sequential Intersection Point Enumeration Algorithm*

As the name suggests, this algorithm basically visits the intersection points defining the feasible region sequentially and checks for optimality. It begins its search along the edge of the feasible region defined by the last constraint. At any particular iteration, it finds the intersection of the current binding constraint with all higher indexed constraints. The next binding constraint is the one corresponding to the minimum value of the intersection point $x_{1,1}$. Next, it compares the minimum of the previous binding constraint to the above computed minimum value of intersection point. If the x_1 value corresponding to the minimum of the previous binding constraint is higher than the x_1 value corresponding to the intersection point, then the current minimum value of x_1 is replaced by the latter. Further, it evaluates the slope of the tangent at the intersection point, and if it is greater or equal to zero, then the algorithm terminates since this must be the first point where the objective function begins to increase.

However, if the x_1 value corresponding to the minimum of the previous binding constraint is lower than that corresponding to the intersection point, then the minimum of the previous binding constraint must correspond to the optimal solution, and hence, the algorithm stops.

This algorithm has a complexity of $O(m^2)$. It is formally stated below and is followed by an illustrative example.

Sequential Intersection Point Enumeration Algorithm

Step 1. Initialize $k = 1$.

Step 2a. Let x_1 correspond to the minimum value of the current binding constraint, i.e.,

$$x_1 = \frac{-b_k}{2a_k},$$

where

$$a_k = \left(2\sum_{l=k}^{m} p_l\right) - p_k \quad \text{and} \quad b_k = \left(\sum_{l=1}^{k} p_l - 2\sum_{l=k}^{m} p_l\right).$$

Step 2b. Find all the intersection points of the current binding constraint and all other higher indexed constraints and pick the one giving the minimum value of x_1, i.e.,

$$\ell = \arg\left[\min_{k<i\leq m}\left(x_{11}^{ki}\right)\right],$$

where

$$x_{11}^{ki} = \frac{c_k - c_i}{a_k - a_i}.$$

Define $x_1^{\text{INTER}} = x_{11}^{k\ell}$.

Step 3a. If $x_1^{\text{INTER}} < x_1$, then define ℓ to be the new binding constraint and update x_1 as below:

$$x_1 = x_1^{\text{INTER}},$$
$$k = \ell.$$

Step 3b. If the slope of the new binding constraint at x_1 is nonnegative, i.e., $2a_k x_1 + b_k \geq 0$, then let $x_1^* = x_1$ and STOP; else go to Step 2a.

Step 4. If $x_1^{\text{INTER}} > x_1$, then x_1, the minimum of the current binding constraint, is optimum, i.e., $x_1^* = x_1$ and we STOP.

Example 5.12 Consider a four-machine flow shop with processing times per unit of 3, 4, 2, and 5 on machines 1, 2, 3, and 4, respectively.

The coefficients a_k, b_k, and c_k are calculated as shown below:

Coefficient	k			
	1	2	3	4
a_k	25	18	12	5
b_k	–25	–15	–5	4
c_k	14	11	7	5

Step 1. Let $k = 1$.

Step 2a. The minimum value of the current binding constraint is given by $x_1 = 25/(2 \times 25) = 0.5$.

Step 2b. The intersection points of constraint 1 with constraints 2, 3, and 4 are 0.429, 0.538, and 0.45, respectively. The minimum of these is 0.429. Hence, we set $\ell = 2$ and $x_1^{\text{INTER}} = 0.429$.

Step 3a. Since $x_1^{\text{INTER}} < x_1$, we define constraint 2 to be the new defining constraint by letting $x_1 = 0.429$ and $k = 2$.

Step 3b. The slope of the new binding constraint is $2(18)(0.429) + (-15) = 0.429 > 0$, we set $x_1^* = 0.429$, and STOP.

The optimal flow time value is obtained by substituting $x_1^* = 0.429$ in the $F(x_1)$ expression for $m = 1$ or 2, which gives the optimal flow time of 7.876. Note that the flow time values corresponding to $m = 3$ and 4 are 7.063 and 7.636, respectively.

5.6.2 *Interval Bisection Search Algorithm*

Unlike the previous algorithm, which sequentially visits the intersection points, this algorithm uses a bisection search which decreases the intervals on the x_1 axis successively. At any iteration, it calculates the midpoint x_1^{M} of the current interval and finds the binding constraint d, which corresponds to the constraint giving the maximum flow time at x_1^{M}. It then evaluates the slope of the tangent to the constraint d at x_1^{M}. If positive, then it implies that we are beyond the minimum. Hence, we then calculate the intersection point of constraint d with other constraints which lie to the left of x_1^{M} and pick the maximum of these. We now check for the possibility of the minimum of constraint d being between this intersection point and x_1^{M}. If this is the case, then we stop since it must be the optimal solution; else we reduce the interval by letting the right end point equal to the intersection point calculated above. A similar analysis holds for the case when the slope of the tangent of constraint d at x_1^{M} is negative. The algorithm continues to repeat this until the left and right end points coincide. This algorithm also has a complexity of $O(m^2)$. It is formally stated below and is also illustrated by an example.

Interval Bisection Search Algorithm

Step 1. Initialize the left and right end points of the interval as follows:

$$x_1^{\mathrm{L}} = 0 \quad \text{and} \quad x_1^{\mathrm{R}} = 1.$$

Step 2a. Calculate the midpoint x_1^{M} of the interval as follows:

$$x_1^{\mathrm{M}} = \frac{x_1^{\mathrm{L}} + x_1^{\mathrm{R}}}{2}.$$

Step 2b. Find the constraint which is binding at x_1^{M} as follows:

$$d = \arg\left(\max_{1 \le k \le m} \{F^k(x_1)\}\right).$$

Step 3a. If the slope of the tangent to constraint d at x_1^{M} is positive, i.e., $2a_d x_1^{\mathrm{M}} + b_d > 0$, then find the closest intersection point to the left of x_1^{M}, i.e.,

$$x_1 = \max_{\{k: d \neq k, x_{11}^{dk} < x_1^{\mathrm{M}}\}} \left\{x_{11}^{dk}\right\},$$

where

$$x_{11}^{dk} = \frac{c_d - c_k}{a_d - a_k}.$$

Step 3b. If the x_1 value corresponding to the minimum of curve d is greater than x_1, i.e., $x_1 < -b_d/2a_d$, then we STOP and $x_1^* = -b_d/2a_d$. Otherwise, we update the interval by letting $x_1^{\mathrm{R}} = x_1$ and then go to Step 2a.

Step 4a. If the slope of the tangent to constraint d at x_1^{M} is negative, i.e., $2a_d x_1^{\mathrm{M}} + b_d < 0$, then find the closest intersection point to the right of x_1^{M}, i.e.,

$$x = \min_{\{k: d \neq k, x_{11}^{dk} > x_1^{\mathrm{M}}\}} \left\{x_{11}^{dk}\right\},$$

where

$$x_{11}^{dk} = \frac{c_d - c_k}{a_d - a_k}.$$

Step 4b. If the x_1 value corresponding to the minimum of constraint d is greater than x_1, i.e., $x_1 < -b_d/2a_d$, then we STOP and $x_1^* = x_1$. Otherwise, we update the interval by letting $x_1^{\mathrm{L}} = x_1$ and then go to Step 2a.

Step 5. If the left and right end points coincide, i.e., $x_1^{\mathrm{L}} = x_1^{\mathrm{R}}$, we STOP and $x_1^* = x_1^{\mathrm{R}}$; else we go to Step 2a.

Example 5.13 Consider the same example used for the illustration of the Sequential Intersection Point Enumeration Algorithm.

Step 1. Let $x_1^{\mathrm{L}} = 0$ and $x_1^{\mathrm{R}} = 1$.
Step 2a. The midpoint $x_1^{\mathrm{M}} = 0.5$.
Step 2b. The calculations are shown in the table below:

Constraint	Equation	$F^k x_1^M = 0.231$	$\max\left\{F^k\left(x_1^M\right)\right\}$
1	$25x_1^2 - 25x_1 + 14$	7.75	8.25
2	$18x_1^2 - 15x_1 + 11$	8.0	
3	$12x_1^2 - 5x_1 + 7$	7.5	
4	$5x_1^2 + 4x_1 + 5$	8.25	
Binding constraint			4

Step 3a. The slope of the tangent to constraint 4 at $x_1^M = 0.5$ is $2(5)(0.5) + (4) = 9 > 0$. Hence, we find the closest intersection point to the left of x_1^M. The intersection points of constraint 4 with constraints 1, 2, and 3 are 0.45, 0.462, and 0.286, respectively, the maximum of which is 0.462, which occurs for constraint 2.

Step 3b. We now compare this value with the minimum of constraint 4, which is $-4/(2 \times 5) = -0.4$. Since $x_1 > -0.4$, we update the right end point to 0.462. The new midpoint is $(0 + 0.462)/2 = 0.231$. The calculation of the constraint binding at this value of x_1 is shown in the table below:

Constraint	Equation	$F^k\left(x_1^M = 0.231\right)$	$\max\left\{F^k\left(x_1^M\right)\right\}$
1	$25x_1^2 - 25x_1 + 14$	9.562	9.562
2	$18x_1^2 - 15x_1 + 11$	8.497	
3	$12x_1^2 - 5x_1 + 7$	6.485	
4	$5x_1^2 + 4x_1 + 5$	6.189	
Binding constraint			1

The slope of the tangent to constraint 1 at $x_1^M = 0.231$ is $2(25)(0.231) + (-25) = -13.462 < 0$. Hence, we calculate the closest intersection point to the right of x_1^M as $x_1 = \min\left\{x_{11}^{12}, x_{11}^{13}, x_{11}^{14}\right\} = \min\{0.429, 0.538, 0.45\} = 0.429$. Since 0.5, the minimum of constraint 1, is greater than 0.429, we STOP. The optimal proportion of work allocated to the first sublot is 0.429 and that to the second is 0.571. The solution is depicted in Fig. 5.8.

5.7 *m*/1/E/II/CV

We now consider a hybrid objective function consisting of a weighted sum of the makespan, (sublot) mean flow time, average WIP, sublot-attached setup time, and transfer time, for an m-machine flow shop with a single lot with continuous and equal sublots.

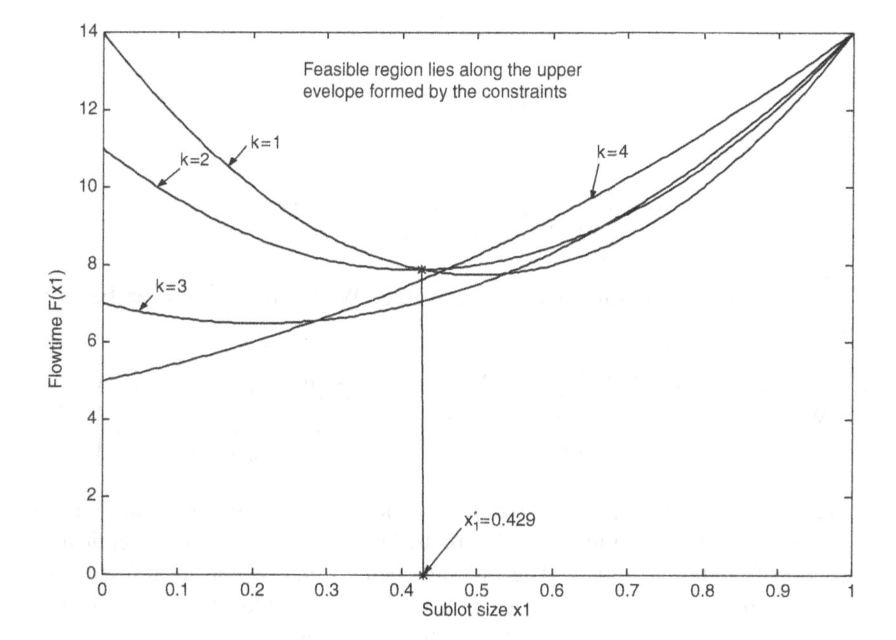

FIGURE 5.8. Graphical solution for Example 5.13

We use the following additional notation:

MFT	Mean flow time of sublots 1 to n
WIP	Average WIP in the system
TT_i	Transfer time for sublot i

We also assume that the sublot-attached setup on any machine is less than the time required to process the entire lot on that machine, i.e., $\tau_k < Up_k, k = 1, \ldots, m$.

In the absence of setup and transfer times, clearly, it is optimal to have unit-size sublots, i.e., $n^* = U$, for minimizing the makespan. This, however, might not be true in the presence of setup and/or transfer times and a different objective function.

As in (5.6), the makespan expression is given by

$$C_{\max}(n) = \left\{ \frac{U}{n} \sum_{k=1}^{m} p_k + \sum_{k=1}^{m} \tau_k \right\} + (n-1) \max_{1 \le k \le m} \left\{ \frac{U}{n} p_k + \tau_k \right\}.$$

Similarly, the flow time of a sublot i can be written as

$$\mathrm{FT}_i = \left\{ \frac{U}{n} \sum_{k=1}^{m} p_k + \sum_{k=1}^{m} \tau_k \right\} + (i-1)\tau_v + (i-1)\frac{U}{n} p_v,$$

where $\upsilon = \arg\left\{\max_{1\leq i\leq m}(p_i)\right\}$. Therefore,

$$\text{MFT}(n) = \frac{\sum_{i=1}^{n} \text{FT}_i}{n}$$

and

$$\text{MFT} = \frac{U}{n}\sum_{k=1}^{m} p_k + \sum_{k=1}^{m} \tau_k + \frac{n-1}{2} \max_{1\leq k\leq m} \left\{\frac{U}{n} p_k + \tau_k\right\}.$$

In accordance with the Little's Law, the average WIP in the system can be written as

$$\text{WIP}(n) = U \left\{ \frac{\frac{U}{n}\sum_{k=1}^{m} p_k + \sum_{k=1}^{m} \tau_k + \frac{n-1}{2} \max\limits_{1\leq k\leq m} \left\{\frac{U}{n} p_k + \tau_k\right\}}{\frac{U}{n}\sum_{k=1}^{m} p_k + \sum_{k=1}^{m} \tau_k + (n-1) \max\limits_{1\leq k\leq m} \left\{\frac{U}{n} p_k + \tau_k\right\}} \right\}.$$

As can be seen from the above expression, the average WIP can be minimized by having the maximum number of sublots, i.e., $n^* = U$, which is identical to the case of negligible setups. The rationale behind this is that, by utilizing the smallest sublot sizes, the sublots leave the system as soon as possible, which in turn minimizes the average WIP. As mentioned earlier, this policy might not be optimal for the objective function under consideration.

We consider a weighted sum of the following performance measures:

1. Makespan
2. Mean flow time of the sublots
3. Average WIP
4. Sublot-attached setup times
5. Sublot transfer times

Therefore, the problem under consideration can be formulated as an integer program as follows:

Minimize: $Z(n) = c_1 C_{\max}(n) + c_2\text{MFT}(n) + c_3\text{WIP}(n) + c_4\text{SA}(n) + c_5\text{TT}(n)$

Subject to:

$$C_{\max}(n) = \left\{\frac{U}{n}\sum_{k=1}^{m} p_k + \sum_{k=1}^{m} \tau_k\right\} + (n-1) \max_{1\leq k\leq m} \left\{\frac{U}{n} p_k + \tau_k\right\},$$

$$\text{MFT}(n) = \frac{U}{n}\sum_{k=1}^{m} p_k + \sum_{k=1}^{m} \tau_k \frac{n-1}{2} \max_{1\leq k\leq m} \left\{\frac{U}{n} p_k + \tau_k\right\},$$

$$\text{WIP}(n) = U \left\{ \frac{\frac{U}{n}\sum_{k=1}^{m} p_k + \sum_{k=1}^{m} \tau_k + \frac{n-1}{2} \max\limits_{1\leq k\leq m} \left\{\frac{U}{n} p_k + \tau_k\right\}}{\frac{U}{n}\sum_{k=1}^{m} p_k + \sum_{k=1}^{m} \tau_k + (n-1) \max\limits_{1\leq k\leq m} \left\{\frac{U}{n} p_k + \tau_k\right\}} \right\},$$

$$\text{SA}(n) = n\sum_{k=1}^{m} \tau_k,$$

$$\mathrm{TT}(n) = n(m-1)\mathrm{TT},$$

$$1 \le n \le U, \text{ and integer.}$$

We assume that $\mathrm{TT} = \mathrm{TT}_i, i = 1, \ldots, n$. $C_{\max}(n)$ and MFT(n) can be shown to be strictly convex functions [20]. Since SA(n) and TT(n) are linear functions, WIP(n) is the only function which requires further analysis.

Let

$$\mathrm{WIP}_k(n) = U \left\{ \frac{\frac{U}{n} \sum_{k=1}^{m} p_k + \sum_{k=1}^{m} \tau_k + \frac{n-1}{2} \left\{ \frac{U}{n} p_k + \tau_k \right\}}{\frac{U}{n} \sum_{k=1}^{m} p_k + \sum_{k=1}^{m} \tau_k + (n-1) \left\{ \frac{U}{n} p_k + \tau_k \right\}} \right\}.$$

be the WIP level dictated by machine k. Observe that, for a given value of n, this WIP level for any two machines differs only by the last term appearing in the numerator and denominator of the above expression, since the other terms are independent of the machine index. Hence, as discussed in Sect. 5.4, for a given value of n, the minimum WIP will be obtained for a machine υ, where

$$\upsilon = \arg \left\{ \max_{1 \le k \le m} \left\{ \frac{U}{n} p_k + \tau_k \right\} \right\}.$$

This implies that WIP(n) is composed of segments of $\mathrm{WIP}_k(n)$. In each segment, $\mathrm{WIP}(n) = \mathrm{WIP}_k(n)$, $1 \le k \le m$. Thus, the average WIP function can be written as

$$\mathrm{WIP}(n) = \min_{1 \le k \le m} \{\mathrm{WIP}_k(n)\}.$$

The nature of the WIP(n) function is shown in Fig. 5.9. The figure indicates that WIP(n) is a segmental strictly convex function, composed of $\mathrm{WIP}_k(n)$ functions, which take the smallest values in each segment.

Note that the intersection points of the above function coincide with that of the $C_{\max}(n)$ and MFT(n) functions, since in all three cases the functions are dictated by the machine that maximizes $((U/n)p_k + \tau_k)$.

The objective function $Z(n)$ is twice differentiable between any two intersection points. The second derivative of $Z(n)$ over each segment is nonnegative, since it is always nonnegative for each of its first three terms. Hence, $Z(n)$ is strictly convex over each segment. Hence, it is a segmental strictly convex function if the function WIP(n) is a segmental strictly convex function. Further, it can also be shown [25] that $Z(n)$ has a unique global minimum n^*, if $c_1 > 0$ or $c_2 > 0$.

We now discuss the issue of the first derivative of $Z(n)$, since it is required by the algorithm presented later. The first derivative of $Z(n)$ (for each segment in which k is the maximizing index) can be written as

$$\frac{\mathrm{d}Z_k(n)}{\mathrm{d}n} = c_1 \frac{\mathrm{d}C_{\max_k}(n)}{\mathrm{d}n} + c_2 \frac{\mathrm{dMFT}_k(n)}{\mathrm{d}n} + c_3 \frac{\mathrm{dWIP}_k(n)}{\mathrm{d}n} + c_4 \frac{\mathrm{dSA}_k(n)}{\mathrm{d}n} + c_5 \frac{\mathrm{dTT}_k(n)}{\mathrm{d}n}.$$

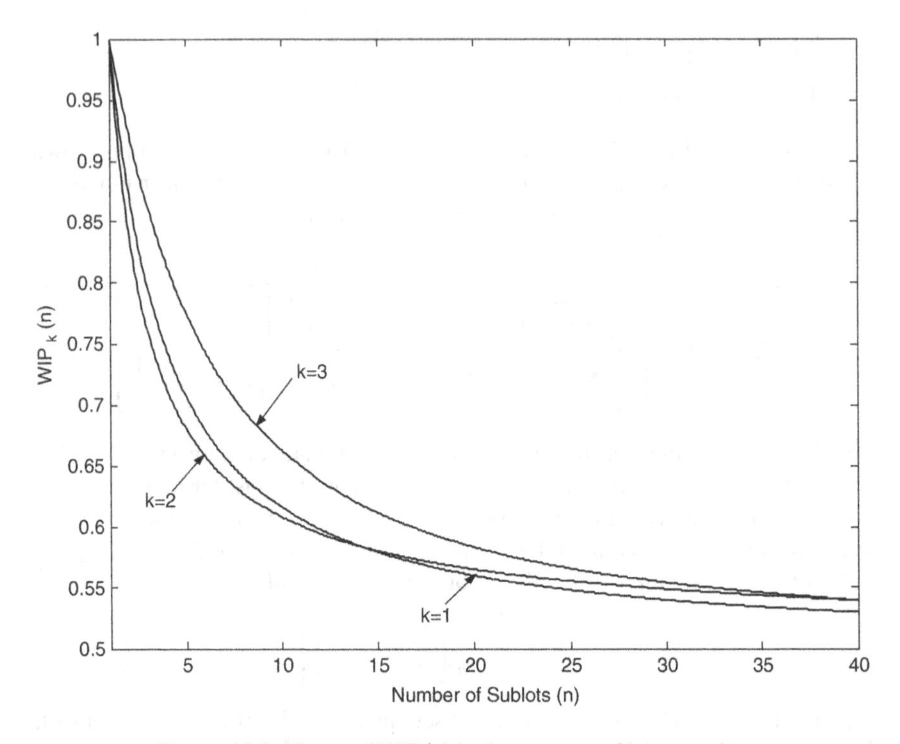

FIGURE 5.9. Nature of WIP(n) in the presence of lot streaming

Therefore,

$$\frac{\mathrm{d}Z_k(n)}{\mathrm{d}n} = \frac{1}{n^2}\left(c_1 U\left(\sum_{\ell=1}^{m} p_\ell - p_k\right) + c_2 U\left(\sum_{\ell=1}^{m} p_\ell - \frac{p_k}{2}\right)\right) + \left(\tau_k(c_1 + (c_2/2)) + c_4 \sum_{\ell=1}^{m} \tau_\ell + c_5(m-1)\mathrm{TT}\right) + c_3 \frac{\mathrm{dWIP}_k(n)}{\mathrm{d}n}.$$

Note that

$$\frac{1}{U}\frac{\mathrm{dWIP}_k(n)}{\mathrm{d}n} = -\frac{n^2\ell_{12} + 2n\ell_{14} + (\ell_{23} + \ell_{34} - \ell_{14})}{\left[2n^2\ell_1 + n(\ell_2 + 2\ell_3 - 2\ell_1) + (\ell_4 - 2\ell_3)\right]^2},$$

where

$$\ell_1 = (t_k/2), \ell_2 = \sum_{\ell=1}^{m} t_\ell, \ell_3 = (Up_k)/2, \ell_4 = U\sum_{\ell=1}^{m} p_\ell, \text{ and } \ell_{ij} = \ell_i \ell_j.$$

Plugging the above formula for $\mathrm{dWIP}_k(n)/\mathrm{d}n$ into $\mathrm{d}Z_k(n)/\mathrm{d}n$ and setting it equal to zero leads to a fourth-order equation, the solution for which can only be obtained via numerical techniques. Instead, we use an approximation for $\mathrm{dWIP}_k(n)/\mathrm{d}n$, which essentially considers only the higher order terms. Therefore,

$$\frac{\mathrm{dWIP}_k(n)}{\mathrm{d}n} \approx -\frac{n^2\ell_{12}}{[2n^2\ell_1]^2}\frac{1}{U} = -\frac{1}{U}\frac{n^2\ell_{12}}{4n^4\ell_1^2} = -\frac{1}{n^2}U\frac{\sum_{\ell=1}^m \tau_\ell}{\tau_k}.$$

After using this approximation for $\mathrm{dWIP}_k(n)/\mathrm{d}n$ in $\mathrm{d}Z_k(n)/\mathrm{d}n$, and equating it to zero, we get the following closed-form expression for the optimal number of sublots n^*:

$$n_k^* = \sqrt{\frac{c_1 U\left(\sum_{\ell=1}^m p_\ell - p_k\right) + c_2 U\left(\sum_{\ell=1}^m p_\ell - \frac{p_k}{2}\right) + c_3 U\frac{\sum_{\ell=1}^m \tau_\ell}{\tau_k}}{\tau_k(c_1 + (c_2/2)) + c_4\sum_{\ell=1}^m \tau_\ell + c_5(m-1)\mathrm{TT}}}.$$

To summarize, we note that $Z(n)$ is a segmental strictly convex function. Each segment, being a part of $Z(n)$ lying between two intersection points, is defined by some machine, which achieves a minimum, i.e., lies at the bottom of all segments of other machines, corresponding to the maximizing index of $((U/n)p_k + \tau_k)$.

The proposed solution algorithm searches for the minimizing solution over each segment of the $Z(n)$ function, keeping a record of the best solution so far. Within each segment, it checks for an optimal solution using the first derivative test above. If the test fails, then the end point of the segment is checked. The algorithm repeats this process until it finds an increase in the objective function value. At this point, it terminates and declares the optimality of the solution found. Note that this algorithm will find the optimum solution if it occurs at an intersection point. However, if this is not the case, then it finds an approximate solution based on the above formula for n^*, developed using the first derivative test. This approximation will be more accurate when the optimum occurs for large values of n.

It is evident that the efficiency of this algorithm can be enhanced significantly if the number of intersection points, i.e., number of machines, can be reduced. Toward that end, we use a *dominance property*, proved by Kalir and Sarin [19], which essentially states that a machine ℓ can be ignored for optimality if its processing time and setup time are less than another machine k, i.e.,

$$\tau_k \geq \tau_\ell \quad \text{and} \quad p_k \geq p_\ell.$$

The algorithm is identical to that presented in Sect. 5.4 except that the objective function (Z) now is different, and also, in Step 2, n_j^* is obtained using the above expression.

5.8 Chapter Summary

The algorithms that we have presented in this chapter for m-machine lot streaming problems, typically, involve a single lot and consistent (or equal) sublot sizes. In the absence of any setup, different heuristic procedures are described. These include the use of equal sublot sizes, a procedure based on the method of Campbell et al. [6] for the m-machine flow shop problem, which determines sublot sizes by considering several pseudo-two-machine problems with processing times on the first of these two machines being equal to the sum of the processing times on the first k machines and that on the other machine being equal to the sum of the processing times on the last k machines, for $k = 1, \ldots, m - 1$, and a two-sublot-based heuristic procedure in which the bottleneck machine plays a central role. This problem can also be addressed by using its network representation like that for the three-machine problem discussed in Chap. 4. An algorithm is presented to determine optimal, continuous, and consistent sublot sizes. This algorithm makes use of the concept of dominant machines, which helps in reducing the problem size. The discrete version of this problem has also been discussed and two algorithms are presented in that regard, which obtain integer sublot sizes by appropriately rounding the optimal, continuous sublot sizes. The worst-case bounds on the performances of these truncation procedures are also given.

We have also addressed m-machine lot streaming problems for an objective function different from that of minimizing the makespan. One such problem involves minimization of completion time. A special instance of this problem is analyzed that involves only two sublots. However, we also consider a more general objective function that involves minimization of a weighted sum of the makespan, mean flow time of the sublots, average WIP, sublot-attached setup time, and transfer time, but for the case of equal sublot sizes. Thus, the problem essentially involves determination of optimal number of sublots in order to minimize this unified objective function. The algorithm that we present builds on the algorithm for the m-machine lot streaming sublot-attached setup problem, which has also been addressed in this chapter.

We also consider a multiple-lot m-machine lot streaming problem for the case of equal sublot sizes. This problem involves determination of both optimal number of sublots for each lot as well as the sequence in which to process the lots. The single-lot m-machine lot streaming problem is solved to determine optimal number of sublots for each lot, and then, a heuristic procedure, termed *bottleneck minimal idleness heuristic* (BMI), is developed that sequences lots so as to minimize the idleness caused at the bottleneck due to the processing of the lots. The number of sublots of different lots is adjusted iteratively with the corresponding sequence to process the lots determined using the BMI heuristic each time, in order to obtain the best possible solution.

6 Concluding Remarks

In this book, we have expounded on the concept of lot streaming in which a lot of the jobs is split into small-size sublots to enable their processing in an overlapping fashion over the machines of a production facility. This is a useful concept for implementation in a batch production environment since it can aid in reducing the makespan, work-in-process, and mean flow times of the lot and its sublots.

The key features of a lot streaming problem are the number of lots, configuration of the machines over which the lots are processed, types of sublot sizes in which to process the lots on different machines (i.e., equal, consistent, or variable), types of the setup times involved (which can be lot and/or sublot-attached or detached), and the removal and transfer times of the sublots. We have presented the underlying analyses, algorithms, and relevant results for a variety of lot streaming problems encountered while processing the lots in the flow shop machine environment. These constitute the body of knowledge on lot streaming. For the ease of understanding, the algorithms are illustrated through numerical examples.

We have, first, presented various models and analyses for the two-machine problems followed by a discussion of the results, analyses, and algorithms for the three-machine and m-machine lot streaming problems. We have also presented a generic model for the streaming of the lots in a flow shop environment as well as several special models that are suitable for specific problem scenarios in consideration. For continuous sublot sizes, these models are, typically, linear programs, which are easy to solve. The requirement of integer sublot sizes results in a linear integer program. Because of the special nature of several problems involving continuous sublot sizes and a few problems involving integer sublot sizes, they can be solved to optimality using simple and polynomial time algorithms. Another approach that has been used for the solution of lot streaming problems relies on their network representation, and we have discussed the use of this approach for the solution of several two-machine, three-machine, and m-machine problems.

The single lot, lot streaming problem involves determination of both the number of sublots, in which to split a lot, and sublot sizes. However, in the presence of multiple lots, an additional issue of determining the sequence in which to process the lots also needs to be addressed. Nevertheless, in the vast majority of the cases discussed, the sublot sizing and lot sequencing problems are independent. That is, the optimal sublot sizes of a lot are independent of the position of the lot in the sequence. It is not uncommon, for ease of analysis, to assume a priori the number of sublots in which to split a lot, thereby resulting in the problem of determining only the sizes of the sublots for that lot.

The objective function that is, typically, used for lot streaming problems is to minimize the makespan. However, for certain instances, sublot completion time and cost-based objective function have also been considered.

Even though, in this book, we have addressed the lot streaming problem in the context of a flow shop, yet it lays a foundation for its further development to incorporate additional features, and thereby, expand the scope of its applicability. Many of the open problems in this regard have been mentioned earlier in the book. Another scenario that seems amenable for exploration based on the analysis presented in this book is the m to one assembly line in which m modules, produced on m separate and parallel assembly/processing lines, are to be assembled to make a product. An effective management of supply chains is essential in today's competitive market place. It offers a fertile ground to explore implementation of the concept of lot streaming. Moreover, development of an efficient solution methodology for the general lot streaming problem still remains a challenging task.

References

1. Adams, J., E. Balas and D. Zawack, "The shifting bottleneck procedure for job shop scheduling," *Management Science*, **34**(3): 391–401, 1988
2. Baker, K. R., "Lot streaming in the two-machine flow shop with setup times," *Annals of Operations Research*, **57**: 1–11, 1995
3. Baker, K. R. and D. Jia, "A comparative study of lot streaming procedures," *Omega: International Journal of Management Science*, **21**(5): 561–566, 1993
4. Baker, K. R. and D. Pyke, "Solution procedures for the lot-streaming problem," *Decision Sciences*, **21**: 475–491, 1990
5. Bukchin, J., M. Tzur and M. Jaffe, "Lot splitting to minimize average flow-time in a 2 machine flow-shop." *IIE Transactions*, **34**: 953–970, 2002
6. Campbell, H. G., R. A. Dudek and M. L. Smith, "A heuristic algorithm for the n-job m-machine sequencing problem," *Management Science*, **16**: B620–B627, 1970
7. Centinkaya, F. C., "Lot streaming in a two-stage flow shop with set-up, processing and removal times separated," *Journal of Operational Research Society*, **45**(12): 1445–1455, 1994
8. Centinkaya, F. C. and M. S. Kayaligil, "Unit sized transfer batch scheduling with setup times," *Computers and Industrial Engineering*, **22**(2): 177–182, 1992
9. Chen, J. and G. Steiner, "Discrete lot streaming in 2 machine flow shops," *Information Systems and Operations Research*, **37**(2): 160–173, 1999
10. Chen, J. and G. Steiner, "Lot streaming with attached setups in three machine flow shops," *IIE Transactions*, **30**: 1075–1084, 1998
11. Chen, J. and G. Steiner, "Approximation methods for discrete lot streaming in flow shops," *Operations Research Letters*, **21**: 139–145, 1997
12. Chen, J. and G. Steiner, "Lot streaming with detached setups in three-machine flow shops," *European Journal of Operational Research*, **96**: 591–611, 1996
13. Dauzere-Peres, S. and J. Lasserre, "Lot streaming in job-shop scheduling," *Operations Research*, **45**(4): 584–595, 1997
14. Dauzere-Peres, S. and J. B. Lasserre, "A modified shifting bottleneck procedure for job shop scheduling," *International Journal of Production Research*, **31**(4): 923–932, 1993
15. Gilmore, P. C. and R. E. Gomory, "Sequencing a one state-variable machine: A solvable case of the traveling salesman problem," *Operations Research*, **12**: 665–679, 1964
16. Glass, C., J. N. D. Gupta and C. N. Potts, "Lot streaming in three-stage production process," *European Journal of Operational Research*, **75**: 378–394, 1994

17. Glass, C. A. and C. N. Potts, "Structural properties of lot streaming in a flow shop," *Mathematics of Operations Research*, **23**(3): 624–639, 1998
18. Goyal, S. K., "Note on manufacturing cycle time determination for a multi stage economic production quantity model," *Management Science*, **23**: 332–333, 1976
19. Johnson, S. M., "Optimal two- and three-stage production schedules with setup times included," *Naval Research Logistics Quarterly*, **1**: 61–67, 1954
20. Kalir, A. A., "Optimal and heuristic solutions for the single and multiple batch flow shop lot streaming problems with equal sublots," Ph.D. Dissertation, Virginia Tech, 2000
21. Kalir, A. and S. C. Sarin, "Constructing near optimal schedules for the flow shop lot streaming problem with sublot-attached setups," *Journal of Combinatorial Optimization*, **7**: 23–44, 2003
22. Kalir, A. and S. C. Sarin, "A near-optimal heuristic for the sequencing problem in multiple-batch flow-shops with small equal sublots," *Omega: International Journal of Management Science*, **29**: 577–584, 2001
23. Kalir, A. and S. C. Sarin, "Optimal solutions for the single batch, flow shop, lot-streaming problem with equal sublots," *Decision Sciences*, **32**(2): 387–397, 2001
24. Kalir, A. and S. C. Sarin, "Evaluation of potential benefits of lot streaming in flow-shop systems," *International Journal of Production Economics*, **66**: 131–142, 2000
25. Kalir, A., S. C. Sarin and M. Chen, "A unified cost based lot streaming problem for the single batch flow shop with equal sublots," Grado Department of Industrial and Systems Engineering, Virginia Tech, 2004
26. Kumar, S., T. Bagchi and C. Sriskandarajah, "Lot streaming and scheduling heuristics for m-machine no-wait flow shops," *Computers and Industrial Engineering*, **38**: 149–172, 2000
27. Pinedo, M., *Scheduling: Theory, algorithms, and systems*, Prentice-Hall, Englewood Cliffs, NJ, 2002
28. Potts, C. N. and K. R. Baker, "Flow shop scheduling with lot streaming," *Operations Research Letters* **8**: 297–303, 1989
29. Ramasesh, R., H. Fu, et al., "Lot streaming in multistage production systems," *International Journal of Production Economics*, **66**: 199–211, 2000
30. Reiter, S., "A system for managing job-shop production," *Journal of Business*, **39**: 371–393, 1966
31. Sen, A., E. Topaloglu and O. S. Benli, "Optimal streaming of a single job in a two stage flow shop," *European Journal of Operational Research*, **110**: 42–62, 1998
32. Sriskandarajah, C. and E. Wagneur, "Lot streaming and scheduling multiple products in 2-machine no-wait flowshops," *IIE Transactions*, **31**: 695–707, 1999
33. Szendrovits, A. Z., "Manufacturing cycle time determination for a multi stage economic production quantity model," *Management Science*, **22**(3): 298–308, 1975
34. Topaloglu, E., A. Sen, and O. S. Benli, "Optimal streaming of a single job in an m stage flow shop with two sublots," Technical Report 06533, Department of Industrial Engineering, Bilkent University, Ankara, 1994
35. Trietsch, D. and K. R. Baker, "Basic techniques for lot streaming," *Operations Research*, **41**(6): 1065–1076, 1993
36. Vickson, R. G., "Optimal lot streaming for multiple products in a two-machine flow shop," *European Journal of Operational Research*, **85**: 556–575, 1995
37. Vickson, R. G. and B. E. Alfredsson, "Two and three machine flow shop scheduling problems with equal sized transfer batches," *International Journal of Production Research*, **30**(7): 1551–1574, 1992

38. Womack, J. P. and D. T. Jones, *Lean thinking*, Simon and Schuster, New York, NY, 1996
39. Yoon, S. H. and J. A. Ventura, "Minimizing the mean weighted absolute deviation from due dates in lot streaming flow shop scheduling," *Computers and Operations Research*, **29**: 1301–1315, 2002
40. Yoon, S. H. and J. A. Ventura, "An application of genetic algorithms to lot streaming flow shop scheduling," *IIE Transactions*, **34**: 779–787, 2002

Index